# Analysing Environmental Discourse

Bloomsbury Advances in Critical Discourse Studies

**Series Editors: David Machin, Michal Krzyzanowski and John Richardson**

Bloomsbury Advances in Critical Discourse Studies is a series that looks at exciting, cutting edge work in critical discourse studies, in terms of its multidisciplinarity of method, theory, and topic of analysis. The series explores how critical discourse studies engages with the social, political and ideological landscape of the modern world, and how these contexts are reflected, (re)produced and mediated through language and other modes of communication.

Titles published in the series:

*Critical Discourse Studies and Technology*, Ian Roderick
*Constructing the Welfare State in the British Press*, Malgorzata Paprota
*Constructions of Migrant Integration in British Public Discourse*, Sam Bennett
*Discourses of Men's Suicide Notes*, Dariusz Galasinski
*The Discursive Construction of Migrant Identities*, edited by Charlotte Taylor, Simon Goodman and Stuart Dunmore
*European Identities in Discourse*, Franco Zappettini
*The Politics and Rhetoric of Commemoration*, Michael Billig and Cristina Marinho
*Popular Music and Multimodal Critical Discourse Studies*, Lyndon C. S. Way

# Analysing Environmental Discourse

## *A Critical Approach*

Scott Burnett

BLOOMSBURY ACADEMIC
LONDON • NEW YORK • OXFORD • NEW DELHI • SYDNEY

BLOOMSBURY ACADEMIC
Bloomsbury Publishing Plc, 50 Bedford Square, London, WC1B 3DP, UK
Bloomsbury Publishing Inc, 1359 Broadway, New York, NY 10018, USA
Bloomsbury Publishing Ireland, 29 Earlsfort Terrace, Dublin 2, D02 AY28, Ireland



First published in Great Britain 2026



Cover design: Jade Barnett
Cover image © Tiverets/ Shutterstock



A catalogue record for this book is available from the British Library.

A catalog record for this book is available from the Library of Congress.

ISBN: HB: 978-1-3503-6029-7
ePDF: 978-1-3503-6030-3
eBook: 978-1-3503-6031-0

Series: Bloomsbury Advances in Critical Discourse Studies

Typeset by Integra Software Services Pvt. Ltd.

For product safety related questions contact productsafety@bloomsbury.com.

# Contents

# Figures

# Tables

# Preface

While writing this book, I struggled with a creeping sense of futility. The period from late 2023 when I started work to early 2025 when the manuscript was submitted saw such catastrophe and upheaval around the world that keeping calm and carrying on from the safety of my office seemed an absurd – even an obscene – thing to do. While environmental decline was on the fringes of my unease, its epicentre was – and still is – in Gaza. Like so many others, I witnessed war crime after war crime through the screen of my phone while being told by respected politicians and media that this is all OK, that this is normal, that the Palestinians had it coming, and that there will be no cessation of the flow of arms to Israel. Any naïve and lingering illusions I might have had about the rules-based liberal order built on universal human rights have been decisively and forever dispelled. The values of the post-war 'West' seem to me now nothing more than a façade constructed to conceal the same calculating cynicism that enabled countless wars, colonial genocides, mass enslavement and the Shoah.

While this realization might seem at first to have little to do with environmental discourse, the argument of this book is that if we think systemically about ecological crisis, we cannot disentangle the harm we do to ecosystems from the harm we do to each other. As Andreas Malm shows in *The Destruction of Palestine Is the Destruction of the Earth*, the dependence of the fossil fuel economy on imperial war is an open secret. Military superiority is built on access to energy, which is in turn used to access more energy. Fossil-powered economies established through violence bring far-flung territories into global networks of combustion and extraction patrolled by gas-guzzling war machines. And the ecological consequences of war do not end when hostilities cease but follow new territorial patterns of extraction, production and exploitation. In the Middle East, Israel is not merely a national project but the eastern outpost of Western fossil capital. Malm cites Joe Biden's assertion that Israel is the best 'investment' the nation makes: 'Were there not an Israel, the United States of America would have to invent an Israel to protect our interests in the region'.[1] He concludes that in the central role it has played in maintaining fossil-fuelled business as usual, the 'steps along the way to the destruction of Palestine were simultaneously steps along the way to that of the earth'.

As I make clear in the first chapters of this book, indifference to the human toll of these geopolitical campaigns is enmeshed in indifference to their ecological costs. This is and has always been the nature of empire, and is as true in Palestine as in Ukraine, Sudan or Afghanistan. And as a university professor earning a salary in the imperial core I cannot exempt myself from complicity in these projects. Higher education in the United States is the bedrock of a militaristic, nationalistic, and capitalist enterprise. If students learn from us the language of critique, this is but one element of a broader skills-set they will use to make their way in life. We can strive to equip them with tools to help the world change course, but they will have to navigate systems that reward thoughtlessness and disincentivize doubt. There is thus no guarantee that our work as educators is not merely reinforcing the death spiral of the world.

Breaking out of the narrative of futility and inevitability is, however, indispensable for any kind of resistance. As I argue in the conclusion, despair has a class character that eats away at coalitions like an acid from the inside, poisoning relations between social groups. We simply have no choice but to operate from a sense that our actions matter. And I would suggest that the global resurgence of right-wing authoritarianism, as dangerous and frightening as it is, reveals to us that we're more impactful than we habitually claim. While the liberal establishment treated students who dared to speak up for Gaza with disproportionate and vindictive brutality, the right is making significant headway in eroding the viability and authority of higher education *tout court*. Any kind of expertise not directly accountable to a strongman, a corporation, or an oligarch is seen as a threat that must be neutralized. We learn from these initiatives that some knowledge (about gender, race, or coloniality, for example) is so dangerous that it must be abolished completely, and that other knowledge (such as about history, climate or communication) is valued only to the extent that it bolsters national feeling, turns a profit or helps win wars. What is it about universities that scares these people so? Though it is easy to ridicule the egomania of a professoriate that imagines publishing an article as equivalent to being on the barricades, it is also worth remembering that pursuing knowledge for its own sake will always undermine the ambitions of those who seek to monopolize the power to influence what people think.

This does not mean of course that we are excused from putting ourselves on the line in more mundane ways, such as standing side-by-side with protesting students or participating in local organizing work that builds good relations and collective power. We should be working towards the day when the fascists are defeated (as they tend to be) in such a way that we do not revert to the late liberal complacency that created the conditions for their rise in the first place. When that day comes, whatever collective future we have will need to be built on solidarity not only with each other but with the more-than-human world.

The community that helped me write this book is a vibrant, critical and intellectually nourishing one to which I am forever indebted. Nancy Tuana, whose philosophical work on ecointersectionality inspired the approach to environmental discourse I present here, has been a decisive force in helping this project come to fruition. Nancy has been a consistent source of good advice, a careful and constructive reader of the manuscript, and has challenged me on a number of occasions to unpick my own reproduction of insidious nature/human dichotomies. I am truly grateful to her for her intellectual generosity and greatness of spirit. My colleague Aparna Parikh offered razor-sharp insights on a draft of Chapter 6, which includes some elements of a literature review conducted by our outstanding graduate student Hazel Velasco Palacios. Together, Nancy, Aparna, Hazel and I have been running a Mellon-funded Sawyer Seminar series at Penn State on gender and sexuality in right-wing discourse. I am very thankful to Aparna for taking the lead on running the seminar while I was doing the bulk of writing on this book.

My thanks also to another Penn State geographer, Tiza Mfuni, whose deep expertise on African conservation made his detailed critiques of chapter five all the more important. The collaborative colloquium 'Thinking Indigeneity from the South' that my dear friend and colleague Gabeba Baderoon and I ran in 2023–4 with support from the Humanities Institute at Penn State offered me a valuable opportunity to explore Indigenous environmental justice scholarship with a group of highly engaged and sophisticated thinkers. The introductory chapters were also shared with members of an international network of scholars that offers critical and constructive feedback on decolonial work-in-progress. I am thankful to all members of the 'Circle' (Angeliki Sifaki, Bineta Thiam, Lucia Sarmiento, Marilena Indelicato, Onni Ahvonen, Sadia Habib, Sudip Sen and the Toms – Houseman, Pettinger and Six) for their feedback – but especially to Marilena for being such a tireless force behind this and many other important initiatives. And then, to go back to the start, I would like to thank Bram Büscher for his ongoing enthusiasm about discursive approaches to political ecology, and for his much-appreciated encouragement of the outlines of this book during a visit to Gothenburg in 2022.

I have had other opportunities to test the ideas in this book. Thanks to Ari Sherris at Texas A&M Kingsville for inviting me to teach a guest lecture on environmental justice in September 2024, where I enjoyed engaging with his students on the pitfalls of green capitalism. Thank you to Michał Krzyżanowski for inviting me to present the nascent argument at the Critical Discourse Analysis meeting in Uppsala in June 2023, and to John Richardson and Sofia Lampropoulou for the opportunity to present an overview of the completed book at the June 2025 instalment of that meeting in Liverpool. It was John who initially encouraged me to submit the proposal to *Advances in Critical Discourse Analysis* at Bloomsbury, the series he edits with Michał and David Machin, and I am profoundly grateful to him for his encouragement and friendship.

The writing process was made all the more pleasant and bearable through weekly check-ins with Penn State's College of the Liberal Arts Writing Program. Under the direction of Cyanne Loyle with help from Jessica Vantine Birkenholtz, Chris Skurka and Maria Truglio, the programme has been a wonderful entry-point into the intellectual life of State College. Thanks to all of my writing buddies, not least among whom is Amy Allen, who not only writes with us but makes the resources available at the College level for the programme to continue. I am also fortunate to be able to rely on the mentorship and support of Sinfree Makoni and Alicia Decker, who have created a wonderfully enabling environment for me in African Studies and Women's, Gender, and Sexuality Studies at Penn State. I am very grateful to them, and also to our College Dean Clarence Lang, for the role they together played in finding a home for the kind of inter/trans/non-disciplinary work I am interested in pursuing within the College of the Liberal Arts.

Writing does not only happen at work, of course, and I would like to thank the friends and family who opened their homes and fed and watered me while I was stuck in laptop-land. My father in law Carlo Milani supported the writing process with infectious enthusiasm and lunchtime wine (which sometimes slowed things down to be honest). Thank you to Silvia Milani and Roberto Pin for their hospitality, and for letting Tommaso and I treat half of their home as an office. Though I did not get to spend as much time as I would have liked with my mom and dad in South Africa while writing this book, I would like to thank them for their love of nature and their hatred of oppression, values I hope I do justice to in my work. And finally, and indeed most of all, I owe more than I can express in these few lines to my husband Tommaso, who seems always to be finding new ways to make my life better and more fulfilling.

Living as I do in a state of privilege and comfort, the gap between my life and the unfathomable suffering elsewhere in the world opens up like a yawning void. If I am able to do anything in my life, I hope it will be to build the kinds of relations that may one day close the abyss between a centre that matters and a periphery where entire worlds and populations are constructed as disposable. Gaza is very much the microcosm of the world system as it exists today. For that reason, I dedicate this book to the people of Gaza, in fervent hope of a free Palestine and an end to injustice and oppression everywhere.

# Introduction: Intersectionality and ecological crisis

In 2010, *Climate Policy* published an article entitled 'How to Eat an Elephant: A Bottom-up Approach to Climate Policy' by Steve Rayner, then director of Oxford University's Institute for Science, Innovation and Society.[1] The adage in the title encodes a gem of folksy wisdom: When faced with a problem as daunting as global warming, which may leave you feeling overwhelmed or discouraged, remember that you don't have to swallow it whole. Cut it up into bite-sized chunks instead, and then you can chew on them one at a time.

Climate change is elephantine. And it is even bigger than we might think. Ecological crisis is intimately enmeshed in economic, social and political problems. Our relations with the planet and our relations with each other are hard to separate. We need only look to inequalities in how people relate to nature as extractors, producers or consumers, conflicts between groups and between nations over territory and resources, and heated political struggles over how to envisage a liveable future, to see that doing anything about the environment requires thinking about the world system as a whole. How to carve up this expanded elephant is tricky, to say the very least. 'One bite at a time' could work to motivate a conception of the world where specific dimensions of ecological harm, social inequality, marginalization or conflict are imagined as existing in isolation, as single issues.[2] As intersectionality theorists have been saying for decades, when women of colour are told by the women's movement that ending racism is not their mission, or by anti-racist movements that misogyny is not their concern, not only are they failed by their comrades, but the *movements themselves fail* to properly articulate the social problems they care about.[3] Because racism and sexism intersect, dealing with one without naming the other is a mistake of both theory and praxis. In ecological questions, racism and sexism (as well as ableism, class oppression, heterosexism and cisgenderism) intersect both with culpability in causing and vulnerability to the effects of environmental devastation. To articulate the 'full dimensions' of any one of these phenomena requires understanding how they intersect with each other, an idea frequently encapsulated in the mantra 'all oppressions are linked'.

The core argument of this book is that the environmental movement fails to apprehend the full dimensions of ecological crisis if it does not approach climate, biodiversity and indeed all life on earth as interwoven with human social life, and specifically with the systems of extraction, exploitation and oppression wound up in colonialism and racial capitalism. It presents the case for a critical approach to environmental discourse that would meet the requirements of what feminist philosopher Nancy Tuana calls 'ecointersectional' analysis.[4] Race, gender, class and disability are not 'additional' to ecological questions, but co-constitutive dynamics that can be traced from the causes of environmental devastation to how its effects are experienced. It ought to be noted upfront that this is by no means a novel argument to make to Indigenous land stewards, environmental justice activists, political ecologists or indeed many environmental humanists, who have, some for years and others for centuries, been drawing the world's attention to the impossibility of separating ecological harm from social injustice. The implications of these insights for critical discourse studies (CDS) have however, yet to be comprehensively thought through, and it is to this task that I apply myself in this book.

My aim in this introductory chapter is to show just how enmeshed in histories of oppression and subjugation the environmental crisis is, and to establish the need for an approach to discourse analysis that understands the profoundly political nature of ecological questions. I argue that the supposedly scientific, impartial framing of catastrophic climate change as 'anthropogenic' does not go nearly far enough in correctly identifying the roots of the total, systemic crisis that humanity faces. Not only does one-dimensional critique place obstacles in the way of understanding the causes of our predicament, but it also frustrates the identification of preventive or mitigatory actions that do not make things worse. Instead of single-issue ecological analysis, we need a broader and more systemic view able to account for the persistent interpenetration of environmental degradation and human social hierarchies. This requires a focus on their shared origin stories. Understanding the historical forces that shaped the systems whose effects are all around us improves both our theorization of how and why they work, and a socio-ecological praxis that seeks to change them. This recognition of the past in the present is a key element of ecointersectionality. I argue below that discourse analysis informed by ecointersectionality and a consciously *political* ecology is an especially crucial project in the current moment.

To return for a moment to Steve Rayner eating the elephant: his version of 'one bite at a time' was not so bad, actually. He did not propose that we first solve problem A before chewing on problem B, in serial fashion. His vision was of communities taking charge of their ecological futures based on contextual knowledge that might yield solutions that made sense from their locations, thus enabling the world to nibble on the elephant from all sides in parallel. Though he does not mention history, gender or race (or even name the 'social') he was clearly onto something. To understand the global enmeshment of social and ecological pasts in present catastrophes, an example from a local contact zone where these systems meet is helpful. We turn now to such a zone, an island 'paradise' reduced to ashes.

## A world on fire

In 2023, the world watched in horror as Hawai'i burned. The fire started early in the morning of 8 August when strong Kaua'ula winds[5] blew power lines onto a three-acre plot of tinder-dry grass near the town of Lāhainā on Maui, the archipelago's second-largest island. Local fire crews initially thought they'd contained it, but strong gusts of wind rekindled the flames. On the back of the hot, dry air, a 'tsunami of smoke'[6] engulfed the land. Within hours, thousands of acres were burning, destroying homes and livelihoods, completely razing the town. Over one hundred people lost their lives, making the wildfire the United States' deadliest since 1918.[7] It would not be fully contained until late September, fifty-one days after it started.[8]

On the heels of catastrophic blazes from Greece to Algeria in 2023, and Canada's worst wildfire season in recorded history, Maui's disaster emblematized a growing sense that 'the world is on fire'.[9] The scientific journal *Nature* reported that a total of 384 million hectares of 'wild' land had been consumed by flames that year, underscoring the 'impact of climate change on extreme wildfire potential and [highlighting] the vast scale and effect of wildfires on society'.[10] The origins of the Maui fire in anthropogenic climate change were likewise affirmed by reputable news outlets, which quoted atmospheric scientists, meteorologists and ecologists as authorities.[11] By mid-August, a fairly consistent picture of the causes of the disaster had come into focus in the media: it was the result of complex interacting factors, and had been a long time coming. *The New York Times*, for example, reported that the 'story of last week's blaze arguably began decades ago, when Hawaii started experiencing a long-term decline in average rainfall'.[12] Climate change increases the likelihood of 'flash droughts'[13] according to a widely syndicated story from the Associated Press. The La Niña weather pattern meant that the 'clouds over Hawaii are thinner'.[14] An acute drought combined with 'land-use change [...] set the scene' for disaster. Because sugar cane and pineapple farms had been abandoned in the 1990s, and because less land was irrigated, 'dry and invasive grasses spread'.[15] When these grasses burn, they take indigenous forests out with them. The grasses have 'adapted to thrive with fire'[16] while the native vegetation has not, and so the grasses take up more and more space such that '26% of our state is now invaded'.[17] The drought had desiccated the grass, but a third ingredient was needed: wind. Hurricane Dora was 1,000 kilometres away from Maui, but the low-pressure system it created to the south combined with 'an anomalous area and oddly dry air to the north'[18] to create 'unusually strong trade winds'.[19] These winds were the bellows of the inferno.

Maui was thus presented as 'a textbook case' of how a changing climate produces 'multiple, interactive challenges that really lead to a disaster'.[20] As a professor at the University of Hawai'i, Mānoa, told *The Guardian*, 'Yes, the grasses were a disaster waiting to happen, but ... the silver bullet is reducing GHG emissions'.[21] As 'the planet heats up, no place is protected from disasters' stated another.[22] Similarly, what 'these ... catastrophic wildfire disasters are revealing is that nowhere is immune to the issue'.[23] The events in Hawai'i were presented as one instance of a global phenomenon that included the drying out of the Amazon, a deadly 2017 wildfire in Portugal and another in Tennessee. Experts sounded the alarm: the risk of catastrophic fires will probably increase with the frequency of flash droughts, and some tropical rainforests are approaching 'a point of no return when the humid ecosystem would forever change'.[24] *The Economist*, a centre-right publication that has the ear of policymakers, warned that 'even cities that have survived for centuries should not feel safe'.[25]

As it stands, this way of telling the story draws attention to three important factors: the systemic and global nature of anthropogenic climate change, its devastating effects and the threat of imminent and cascading disaster unless immediate action is taken. While these claims might seem like a clarion call to humanity to wake up and do something, they fall short for the simple reason that they leave out important parts of the story. As we shall see, the narrative presented by scientists in reliable and respectable publications is selective, incomplete and deeply ideological. It erases the part of the picture we most need to see, obstructing from view the elements that might help humanity survive the coming disasters. To see why this is the case, it is useful to listen to people who know Maui as more than just a local example of a global problem.

## The fire as 'exclamation point'

The voices of Kānaka Maoli (Native Hawai'ians) were present in mainstream publications in August 2023, though mostly as distraught community members, or in the opinion pages, and only rarely with the imprimatur of 'expert' or 'scientist'. In the 'Ideas' section of *TIME* magazine, Kaniela Ing, national director of the Green New Deal Network and Kānaka Maoli from Maui, wrote that the fire resulted from a 'dangerous dance between climate change and centuries of colonial greed'. For Keala Kelly, a filmmaker and Hawai'ian sovereignty activist, 'Lahaina is not America. But that fire is American. And it was really lit as far back as 1893'.[26] According to Ing, in the long history of injustice in the plantation and tourism economies, the 'fire is an exclamation point'.[27]

What these narratives tend to have in common is that they trace the origins of the crisis not to the demise of the sugar plantations in the 1990s, but to their origins in the 1800s. The first sugar planter, on Kauai in 1835, was a Bostonian named William Hooper.[28] By the 1890s the economic interests of the US sugar magnates were at odds with the Hawai'ian monarchy, and so in 1893 they fomented the coup that deposed Queen Liliuokalani, leading to the annexation by the United States of the islands in 1898.[29] This was a key event in the 'long and ongoing history of settler colonial resource theft and trickery'[30] based on what many Hawai'ians consider an illegal occupation, as the law which annexed the islands was made in Washington, DC.[31] Pioneer Mills, a US sugar company, dominated the area around Lāhainā, which had previously been known as the 'Venice of the Pacific' due to its extensive waterways. It was Pioneer Mills plantations that 'sucked these water resources dry'[32] according to University of Hawai'i Mānoa law professor Kapua'ala Sproat. The Kānaka Maoli political scientist Uahikea Maile argues that the local ecosystem was 'strategically and purposefully altered to feed colonial forms of profiteering and wealth accumulation and greed'.[33] A particularly meaningful resource was the large pond, Mokuhinia, which surrounded Moku'ula, the seat of the royal family when Lāhainā was capital of Hawai'i between 1820 and 1845.[34] The pond was drained by the plantations, who filled it in with dirt in the early twentieth century. Later, they built a baseball field and parking lot on top of it.[35]

While it was US-owned plantations that first transformed the 'once-wetland haven [...] into a vulnerable tinderbox',[36] tourism exacerbated the situation. Hawai'i had become the fiftieth US state in 1959; during the 1960s, annual visitors to the islands increased from 250,000 to 1,7 million, and by 2022 the number was over 9 million.[37] When globalization made sugar less profitable in the 1990s, property developers such as the West Maui Land Co (WML) took over huge tracts of land and irrigation systems from companies such as Pioneer Mills.[38] Migrants and holidaymakers from the continental United States bought themselves luxury homes, and hotels mushroomed across the island. The former royal capital became known as 'a resort paradise with immaculate surf breaks and snorkeling sites'.[39] Tourism, then, was the new sugar. With the influx of new settlers, Kānaka Maoli were reduced to around 10 per cent of the population.[40]

It is against this historical context that we can read what happened next. In August 2023, while most people watched the conflagration in horror, some saw an opportunity. Less than a week after the fire started, Lāhainā locals reported they'd been approached by realtors interested in buying their land.[41] Because many people inherited their family homes, they'd never had mortgages and were not required to have homeowners' insurance. The fire had destroyed some of the only low-income housing on the island, and these families would need money to rebuild. With the median cost of a home on Maui at over $1m, new construction was bound to be out-of-reach for most people.[42] As developers circled desperate families, activists petitioned Democratic governor Josh Green to place a moratorium on land sales.[43] While he at first said that he would consider it, his administration eventually baulked, saying it was a 'tricky legal question'[44] that would not be compatible with owner's or buyer's rights under US law.

Land was not the only resource that was vulnerable. In June 2022, West Maui had been designated a surface and groundwater management zone, limiting what property developers could do with the water flowing through the irrigation systems, which were owned by a WML subsidiary.[45] This 'designation' had been the outcome of years of community activism to prevent holidaymakers and wealthy landowners from being prioritized in water allocations, leaving streams and taro paddies dry.[46] With the fire still raging, the WML successfully pressured the governor to suspend the state water code on the basis that people were supposedly 'fighting against the release of water to fight fires'.[47] The avowed goal of the WML was to do away with designation.[48] The Kānaka Maoli deputy-director of the state water commission, Kaleo Manuel, was made a scapegoat for this supposed lack of water, and redeployed.[49] The Hawai'ian supreme court, in their April 2024 ruling on a related case, found that the state had 'mustered nothing, even scantly, to support its instigative claims that "there was not enough permitted water to battle the wildfires this morning"'.[50] Claiming that fair public water allocation prevented firefighting turned out to be nothing more than a ploy to restore control of water to private actors, giving large landowners the right to work alone with the government 'behind closed doors'.[51]

Sproat terms the strategy that played out on Maui, where profiteers take advantage of tragedy to privatize public goods, 'plantation disaster capitalism'.[52] In Naomi Klein's definition, disaster capitalism 'uses the desperation and fear created by catastrophe to engage in radical social and economic engineering'.[53] Klein and Sproat point to attempts to do away with public schools and housing projects in New Orleans after Hurricane Katrina in 2005, to privatize the electricity grid in Puerto Rico during and after Hurricane Maria in 2017, and to seize land from small-scale fishers and farmers in Thailand and Sri Lanka after the 2004 tsunami.[54] The burnt-out ruins of West Maui become an irresistible blank slate just waiting to be filled with 'the most perfect, beautiful plans'[55] of developers – beachfront McMansions, hillside developments and perfectly green golf courses.[56] Governor Green meanwhile was quick to go on a campaign to convince tourists to come back to Maui, to revive the economy and reduce unemployment, 'helping our people heal'.[57] The activist position was quite the opposite. Kaniela Ing, among many others, called for an indefinite halt to tourism while locals take 'time to grieve and heal' saying that they 'want to organize the "Rebuild Maui" campaign before disaster capitalists do'.[58]

## Competing narratives?

It would be reductive to characterize this latter account as *the* Kānakā Maoli version. Many Native Hawai'ians are invested in and profit from the tourist economy, and whether or not to revive the economy as it was remains a divisive issue.[59] The narrative presented above does, however, contrast markedly with the story about how the Lāhainā fire can be attributed to climate change alone. It maps onto temporal and spatial scales not mentioned by scientists, apportions agency differently and analyses systems in a distinctive way. To use for a moment a concept from media studies: the latter narrative 'frames'[60] the fire as a story about US imperialism and colonial greed, while the former frames it as the result of objectively measurable and impersonal ecological and climate changes. In the climate crisis frame, anthropogenic greenhouse gas emissions trap heat in the atmosphere, changing climate patterns such that Maui's rainfall has been on the decline since the 1990s, while making faraway hurricanes unusually strong, and at the same time as invasive grasses have taken over from irrigated sugar and pineapple plantations. These are the 'multiple, interactive challenges that really lead to a disaster'.[61] The colonial frame shifts attention from the vagaries of 'anthropogenic climate change' to specific people: colonizing agriculturalists, capitalists and property developers who 'sucked these water resources dry'.[62] We learn that the reason invasive grasses were able to cover 26 per cent of the island is the *start* of the plantation system in the nineteenth century, not its *end* in the twentieth. We learn that historical and current encroachment on land and water rights is what set the scene for the incineration of Lāhainā. And we understand that the return of tourists might not be 'healing' at all – that they are part of the larger problem in the first place.

It is important to note that the colonial frame does not question the reality of climate change. None of the people cited above challenged the proximate causes of the Maui wildfires identified by meteorologists or other climate scientists. They just see the bigger picture: Ing's 'dangerous dance between climate change and colonialism'.[63] Similarly, though we cannot be sure, the scientists cited in the first set of articles may well have affirmed elements of the colonial frame had they been asked. The point, however, is that they were not asked. From an editorial standpoint, the media's duty is to report how scientists reason through the quantifiable biogeophysical causes of disaster, while activists and Indigenous people are useful for context; if they want to make specific claims about causation, these are presented as opinion.[64]

To be clear, presenting these contrasting frames as equally valid perspectives on the same event is both an ecological cop-out and a moral failure. The colonial frame not only has the more sophisticated analysis of the 'multiple, interactive challenges' that caused the fire, but it also holds specific groups of people responsible and proposes solutions that undermine business-as-usual. Rather than vaguely calling for 'reducing GHG emissions', the colonial frame militates for fundamental economic and political change. It is thus a far more threatening frame to the capitalist status quo than the climate frame, which seems to make the crisis the fault of everybody in general and thus nobody in particular. The shift in temporal scene from the 1800s to the 1990s is perhaps the most egregious erasure in the climate frame, as it posits as the cause of the catastrophe the *end* of a mode of production whose *beginning* in fact set the scene for disaster. It does nothing to critique the relentless commodification of Hawai'ian land and resources, the reduction of its Indigenous caretakers to a racialized minority[65] whose presence bolsters the indigeneity claims of white settlers,[66] nor the system of racial capitalism which funnels material wealth to the beneficiaries of colonialism. In thinking of the climate as somehow separable from capitalism and colonialism, the climate frame thus obscures both the causes of and the solutions to the global environmental crisis.

## Converging calamities

How then are we to account for the enmeshment of these world systems, and move towards analysis that enables a better response to interlocking crises? If colonialism and racism are implicated in ecological breakdown, does this mean that we have to solve all of these problems at once, or admit defeat? As political theorist Nancy Fraser observes, we cannot underplay the scale of the crisis:

> What we face, thanks to decades of financialization, is not 'only' a crisis of rampaging inequality and low-waged precarious work; nor 'merely' one of care or social reproduction; nor 'just' a crisis of migration and racialized violence. Neither is it 'simply' an ecological crisis in which a heating planet disgorges lethal plagues, nor 'only' a political crisis featuring hollowed-out infrastructure, ramped-up militarism, and a proliferation of strongmen. Oh no, it's something worse: a general crisis of the entire societal order in which all those calamities converge, exacerbating one another and threatening to swallow us whole.[67]

It is thus not we who are eating the elephant, but the elephant that is eating us. In a time of mounting emergencies, the sheer weight of all of these problems may seem overwhelming. As complex and numerous as these overlapping crises appear, however, analysis of their linkages reveals shared roots in specific historical processes and relations of power. In understanding the interconnectedness of coloniality, heterosexism, racism, ableism, class domination and ecological devastation we may thus move towards an aetiological analysis of their shared origins in colonial racial capitalism.

As it becomes less feasible to deny that the social order is in crisis, explaining exactly what has gone wrong is an important critical project. In the case of Lāhainā, it was clear that crisis spelled opportunity for property speculators swooping in to buy up cheap land from desperate families. After the fire, as community leader Keʻeaumoku Kapu put it, 'our island is now turned into a cheaper commodity'.[68] In the face of material disruptions and disasters, discursive struggles seek to fix their meaning in ways that renew the licence for hegemonic power. As a number of scholars have pointed out, capitalist 'catastrophism' can exacerbate inequality and make alternative social visions harder to achieve.[69] The urgency of a particular problem is constructed as presenting stark choices, where questions of justice must 'temporarily' be set aside as some semblance of order is restored.[70] Describing a society as in an 'emergency' can work subtly to protect the institutional arrangements that produced the problem in the first place.[71] Disasters, as we have seen, open up paths for the reproduction of colonial exploitation and appropriation.[72] Though the colonial frame clarifies origins, specific catastrophes are always complex outcomes of specific dynamics that will shift from place to place. Reading these complexes through a colonial frame revokes the free pass given to capitalist extraction and exploitation to present itself as the solution to problems it caused in the first place.

The twentieth-century environmental movement has been characterized by many people as breaking decisively with the voracity of the commodification and exploitation of the earth. However, as I discuss in the next two sections, a distinction must be made between the kind of environmentalism that goes to the roots of converging calamities, and that which addresses only particular fruits, such as the extinction of a charismatic animal, or the pollution of a favourite hiking spot. The former category is associated with anti-colonial struggle and environmental justice, often associated with 'the environmentalism of the poor'.[73] The latter category covers environmentalism 'of the rich' which – unfortunately – characterizes much of what is thought of as mainstream environmentalism. In my discussion below, I distinguish the type of environmentalism that seeks to universalize a particular ecological subject while obscuring its race and class interests (what I will call 'imperial environmentalism') from subaltern decolonial and anti-racist environmentalisms concerned with root causes (which I shall call 'radical environmentalism').

## Imperial environmentalism

In his 1995 book *Green Imperialism*, Richard Grove traces the origins of Western environmental discourse to European anxiety about the productivity of their new colonial possessions. Arguing against the notion that the environmental movement is part of a 'new and revolutionary programme'[74] he shows how in British, French and Dutch colonies ecological thinking proved useful to colonial administrators in 'ensuring sustainable timber and water supplies and in using the structures of forest protection to control their unruly marginal subjects'.[75] Scientific and religious discourses constructed knowledge of Nature as knowledge of God, and administrators took their impact on ecosystems quite seriously.[76] 'Sustainable development' has thus arguably been around since at least the eighteenth century. Edenic, Romantic and 'physiocratic'[77] thinking further infused the scientific and religious roots of environmentalism. Argyrou Vassos shows how Western environmental thought developed as an elite cultural project working to reproduce a justificatory epistemology for existing power structures.[78] Instead of representing a break with modernity, mainstream environmentalism worked to reproduce racist and colonial notions of who can truly be human, in that European 'man' (whether as scientist, politician or poet) is positioned at the centre of discourse, uniquely able to see the totality of the entangled social and natural worlds, while lesser beings are defined by their failure to appreciate sublime ecological harmony, a move to 'innocence' that works to cleanse the imperial conscience of complicity in oppression and ecological harm.[79]

In the North American context, the preservation of wilderness was intimately involved with anxieties about settler masculinity. Nineteenth-century environmentalists drew connections between the health of the environment and the vigour of white American men. As Miles Powell has shown, the construction of 'frontier sportsmen' as masculine icons was rooted in mainstream scientific concerns about the debilitating effects of urbanization and industrial modernity.[80] The establishment of wilderness areas in the United States was explicitly justified by key advocates as a racial project that would return a pioneer spirit to old-stock white men, helping them return to potency and vitality through exposure to a preserved 'wild frontier'. Nineteenth- and early-twentieth-century preservationism was energized by fears of being overwhelmed by racial others while connecting the ability of white settlers to stave off this threat with their having a healthy connection to a healthy land. The history of US environmentalism is thus enmeshed with 'scientific racism, eugenics, immigration restriction, and population control'.[81]

Dorceta Taylor proposes a periodization of the (white, middle-class) environmental movement in the United States that marks a shift from a 'Romantic' to a 'New' environmental paradigm in the 1960s (in the wake of Rachel Carson's *Silent Spring*[82]) and a 'Reform' paradigm from the 1980s which saw environmental organizations grow 'increasingly big, bureaucratized, hierarchical and distant from local concerns and politics'.[83] This institutionalization of the environmental movement[84] by the 1990s established a more-or-less mainstream green politics that was (unsurprisingly) predominantly middle class and white. A similar trend was noticeable throughout the postcolonial world. In the years leading up to the transition to democracy in South Africa, for example, prominent liberation leader Gomolemo Mokae remarked that the 'green movement agitates on behalf of flora, of fauna, of the entire animal kingdom, with but one exception: black mankind'.[85]

Far from representing a radical break with the patriarchal, white supremacist and colonial-capitalist order, then, mainstream environmentalist discourse originating in imperial centres worked historically to shore up and obscure the workings of existing gendered, raced and class interests. White domination and Eurocentrism are furthermore no less of an issue in contemporary climate crisis discourse. In interviews with fourteen climate policy experts based in the Global North, the industry insider Chris Shaw identifies 'discursive guardrails' which anchor dominant representations of the climate crisis to established Western liberal institutions.[86] Their discourse renovates individual choice as an organizing value, constructs Western science as above ideology and national interests, and disconnects climate change from its historical causes. These discursive moves depoliticize the climate emergency, affirming the idea that capitalism and Western supremacy are solutions to problems they caused in the first place. As Shaw writes, new 'stories' about sustainable lifestyles 'will show how the whole of humanity can align with the norms of western liberal democracies, and live clean, smart, prosperous and fulfilling lifestyles without dangerous emissions of greenhouse gases'.[87]

There is also plenty of blatant collusion between what counts as environmentalism and the guiltiest industrial players. In the halls of power, conservation NGOs share lobbyists with oil companies and talk of the 'unexpected benefits' of aligning their projects.[88] Leading universities take money from the petrochemical industry to open 'sustainability' research centres.[89] These efforts tend to be dismissed as 'greenwashing' (a term I discuss in Chapter 3) by other mainstream players. What comes in for less critique, however, at least in the imperial core, is the construction of an undifferentiated, dehistoricized humanity that shares ultimate responsibility for our ecological breakdown. A key principle of the mainstream environmental movement around the world has been insisting, against the protests of fossil fuel industry denials, that climate change is anthropogenic: *caused by humans*. Scientists have rallied behind the idea of naming our current geological age the 'Anthropocene' to signal the extent of our impact. But who is the '*anthropos*' who caused this crisis? Are we all equally to blame?

Scholars trying to get more specific have suggested other names. 'Capitalocene,'[90] for example, attributes the ecological crisis to the rise of capitalism.[91] Another term is 'Plantationocene' which, as Donna Haraway points out, usefully picks out how 'the slave plantation system was the model and motor for the carbon-greedy machine-based factory system'.[92] Nicholas Mirzoeff argues that because 'the anthropos in the Anthropocene turns out to be our old friend the (imperialist) white male, my mantra has become: it's not the Anthropocene, it's the white supremacy scene'.[93] The move to represent humanity as a unitary and universal subject, what Sylvia Wynter describes as 'the incorporation of all forms of human being into a single homogenized descriptive statement that is based on the figure of the West's liberal monohumanist *Man*'[94] lies at the heart of imperial environmentalism. It thus focuses in a very narrow fashion on 'fixing' human–nature relations without ever recognizing that the problem never was with all of humanity, but with the narrow and colonial conception of *Man* as *Homo economicus* in the imperial core. This failure of recognition is a failure to correctly understand the roots of the crisis in monohumanism, and a recommitment to the same patterns of thought and action that brought about our intersecting calamities.

## Radical environmentalism

My use of 'radical' here cleaves to the word's etymology: it has to do with *roots*. Radical environmentalists address the historical and material roots of the entwined processes of social and ecological devastation. Because these processes are connected with the attainment of wealth and prosperity for some and poverty, death and disease for others, radical environmentalism is intrinsically a struggle for justice. It encompasses calls for reparation, for the return of stolen land and for reckoning with historical and on-going genocides. It considers a restructuring of the world system away from white supremacist capitalist exploitation as the sine qua non of ecological struggle. Where mainstream environmentalism sees protecting 'nature' as the world's single most pressing issue, radical environmentalism follows Audre Lorde in understanding that there is 'no such thing as a single-issue struggle because we do not live single-issue lives'.[95]

Historical and ongoing Indigenous resistance to colonial processes of dispossession and extraction is the paradigm case of radical environmentalism. In contrast to mainstream environmentalism, radical environmentalism asserts that 'the current environmental crises which are named through the designation of the Anthropocene, can be viewed as a continuation of, rather than a break from, previous eras that begin with colonialism and extend through advanced capitalism'.[96] The reorientation of the timescale to take in the colonial-era devastation of lands and relations is aptly encapsulated in publications such as Nick Estes's *Our History Is the Future* and Kyle P. Whyte's 'Our Ancestors' Dystopia Now'. Another paradigm case of radical environmentalism is the environmental justice movement. Environmental justice critiques how exposure to environmental harm disproportionately affects racialized and subalternized people, while enjoyment of its environmental benefits moves upwards through racial and class hierarchies to disproportionately accrue to those at the apex. In a not dissimilar fashion to how race 'is the modality in which class is lived'[97] Laura Pulido has argued that environmental injustice is the context in which racial capitalism is lived.[98] While people around the world have been fighting for environmental justice in different ways for centuries, the emergence of a distinct movement that interpreted racial justice as aligned with environmental quality is typically dated to North America in the 1980s, when activists started to map out the unequal distribution of environmental harms and benefits along racial lines.[99] The idea that it is acceptable to expose poor people to environmental toxins goes deep within an insidious form of market fundamentalism that holds sway in the corridors of power. Rob Nixon opens *Slow Violence and the Environmentalism of the Poor* with an epigraph from erstwhile World Bank president Lawrence Summers, who observed that 'the economic logic behind dumping a load of toxic waste in the lowest-wage country is impeccable' and that 'countries in Africa are vastly under polluted'.[100] Multinational corporations in rich Western countries may thus happily send their pollution offshore, back along the routes scored by the transatlantic slave trade, and the comprador classes of developing nations will obligingly collude in suppressing labour, environmental and land rights to make room for the poisons of the West.[101]

The deadly effects of the rich world's economic development are furthermore experienced in the Global South not only through differential exposure, but through differential vulnerability. In Ruth Wilson Gilmore's classic definition, racism is the 'production and exploitation of group-differentiated vulnerability to premature death'.[102] Racialized and poor people around the world are bearing the brunt of climate change while being the least responsible for it.[103] The impact of environmental degradation is differentially experienced along gendered lines, with women shouldering a disproportionate share of social reproduction in the context of mounting risks.[104] Part of the project of environmental justice is to point to the contingent nature of these distributions. The risks and benefits of modernity are distributed unequally along social hierarchies because of historical domination and greed, but if seen at a global level there is no inherent principle of scarcity that necessitates that anybody should suffer. In terms of energy usage, for example, while the global amount of energy consumed would be enough to secure a decent standard of living for everybody on earth, the rich world consumes the lion's share.[105] Estimates suggest that the wealthiest 1 per cent of the world's population consume the same amount of energy as would 1.7-billion people at a decent standard of living.[106] And yet, as I will discuss in Chapter 3, poor nations are frequently upbraided by rich ones to accelerate their green transitions, or paid to reduce their carbon emissions even as the wealthy in the West buy themselves superyachts or go on trips to space.

The critical logics of Indigenous, anti-colonial and anti-racist environmentalisms are increasingly converging in theory and praxis. As the world heats up, differential exposure and vulnerability to the harm this will bring that cleave to historical patterns of domination have been theorized by Farhana Sultana as 'climate coloniality'.[107] In a similar vein, the geographer Max Liboiron shows how pollution that is allowed to disrupt ecosystems recapitulates the 'bad relations' of colonial 'entitlement to the Land to assimilate that pollution' – concluding that pollution itself 'is colonialism'.[108] Calls from a number of scholars[109] have been made for the need to 'decolonize' environmental justice, which can treat the non-human world merely as a resource to be shared equally, rather than as part of the more-than-human world in which we are all implicated and where we strive to be in good relation. My inclusion of both traditions in a broad-church 'radical' category is motivated by the potential for coalition between them. Environmental justice activism and scholarship have effectively and powerfully critiqued capitalism and racism as the causes of ecological harm, and the door is open for more robust engagements with coloniality and the problems of human/non-human dualisms.

To advance an anti-colonial, anti-racist and anti-capitalist environmental politics requires a reorientation, as we have seen above, to timescales, histories and origins. It also requires a reorientation to each other. Kyle Whyte argues that while settler environmentalists fret about *environmental* 'tipping points', Indigenous nations understand the dependence of ecological balance on *relational* tipping points, which have long since been exceeded. Observing that '[consent,] trust, accountability, and reciprocity are qualities of relationships that are critical for justice-oriented coordination across societal institutions on any urgent matter'[110] Whyte argues that the absence of these relations is what stands in the way of addressing the ecological crisis in a just fashion. One way to think through this principle of relationality is ecointersectionality, which I will now lay out.

## Ecointersectionality

'Intersectionality' has become a ubiquitous term applied to a wide variety of social problems, such that (as the intersectional feminist Patricia Hill Collins observes) 'its very popularity causes it to lose meaning'.[111] Intersectionality theorizes interlocking structures of social oppression. But in order to sustain its 'critical edge'[112] Collins argues that it ought not merely to be seen as an analytical strategy to understand these problems, but a 'critical praxis [that] sheds light on the doing of social justice work'.[113] The orientation of radical environmentalist activism and scholarship to the nexus of extraction, environmental despoliation, racism, sexism and colonialism, and how they are experienced and resisted by marginalized groups, would seem to make it the field a paradigm of intersectional praxis. So what value, if any, does prefixing intersectionality with 'eco-' have?

The decolonial political theorist María José Méndez argues that if intersectionality is to make sense of the enmeshment of social and ecological crises, it will need to address not only coloniality, patriarchy and capitalism but the Western modes of thinking about agency and human–nature relations that currently inform even critical scholarship. Nancy Tuana diagnoses this orientation as 'ecological indifference' – a sensibility that 'separates the "human" from the "nonhuman" and views the latter as essentially lacking in worth apart from what it might offer the former domain'.[114] When 'nature' is an inanimate resource separable from humanity and open to exploitation, the unequal distribution of its harms and benefits engenders what Tuana calls the 'racial climate'. This climate is set in motion through, for example, the colonial scorched earth and overhunting practices that resulted in the genocide of Indigenous peoples[115] and the appropriation of their lands. The ruination of ecosystems and the subjugation of certain groups make people and environment 'disposable' in the interests of the powerful few, whose domination is thereby sustained. Resisting this constructed disposability becomes a matter not merely of justice for humanity but of just relations throughout the more-than-human world, the bedrock of ecointersectional critique. Ecointersectionality, put simply, is intersectional praxis that rejects human–non-human dualism and the ecological indifference this engenders.

Because each ecointersection is constituted by myriad processes, Tuana argues that trying to extract an overarching principle or generalization across cases risks obscuring 'the complex histories and multiple vectors of power' involved.[116] Each location thus requires a 'genealogical sensibility' where origins may be traced not as the source of an inviolable constancy but as the emergence of an identifiable dynamic of struggle, discord and friction. Because the idea of the ecointersection resists the reduction to any oppression to a fundamental vector or category, she argues that 'it is crucial to cultivate our commitment to seeing complexities, to situating our intersectional analyses both spatially and temporally, as well as working to understand the many interactions among beliefs, values, institutions, comportments. It means accepting that our analyses will always be limited, that there are always more facets of the problem than any one analysis is able to comprehend'.[117]

To move to the project at the heart of this book, the complexity of ecointersectional discourse analysis – i.e. analysis both that is informed by a case-specific 'genealogical sensibility' and that grapples with analysis of interlocking social structures without lapsing into ecological indifference – may seem daunting at first. But it is brought within reach if we forge a methodological partnership between CDS and political ecology. To explain why this is the case, I outline the strengths of these distinct traditions below.

## Why CDS should engage with political ecology

Political ecology is a transdisciplinary approach to the study of the enmeshment of human social life in ecosystems at various scales. Sometimes classified as a subfield of human geography, it differs (in the political ecologist Paul Robbins's popular definition) from other approaches to the natural world in that it views 'ecological systems as power-laden rather than politically inert [and takes] an explicitly normative approach rather than one that claims the objectivity of disinterest'.[118] Borrowing its methods from both the natural and the social sciences, and occasionally the humanities, much political ecology departs from a critical realist philosophical perspective, which, according to the geographer Roderick Neumann, recognizes 'the materiality of nature while maintaining that our understanding of nature is discursively constructed'.[119] The field has advanced Marxist critique of the political economy of environmental dynamics, and was subsequently informed by post-structuralist social theory, as well as more recently by decolonial and intersectional approaches.[120] As such, a thoroughgoing concern with semiosis, mediation, ideology and representation is evident in the work of prominent political ecologists.[121]

What makes political ecology so valuable for CDS is its combination of a recognition of the importance of discourse with structural analysis of the spatial and historical-material dynamics of biodiversity conservation, climate change, the energy transition and other important environmental issues. An approach that is both historically aware and empirical (whether qualitative or quantitative) is indispensable in properly analysing how interlocking systems of oppression shape life across locations in space and time. Political ecology enables properly ecointersectional analysis, offering the analyst an account of how exposure and vulnerability to risk and harm are distributed along social hierarchies.

A more robust engagement with political ecology would furthermore respond to the challenge of Nancy Fraser's dismissive remark that some scholars are 'sophisticated practitioners of discourse analysis while remaining utterly innocent of the traditions of *Kapitalkritik*'.[122] With its grounding in Marxist political economy, political ecology makes good on the commitment of critical discourse analysis to conduct dialectical analysis of how the symbolic and the material co-construct one another.

A dialectical approach would also address a significant problem in CDS with the way that environmental discourses tend to be analysed. Too often, 'green' discourses are granted a free pass. Climate scientists or conservationists are given the benefit of doubt, as they seem to be against extraction, pollution and monoculture, and for nature, biodiversity and survival. And these are good things, surely? In one popular approach to the analysis of environmental texts, discourses that are assessed as advancing 'a philosophy of ecological harmony'[123] are not only exempted from critique but advocated for promotion. Without some engagement with the empirical world of political ecology, this kind of pollyannaism is a profoundly risky move.

## Why political ecology should engage with CDS

It takes a lot of discursive work to make theft, murder and exploitation seem like unremarkable and acceptable features of social life, leaving the world careening towards catastrophe with our apparent consent. The embodied materiality of human suffering must be hidden from view behind unobjectionable-sounding abstractions such as 'market forces', 'sustainable development' or the vagaries of 'climate change'. The people who benefit most from inequality must be seen to have earned their elevated status, and those who lose out constructed as lacking the ingenuity or perseverance to better themselves. While it is absolutely the case that people are subjected to coercive violence, power is exercised not only through violence (slow or fast) but through securing the acquiescence of those who might intervene. At its best, discourse analysis explains with precision and in detail how the status quo is maintained through the 'spontaneous consent' of the masses.[124]

If the mechanisms through which consent is manufactured were apparent to participants in discourse, pathways to resistance might open up. These must therefore be hidden from view if hegemonic orders are to be maintained. Textual analysis that deserves to be called 'critical' at all must succeed in unveiling how power relations are reproduced in ways that are not 'normally apparent'[125] to people going about their daily lives. Though this idea is not always popular with scholars who have only a passing acquaintance with discourse analytical techniques, close attention to signification, interaction, and lexical and grammatical structures is absolutely essential for explaining how power infuses social life. In Ben Rampton's account, they provide 'a microscope'[126] that makes good on the Foucauldian project of empirically accounting for how 'power reaches into the very grain of individuals, touches their bodies and inserts itself into their actions and attitudes, their discourses, learning processes and everyday lives'.[127] Discourse analysis by political ecologists can only be strengthened and nuanced by a more robust engagement with the concepts, tools and traditions of CDS.

## Plan of the book

The rest of this book will devote itself to the project of outlining an ecointersectional approach to discourse analysis that brings political ecology into conversation with CDS. I am by no means claiming that the model I outline here is the only workable critical approach to environmental discourse. The first chapter will thus assess other approaches in an attempt to build on their insights and conceptual frameworks. The fields of environmental communication and ecocriticism in particular have produced a wealth of important critical work. My approach is to situate ecointersectional discourse analysis within the 'broad church' of the environmental humanities, and to draw on its scholarship. I also review the track record of CDS in analysing environmental discourse in this chapter, and specifically the prominence of 'ecolinguistics' as a discourse analytical approach. While there is much about ecolinguistics that is admirable, I argue that it is ultimately an unsuitable basis for ecointersectional analysis, as it works to obscure the roots of ecological crisis.

In the second chapter, I present the key theoretical assumptions and the conceptual tools I propose as an approach to ecointersectional discourse analysis. I do not imagine that this is a complete or definitive 'toolkit', and it is certainly possible that other theoretical starting points might also prove productive. In line with the queer theoretical tradition, I act as a methodological 'scavenger',[128] seeking out methods appropriate to the analytical task regardless of their disciplinary home. I do however claim coherence in my ontological commitments, which I make within the constraints of post-Marxist and post-foundational discourse theory, a decision I motivate in detail. In this chapter, I also name and define the key concepts which will be pulled through in the analytical chapters: the notion of the **imaginary**, and detailed definitions of how **subjectivity**, **institutions**, **affect** and **chronotopes** (space-times) are theorized as structuring the imaginary. Other relevant tools from CDS that cut across these categories are also briefly presented here.

The four analytical chapters (3 to 6) are each made up of three movements:

I. An account of the dimensions of the imaginary that combines found textual 'moments' or vignettes in the discourse-scape with scholarly attempts to explain them. This movement aims for breadth and comprehensiveness, feeling for the edges of a particular discourse regime.
II. A focused summary of the elements of the imaginary: its guiding values, subject roles, chronotopes, institutions and affects, built on the scholarship and analysis presented in movement I.
III. Textual analysis of several cases, where ecointersectional discourse analysis is modelled using the elements of the imaginary established in movement II and the conceptual tools presented in Chapter 2.

I propose that there are four dominant imaginaries in contemporary environmental discourse that practitioners of CDS should work to deconstruct:

1. **Green capitalism** (Chapter 3). In the green capitalist imaginary, there is no contradiction between people, planet and profit. The mechanisms of the market have been certified as environmentally friendly. Conscientious consumers are now empowered to make ecologically sound purchasing and investment decisions, as is their moral responsibility. Scientists,

financiers, inventors, benevolent billionaires and inventors work together to deliver ecomodernity. Our hope is invested in sustainable development for all, which requires little sacrifice, only the ethos of living in harmony with the planet, and encouraging other consumers to adopt the behaviours that will deliver our shared 'future-eco-perfect'.

2. **Overpopulation** (Chapter 4). In this imaginary, humanity weighs heavily on the earth, threatening to exceed its carrying capacity. We are likened to a plague, and the more children we have the more responsible we are for disease, hunger and our own suffering. To mention this requires brave truth-tellers and people who are not afraid of being called names, and who assert that it is in fact 'Not Racist™' to state a scientific fact, or to point out the difference between the profligate, fecund 'jungle' and the manicured, abstemious rich-world 'garden'. What people in poor countries need most from the rich world is family planning, and our aid to them should thus focus on educating them about smaller families. This will reduce their suffering, as when disasters come, there will be fewer people around to die.
3. **Colonial conservation** (Chapter 5). In the colonial conservation imaginary, 'wild nature' has an unassailably positive value. Noble European conservationists in reverent communion with wilderness are saddened when they see their Edens spoiled by invading human populations. The locals are either with them or against them, and preservation means keeping these people out. When animals are killed without permission, the savagery of poachers is the target of a violent rage. The exertions of the protectors of enigmatic species and iconic landscapes may be rewarded with the enjoyment of luxury, and they dream of expanding the horizon even further, creating more conservation *Lebensraum* by extending the frontier for nature, using military, scientific, developmental, and moralistic registers to justify forced removals, extrajudicial killings and the fortification of protected areas.
4. **Econationalism** (Chapter 6). In this imaginary, an ethnically defined nation and its bordered territories are a natural whole threatened by invading Others whose lack of organic connection to the land underlies their failure to care for it. They are the sources of pollution, disease and social strife we see reflected in landscapes of devastation and decline. The national landscape and its constitutive ethnoscapes serve as incontrovertible evidence of the fact that this territory belongs to us and to us alone. Our acts of ecological patriotism prove rootedness and autochthony. In settler-colonial contexts in particular, Indigenous people

may be used as hollowed-out avatars in white settler identity work, or 'Ecological Indians' that sanctify the presence of settlers on the land. True ecological patriots believe in strong borders, which prevent pollution and disease from entering the national body.

The final chapter of the book articulates the case for 'critical hope'. Because green-techno-platform-financial capitalism has managed to establish itself as postmodernity's horizon of intelligibility, 'hopeful' tends to be a profoundly reactionary way to feel, especially when such hope is built on yearning for wealth, consumption and economic growth. Critical hope, on the other hand, is the practice of refusing to give in to despair in the face of overwhelming odds, instead recognizing the extensive evidence from the historical record of moments where the unthinkable has happened, and where what once seemed hegemonic was brought crashing down. This is the social vision of radical environmentalism, and one that is only attainable once imperial environmentalism has been superseded.

## Conclusion: A 'renewed sense of urgency'

From the outline above, it should be clear that I am focusing on 'problem' discourses. It is my contention that this is the appropriate role for CDS to play, and that it is also a crucial component of social change for the better. Critique is not pure negativity but the bridge to better praxis. Radical environmentalists have been proving this to the world for centuries. One does not have to read this book in order to conduct correct analysis of environmental catastrophes, as the residents of Lāhainā proved in the aftermath of the fire. As Naomi Klein and Kapua'ala Sproat observe, 'as soon as the real estate developers started circling, local residents began organizing to call out disaster profiteering'.[129] *The Guardian* reported that on Maui the 'Native Hawaiian struggle for sovereignty has gained a renewed sense of urgency'[130] while NBC News reported that people were working towards 'complete independence from U.S. interference'.[131]

At its best, discourse analysis can reveal the particular, socially located truths of communities that have been on the receiving end of colonial and racial capitalism. In bringing ecointersectionality into conversation with CDS, I have two main aims related to two audiences in mind. For scholars who grapple with materiality while recognizing the power of semiotic representation to motivate, legitimate, and perpetuate unjust and ecocidal structures, my aim is to provide a powerful but easy-to-use toolkit that enables detailed and critical textual analysis. For scholars who already work mainly with text, my aim is to provide state-of-the-art reviews of the dynamics that are shaping the ecological crisis, and the discursive structures providing cover for powerful interests. In the former category, I would include political ecologists, geographers and environmental sociologists, while in the latter, I would include critical discourse analysts, environmental communication scholars and environmental humanists more broadly. Much of what I do in this book is not my own invention, but an attempt to bring these two broadly construed groups of scholars into a more robust conversation with each other.

1

# Critique and discourses of the earth

In April 2023, *The New York Times* reported that, despite having come to power three years earlier on a platform that promised a green economic transition, the Biden administration had given the green light to new oil and gas projects in Alaska and elsewhere that would see the equivalent of over a billion barrels of oil extracted.[1] Replying to the account of another activist[2] who shared this news, @SaintPalermo tweeted: 'Not only are we not doing anything to slow the pace of catastrophic #ClimateDisaster – but we are racing to the edge of the cliff.'[3] In their tweet, @SaintPalermo included an image by the Austrian political cartoonist Marian Kamensky (Figure 1).

**Figure 1** Political cartoon by Marian Kamensky. Courtesy of the Artist.

It is hard to fault the factuality of @SaintPalermo's tweet and the image it contained. The sense that the world is rushing towards disaster is borne out in analysis of the progress that has been made in curbing carbon emissions. Between 1750 and 1992, it is estimated that human use of fossil fuels had put 854 gigatons of $CO_2$ into the atmosphere. The 1992 signing by 154 nations of the UN Framework Convention on Climate Change (UNFCCC) in Rio de Janeiro was supposed to be an inflection point, representing 'a significant step forward in meeting what is potentially the greatest global environmental challenge facing mankind'.[4] And yet by 2022, thirty years after Rio, mankind had burned their way through a pile of hydrocarbons so large that an *additional 919 gigatons* of $CO_2$ had entered earth's atmosphere. To put this plainly: after signing a landmark agreement acknowledging the graveness of a crisis hundreds of years in the making, the world's nations accelerated consumption of fossil fuels to the extent that 8 per cent *more carbon* was released into the atmosphere in thirty years than in the 240 years prior. 'Racing to the edge of the cliff' is thus no overstatement.

At the same time, the image @SaintPalermo used to illustrate their argument raises important questions about how environmental activists understand the planetary crisis and their role in fighting it. In the drawing, the activist is glued to the road, indexing a 'non-violent direct action' (NVDA) tactic associated with the youth-driven activist group Extinction Rebellion (XR) in the UK and *Letzte Generation* (Last Generation, LG) in Germany.[5] The activist cuts a brave figure, outnumbered and alone. They are centred in the image, the hero of a struggle against two superior forces: the weight of the world pressing down on their shoulders, and the police, an institution tasked with squashing dissent. The policemen are there to ensure that obstructions to business as usual, such as people glued to roads, are dealt with swiftly and forcefully. These dramatis personae are shown in the midst of a chain of events that presumably has one of two outcomes: either the activist stays glued in place and the world is saved, or they are removed and the planet plunges off the cliff.

What does this image propose, and what does it presume? I will start with the loneliness of the activist, which is polyvalent. Their being only one person affirms that principled individuals can take action in the face of overwhelming odds, in a similar fashion to the popular mantra, 'I am one but still I am one. I cannot do everything, but still I can do something.'[6] It powerfully centres human agency in challenging the status quo while valorizing resistance to state repression through the mythopoetic figure of the lone hero. Every semiotic choice has consequences, however, and one effect of the choice made here is to foreground individual action while pushing groups and coalitions outside of the frame.

Indexing road-gluing as NVDA does hint at a broader organization, but there are no comrades in sight. The protestor and the police have to be represented, furthermore, and these ones happen to have been represented as white men on a grassy clifftop reminiscent of a European or Pacific US coastline. Unavoidably, and almost certainly unintentionally, this image in isolation affirms the centrality of individual white activists in the West fighting alone to save the entire world.

Perhaps even more insidiously, a lone activist 'saving the world' by preventing it from going over the cliff reproduces the idea that the survival of planet earth depends upon humanity, rather than the other way round. The notion that people who care about the world carry it on their shoulders or – more typically – in their hands, is fairly common in environmental communication and education.[7] A Google Images search for the term 'environmentalism', for example, yields multiple instantiations of this trope. In the first image result, multi-coloured human hands hold up a globe. In the second, people's arms grow to hold leaves as if they were the trunks of a tree. In another, cupped hands hold soil from which a small plant grows. The sheer regularity of these images is a testament to a well-worn environmentalist cliché that we hold the world in our hands. What bears repeating is that it is humanity that depends on the world, and not the world that depends on humanity. This can be a tough sell to strains of environmentalism with their roots in European Romanticism. That 'Man's dominion / Has broken Nature's social union'[8] has been a preoccupation of Western cultural production since at least the eighteenth century. We see all around us the damage we do to the natural world and blame our voracious and wasteful culture. And yet this mode of thinking repeats two mistakes of the ostensible target of our critique: it draws a dividing line between 'Man' and 'Nature' and implicitly subjugates the latter to the former.[9] In short, talk of 'Man's dominion' has failed to humble Western societies in how they relate to the web of life which sustains us all. The fact is that, however bad things get, a diminished but protean 'Nature' will survive the current crisis; 'Man's' future is altogether less certain.[10]

My argument is not that everything about this image is harmful. As I have outlined, it makes an important truth about our frenzied descent into an orgy of fossil fuel consumption while critiquing the role of repressive state institutions in protecting business as usual. Arguably, references to the NVDA of XR or LG justify some of the semiotic choices. My argument is that critical textual analysis reveals the underlying assumptions and claims we make about our relation to the world, to each other, and to our own culpability for and complicity in systems of oppression when producing environmentalist texts. The transformation of environmental discourse (and thus of our thinking and our scope of action) is not possible without this revelatory work. The meaning we make as people is both constrained by prevailing discourses and social structures, and productive of them, in that every communicative act is an opportunity to combine the resources we have at our disposal in novel ways. Micro-interactions such as @ SaintPalermo's tweet reveal how individual human agents, within constraints, make sense of ecological, social and moral 'macro-realities' while also producing them. Communicative acts and these 'bigger pictures' are in a dialectical relationship, where patterns in our utterances coalesce into constellations of shared meanings, which in turn enable the production of future utterances. While the existence of these constellations, macro-realities or bigger pictures might seem self-evident, the empirical entry-points we have at our disposal are the countless tiny points of light in everyday text and talk of which they are composed.

The elements of the image, as I have discussed it, correspond, broadly, to the analytical seams of ecointersectional discourse analysis. The notion that the natural world is headed for a precipice with only activists standing in the way is one example of an **imaginary**. The imaginary theorizes the world, explaining how it works. It organizes relations between discursive elements, among which are the **subject roles** (such as policeman or protestor) which are given specific attributes, the **institutions** that are represented (law and order, activist organizations) in specific **chronotopes** (space-times, which place the planet on an incline between an up and a down, a before and an after). A fifth avenue for analysis is **affect**: the feeling of alarm at impending catastrophe, anger at the cops and a sense of being 'confounded' (as communicated in an emoji used in the original tweet) at the stupidity of humankind in accelerating the destruction of the earth. Before laying out these concepts in the next chapter, I will locate the approach I advance in this book in the context of other analytical approaches to environmental discourse.

## Approaches to analysing environmental discourse

The focus of this book on environmental discourse and not *environmentalist* discourse prioritizes examination of the construction of the ecological crisis by social actors over boundary disputes about what truly 'belongs' in the environmental movement. So much of what was 'environmentalist' in the past has been captured by corporate NGOs or the private sector and drained of any radical potential. Care for the environment has been commodified, circulating as an inert, unthreatening sign. I thus count any discourse organized around ecological values as 'environmental'. In John Dryzek's *The Politics of the Earth*, now in its fourth edition, these discourses can be carved up into four categories.[11] In recognition of planetary limits, and against Promethean notions of endless growth, the first type of discourse, 'limits and survival', accepts that urgent coordination and action among elite actors are required in the face of looming tragedy. Environmental problem-solving, the second type, includes discourses that emphasize the role of experts (the sub-category Dryzek names 'administrative rationalism'), laypeople ('democratic pragmatism') or markets ('economic rationalism') in coping with environmental problems. The third type, sustainability discourse, includes a sub-category that holds that global development can be environmentally benign, and another called 'ecological modernization' (or, as it is usually called, 'ecomodernism'), which values technological advancement and green industries. The final category 'green radicalism' includes a range of sometimes contradictory social visions including those that seek to alter human environmental consciousness ('green Romanticism') and those that advance a politics organized around ecology, such as environmental justice, eco-Marxism or garden variety Green Party liberalism, all of which he places in the sub-category 'green rationalism'.

Though Dryzek's taxonomy is a useful starting point, it is not without its problems. As already mentioned, Romanticism is only 'radical' in the sense that it served as a critique of Western industrialization, not in any materialist or systemic sense. The liberatory and redistributive social vision of 'degrowth' (discussed in Chapter 3) is grouped together with the ethnonational chauvinism of the Club of Rome, for example, as both accept the existence of planetary limits (or 'boundaries'). Colonialism is barely mentioned. And while far-right ecologism is name-checked, it does not fit clearly into any of the categories.[12] Dryzek's typology is furthermore Western and imperial in its orientation, mapping the poles of environmental discourse from Walden Pond to Rome. This bias is compounded by a tendency to trace the origins of environmental politics to the late twentieth century. Most crucially, Indigenous and decolonial resistance to colonization and imperialism fall outside of its purview. Plausibly, as discussed in the previous chapter, these movements could be grouped with 'environmental justice', though it is hard to see how historical resistance to imperial genocides and ecocides drawing on Indigenous epistemologies would be counted as either 'rationalist' or 'Romantic' – the two options on offer in the taxonomy.

The tendency to conflate 'discourses of the earth' with Western epistemologies and environmental politics is a usefully revealing one, however. It speaks to a specifically 'green' configuration of the coloniality of knowledge where planetary care is what imperial metropoles say it is. An orientation to the overlapping world systems of colonialism, capitalism and hierarchies of the 'human' enables a split view of environmental discourse. On one view, we see a formation laden with power that seeks to fully determine the meaning of the 'environmental'. On the other, we see what is suppressed, what is left out, and are moved to ask why. This critical project contrasts the environmentalism of the powerful to understandings of humanity's entanglements in the web of life that show up the violences and oppressions of empire while simultaneously advancing liberatory visions of nature–society relations. What is needed, therefore, is an approach to environmental discourse rooted in recognition of histories of domination and oppression both of other humans and the more-than-human world, and of how the two are inextricably entwined.

My project is broadly aligned with the environmental humanities, which as Ursula K. Heise notes pays close attention to language in order to critique the construction of the ecological crisis as 'basically techno-scientific' while drawing attention to 'questions of socioeconomic inequality, cultural difference, and divergent histories, values, and ethical frameworks'.[13] Much of this work is rooted in critical discursive and rhetorical traditions. Pathbreaking environmental humanists Jane Bennett and William Chaloupka argue that if 'one begins to see "rhetoric" where Western social theory had claimed "logic," the privileged form of nature discourse becomes interpretation, reading, disclosure of rhetoric-hiding-as-scientific-fact'.[14] They go on to argue that 'reading' environmental discourse from this perspective

> attacks essentialist and foundational claims, exposing the exclusions and impositions they engender. And it exempts no such claims, even if they ground an otherwise politically attractive agenda, such as ecology. This critical movement throws the idea of nature and its alleged lessons into doubt more thoroughly and relentlessly than any positivism or skepticism before it ever did.[15]

The 'relentlessness' of this critique and its radical potential is however not evenly in evidence across the environmental humanities, which have adopted a variety of approaches to the 'critical' informed by diverse disciplinary traditions as well as theoretical and political positions. It is worth exploring the diversity of this work briefly, as it prepares the way for the analytical approach I propose later in the next chapter.

## Building discursive typologies

The environmental sociologist John Hannigan distinguishes discourses based on the idioms and motifs they use in rhetorics of rectitude and rationality that support claims. Hannigan's typology has three parts, which are even more narrowly rooted in popular North American texts and movements of the twentieth century than Dryzek's. He proceeds chronologically from 'Arcadian' to 'Ecosystem' and thence to 'Justice' discourses.[16] What is progressive in Hannigan's tripartite typology is its grouping of survivalist, environmental problem-solving and most sustainability discourses into the 'ecosystem' category while splitting Dryzek's green radicalism into two major tendencies, one rooted in a Romantic nostalgia for lost Edens, and the other in an ethic addressed to injustice, recognizing the unevenness both in culpability for environmental degradation and in the experience of its effects.

Hannigan also usefully identifies three stages in the social construction of environmental problems. In the first, the problem must be 'assembled' – by which he means it must be named, delimited and backed up with evidence.[17] Though this is typically done by scientists, laypeople from affected communities may also be involved. In the second stage, the problem is 'presented' – i.e. communicated, typically through the mass media. Finally, it is politically 'contested', which may result in action or co-optation by powerful interests, or defeat in the face of countervailing claims. While all three of Hannigan's stages are open to humanistic inquiry, the second is a pivotal one for textual approaches as it is at the interface between the empirical and the political that much of the construction of environmental crisis is done. As I will outline below, four discourse analytical traditions may be mapped onto these stages. Environmental communication, ecocriticism, critical discourse studies (CDS) and ecolinguistics all have the potential to explain plays of power and domination in the construction of environmental problems and the contestation of their solutions.[18]

## Environmental communication

Because of its tight focus on the mediation of knowledge about nature, environmental communication constitutes a 'scholarly habitus' between the humanities and the sciences.[19] It is not only a subfield within communication studies, but, in the words of prominent environmental communications scholar Tema Milstein, 'a metafield that necessarily cuts across disciplines'.[20] Geographers, political ecologists and environmental scientists, as well as communications scholars are interested in deconstructing texts that work to persuade people about environmental problems or explain the natural world to them and these distinct methods of analysis enrich environmental communications scholarship in a mutually constructive fashion.[21] Practitioners thus use a variety of methods of textual analysis. Milstein, for example, has deployed ethnographic and linguistic analysis of common-sense metaphors to critique Western tourism's 'ocularcentrism' in constructing Nature as putting on a 'performance'.[22] While moments of awe when encountering the more-than-human world can provide opportunities for developing a posthuman connection to it, the performer metaphor 'reasserts cultural binaries of self-other, audience–performer, reifying boundaries as they momentarily disassemble'.[23] These metaphors serve as traces of the power of the Western gaze to construct Nature. One might add that 'ocularcentrism' is also a key element of 'white socio-spatial epistemology'.[24]

In a similar fashion to other subfields in communication studies, environmental communication has its eye both on the effectiveness of a specific set of professional practices and on how they relate to the broader socio-political arena. The Habermasian distinction between social research that improves performance and that which orients to human emancipation is salient here.[25] In a popular textbook, for example, students are told that environmental communication 'helps to prepare you to enter many professions in which communication is central to an entity's involvement in environmental affairs'.[26] This focus on effective and ethical communications practice is turbocharged by the context of impending environmental catastrophe. One of the subfield's founding figures, Robert Cox, thus advocates for positioning it as a 'crisis discipline' with a strong bias towards action and against inaction, as the risks of the latter are assessed as outweighing those of the former.[27] Much environmental communications research thus mirrors strategic communication in its focus on how to meet urgent communications objectives effectively and efficiently, an orientation which arguably shifts attention away from socio-political critique and researcher reflexivity and towards 'usefulness' in the world.[28]

As discussed in the introduction, invocations of crisis may provide cover for entrenched power interests, and the notion of the 'crisis discipline' has thus been subjected to a steady stream of objections.[29] Some scholars have accepted this framing while seeking to define crisis in more expansive terms, including advocacy for 'environmental justice climate change activism and research for social impact'.[30] They and others have aligned themselves explicitly with a critical project that draws attention to how environmental discourses 'reflect and reproduce a particular political economy of interests'.[31] There have also been calls to 'decolonize' environmental communication, though a recent review of this scholarship mentions studies that critique ecocidal colonial practices, but none that critique imperial environmentalism, or the sub-discipline itself.[32] That being said, discursive approaches to environmental communication rooted in at least some critique of colonialism, capitalism or hierarchies of the 'human' are plentiful, and critical environmental communication scholars have been navigating the space between effective action in the world and critical distance from it.[33] As I will describe later in this chapter, ecointersectional discourse analysis both builds on these existing projects and aims to facilitate future ones.

## Ecocriticism

A second rich seam of environmental textual analysis is ecocriticism, where socio-politically critical work sits cheek-by-jowl with predominantly aesthetic critique. The distinction between environmental communication and ecocriticism has been characterized as relating to method, with the former mining large datasets using social science methods and the latter focusing on the close reading of texts using various literary theories.[34] As discussed above, however, this does not reflect the variety of work within environmental communication. A more apposite distinction might be drawn between the kinds of texts each field tends to analyse, with ecocritics preferring 'literary' texts (whether fictional or not[35]) and communications scholars oriented towards more mundane and institutional ones. This is also an imprecise distinction, however, as evidenced by overlaps in texts, methods, and the theoretical and political commitments of the subfields, motivating some scholars to pursue a shared research agenda based on the 'positive humanitarian and ecological practices' at the heart of both.[36]

Ecocriticism is defined by the literary scholar Cheryl Lousley as a 'political mode of literary and cultural analysis … [which] aims to understand and intervene in the destruction and diminishment of living worlds'.[37] Not all ecocriticism is grounded in an explicit politicization of human social relations, and some strains of it are associated with Deep Ecology, Romanticism and the cultivation of a Thoreauvian 'deference to the object'.[38] Much ecocriticism is still white and Western, shying away from decolonial praxis in favour of the idealist tendencies of postcolonial theory.[39] These tendencies within the subfield have been robustly contested by anti-colonial, anti-racist, ecofeminist and queer-of-colour ecocritics, whose work I discuss below in laying out the elements of an ecointersectional approach to discourse analysis.[40]

## Ecological themes in critical discourse studies

While communications and literary theory inform environmental communication and ecocriticism, respectively, the third major tradition of environmental discourse analysis is grounded in critical linguistics. I will lay out more of this theory in the next chapter, but suffice it to say for now that critical approaches to discourse analysis tend to be oriented around social problems and, as articulated by two of the tradition's founding figures Ruth Wodak and Norman Fairclough, to declare openly 'the emancipatory interests that motivate' intervening 'on the side of dominated and oppressed groups and against dominating groups'.[41] CDS uses linguistic theory, but linguistics is only one of its disciplinary 'homes'. There is no CDS department; what there is instead is an approach to discourse analysis that is taken up broadly across the humanities and social sciences – including by environmental communications scholars, political ecologists and others. Critical discursive approaches can help to deconstruct text and talk at every stage of Hannigan's model, from the 'assembly' of the evidence for a problem through 'presentation' and into the realm of 'contestation'. Anaïs Augé, for example, has analysed the use of metaphors of climate change in large corpora of technical scientific papers, media texts, political speeches and social media chatter.[42] Her work shows that the use of metaphor in argumentation has specific social effects; for example, in dehumanising climate migrants as threatening waves, plagues or conversely as valuable commodities.[43] Corporate greenwashing has also been a common target of analysis.[44] Practitioners of multimodal critical discourse analysis (MCDA) have shown how the use of charts, infographics, bullet points, icons and tables in the United Nations Sustainable Development Goals (SDGs) create the illusion of harmony and clarity where there is actually confusion and struggle, working to obscure the neoliberal free-market ideology whose hegemony the SDGs work to affirm. At a different institutional scale, analysis of the iconography of environmental justice movements reveals schisms in the mobilization of racial and ethnic identity and in stances to political authority.[45]

While advocacy for a better and more just world is what puts the 'C' in 'CDS', the identity of exactly who it is whose 'emancipatory interests' guide analysis has come in for critique. CDS seems to privilege human social relations at the expense of ecological impact. From an ecointersectional vantage point, the divide between humans and their environments is left unchallenged. There certainly appears to be a thoroughgoing anthropocentrism in models of meaning-making that construct *Homo sapiens* as uniquely capable of sign-making, an idea that the nascent project of ecosemiotics seeks to address.[46] Some attempts to incorporate the ecological into CDS seek to prioritize both human liberation and ecological balance in a way that is human-oriented, though not human-centred.[47] Others have accepted that one depends on the other, as in the linguist Arran Stibbe's observation that 'freedom and democracy do not automatically lead to sustainable levels of consumption, and peace in a society that exceeds environmental limits will be short lived'.[48] The argument that CDS lacks the potential for analysing the construction of human-nature relations has spawned a distinct approach to critical discourse analysis (CDA) known as 'ecolinguistics'.[49] In the next section, I argue that while this approach sounds promising, the pitfalls and problems woven into its theoretical cloth make ecolinguistics an unsatisfactory starting point for ecointersectional discourse analysis.

## Ecolinguistics

At the heart of ecolinguistics lies an ecological spin on the idea, associated with linguists Benjamin Whorf and Edward Sapir, that the language used by a specific culture encodes and thereby constrains its customs and behaviours. The strong version of this thesis, linguistic determinism, holds that a language fully delimits what can be thought. The less rigid interpretation of this notion holds that our perception of the world is influenced by the linguistic forms we have at our disposal. This weaker claim is affirmed in the CDS dictum that the linguistic structures we use do not merely 'reflect or represent social entities and relations, they construct or "constitute" them'.[50] On either version, the disastrous and exploitative relationships between humanity and the planet are understood as having their roots completely or partly in patterns of language use that the ecolinguist is well-suited to deconstruct and change. In a key moment in the project's history, the British linguist Michael Halliday identified the 'lexicogrammatical' roots of the ecological crisis at the International Association of Applied Linguistics's 1990 meeting. He singled out the 'simple message' he saw everywhere in the media at the time, that 'growth is good. Many is better than few, more is better than less, big is better than small, grow is better than shrink'.[51] Arguing that growthism is 'deeply engrammatized'[52] within 'Average Standard European' (ASE) languages, Halliday encouraged linguists to innovate ecologically sound 'language planning'.[53] Ecolinguists have responded to Halliday's call by initiating a project they describe not as the analysis of ecological discourse, but as the *ecological* analysis of discourse.[54]

A key figure in the recent development of this approach is Arran Stibbe. Stibbe has critiqued the legacy of linguistic determinism, arguing that positing as an obstacle to ecological harmony the deep grammar of ASE languages ultimately excuses inaction. Any project trying to change these deep structures would clearly be unfeasible. He also argues against the idea that linguists should prescribe more eco-friendly vocabulary – such as referring to 'food industry captives' instead of 'livestock' – on the basis that these terms are easy to mock and extremely difficult to introduce in contexts where attempts at 'verbal hygiene'[55] are already political lightning rods. Instead, he argues for the centrality of *stories* in shifting social norms, citing the Nigerian novelist Ben Okri: 'Stories are the secret reservoir of values: change the stories that individuals or nations live by and you change the individuals and nations themselves.'[56] Stories for Stibbe are mental models of the way the world works that are shared at a cultural level; they can thus 'be questioned from an ecological perspective: do they encourage people to destroy or protect the ecosystems that life depends on?'.[57] Whether the answer to this question is yes or no determines if they ought to be resisted or reproduced. And because there are bound to be ideological disagreements between ecolinguists as to what counts as destructive or protective, he introduces deep ecologist Arne Naess's term 'ecosophy', or ecological philosophy, to name any set of normative principles aimed at harmony between humanity and nature. The analyst is encouraged to state their own ecosophy upfront, and then to compare this credo to that of the text they are analysing. If the 'story' identified in the text is discordant with their ecosophy, it is judged harmful and subjected to CDA. If, however, the text is aligned with their own beliefs, they are encouraged to choose a different route: positive discourse analysis (PDA). Where CDA is understood to do the useful work of critique and deconstruction with regard to harmful stories, PDA prepares the way for beneficial ones to be reproduced and promoted by scholars.

PDA, which developed around the turn of the millennium, arose from a dissatisfaction with CDA's supposed negativity. 'If discourse analysts are serious about wanting to use their work to enact social change,' wrote Jim Martin, PDA's originator, 'then they will have to broaden their coverage to include discourse [...] that inspires, encourages, heartens; discourse we like, that cheers us along'.[58] CDA and PDA are presented as complementary, in a 'yin/yang perspective [ ... where] deconstructive and constructive activity are both required'.[59] In the face of ecological crisis, we clearly need solutions as well as critiques; Stibbe avers that there is 'little point in exposing the problems with current ways of using language unless there are beneficial alternative forms of language available to move towards'.[60] Stibbe's ecolinguistics is thus conceived of as a two-pronged project: deconstructing and rejecting harmful stories that valorize unnecessary consumption and endless growth or that constitute 'shallow environmentalism', and describing and promoting ecologically sound ones, such as can be found in certain haiku, Native American sayings, or the 'imaginative naturalism' of Rachel Carson or William Wordsworth.[61]

As a subfield, ecolinguistics has carved out space for itself as an academic and activist project. It is organized as an international association which holds regular meetings and publishes its own journal.[62] A growing bibliography of ecolinguistic analysis speaks to the popularity of the approach; work has focused on everything from placenames[63] to large corpora of media and other institutional texts that thematize debates about the environment.[64] Part of its appeal is that it makes room for radically divergent ecosophies from the capitalist ecomodernism of Bjørn Lomborg on the right to the ecosocialism of Murray Bookchin on the left. Ecolinguists frequently assert the importance of their being part of a project that does not merely critique problematic texts but that guides them 'in the building of a better world'.[65] This solution-orientation, combined with the relatively jargon-free framing around 'stories we live by' (as opposed to 'discourses' – or my choice in this book, 'imaginaries') and the provision of a comprehensive and easy-to-use textual analysis toolkit, makes ecolinguistics a formidable and serious approach, with room to grow.[66]

It should already be clear from the discussion in the previous chapter, however, that ecolinguistics cannot offer a solid basis for an ecointersectional approach to CDS. First and foremost, it jettisons analysis of the complexity and historical specificity of discursive locations in favour of comparison with a fixed 'ecosophy'. While the 'critical' in CDS tends to be associated with the Frankfurt School's critique of capitalist modernity, ecolinguistics remains agnostic about the roots of ecological crisis, as evidenced by its embrace of a range of ecosophies from cornucopian ecomodernism to Deep Ecology and ecosocialism. This inclusivity may seem like a strength, making room in the tent for a diversity of opinions, but it obscures the historical roots of planetary catastrophe. Once an ecosophy is chosen, furthermore, it is insulated from critique and treated as a set of principles on which analysis rests unproblematically. It is an object to which the text must be compared, and no further theory-building or estrangement from one's own investments as a researcher is required. This, arguably, can lead to dogmatism instead of critique, as reflexivity is foreclosed.

When it comes to advocating for PDA in the case of discourses 'we like, that [cheer] us along', this may easily 'slip over into complicity in injustice or oppression' as Paul Chilton and Ruth Wodak observed in relation to Jim Martin's work.[67] At one level, this is because when we exempt discourses that give us a warm glow from critique, we make ourselves unwitting accomplices in reproducing the problematic suppositions and hegemonic scripts that even the cosiest-seeming texts may harbour. At another, if, as Stibbe claims, 'beneficial' discourses are more common in marginal and suppressed texts, what guarantee do we have that they will still have positive effects when they break out of the narrow confines of their present production and reception? As scholars of language and politics have observed for decades, discursive change is fundamentally about the de- and recontextualization of existing discursive fragments, the political character of which may shift profoundly depending on where, when and in what combination they are uttered.[68] Ostensibly 'progressive' discourses such as feminism or gay rights may be co-opted from the societal margins into the mainstream by societally dominant groups that advance exclusionary and oppressive social aims.[69] The work of 'promotion' is thus shielded from the work of critique at the expense of contextual analysis.

This problem is brought into even starker relief if we examine the sources of the 'beneficial' stories that are mentioned. Instead of citing political ecologists or contemporary Indigenous climate justice activists and intellectuals for inspiration, Stibbe turns to historical figures such as Chief Seattle and Luther Standing Bear, Japanese nature haikus, and Romantic poetry. Stories about individuals relating to raindrops or the seasons are welcome; analysis of injustice and oppression less so. Colonialism is not mentioned; racism and capitalism only in passing. Stibbe's orientation is primarily towards Deep Ecology, focused in a Romantic[70] vein on correcting an undifferentiated humanity's relationship to earth, rather than on addressing the complex and difficult questions surrounding our relations with each other. Instead of naming the systems and histories that have brought humanity to the brink of disaster, we are told to focus on positivity, and on solutions. It is this aversion to critique that ultimately makes ecolinguistics an unworkable basis for ecointersectional discourse analysis.

## The solution is critique

The political theorist Wendy Brown once pondered what caused otherwise 'progressive' scholars

> to reduce critique, which at its best functions as a means of strengthening our political hand by attempting to know better the cards we are dealt, to a wholesale rejection of those cards? Anxiety about critique, reduction of it to dismissal or

> mere negativity, is ubiquitous [...] It is as if we fear losing any object that we scrutinize too closely or whose ambivalent or corrugated character we expose to the light.[71]

This anxiety is clearly felt by ecolinguists who wish to shield treasured poetic or ecological texts from disintegration under too harsh an examination. But they have misunderstood critique, which is caricatured as 'pure negativity'[72] – the 'yin' to a positive 'yang' that distinguishes what we must resist from what we must promote. For how can we promote decontextualized discursive fragments without also critiquing how they play out in the world? If we do so, we miss the potential to explain how strategies of resistance work in practice, and how resistance produces novelty and discursive change.[73] If done well, as Fairclough argues, critique draws attention to 'the many ways in which the dominant logic and dynamic are tested, challenged and disrupted'.[74]

And yet a critical pose is persistently associated with a failure to be constructive and a perverse resistance to being part of the solution.[75] As one rebuttal of 'ultraleft' environmentalism put it, 'Blaming "Capitalism" is not an alternative to solving problems.'[76] Poking holes in other people's ecological utterances because of a perceived lack of ideological 'purity' might seem to feed into 'discourses of climate delay' such as the sense that it is already too late to do anything, or that change is impossible given the enormity of the problems we face.[77] This is the power of Robert Cox's invocation of the 'crisis discipline'. As shown in the opening vignette of this chapter, since 1992 the world has descended into a fatalistic, frenzied necropolitical orgy of extraction, combustion and jostling for position at the front of the trough, where a sense of disaster just over the horizon has not changed patterns of consumption but occasioned hedonism and hyper-accumulation instead alongside widespread societal mistrust. Irresolution has crept in where focus was needed, fragmentation in place of solidarity and despair has supplanted action. And so when everyone's a critic, critique is considered part of the problem. As critical environmental communication scholar Murdoch Stephens observes, 'when confronted with the urgency of climate change, a critical reading seemed to be at best a form of folly and at worst to be actively complicit with the worst of fossil fuel profiteers'.[78]

The question of complicity is an important one to which I return below. For now, the idea that as academics we should stop carping and just put our shoulders to the wheel is a by-product of a higher education model that has established itself as a key site for delivering the 'solutions' demanded by the hegemonic social group.[79] In the higher education sector in the UK, for example, definitions of research 'excellence' prioritize work that meets industrial needs, and that can clearly be applied to problem-solving. As Fairclough pointed out, over twenty years ago, the academy became the arm of the global economy tasked with building its workforce, where research is driven by business and government priorities and 'talk of the autonomy of universities is largely nostalgia'.[80] The applied linguists John O'Regan and John Gray show that the 'neoliberal education imaginary' – by no means limited to the British context – involves 'the deliberate neutralization of the university as a centre of independent knowledge creation and learning, and hence as a potential locus of intellectual opposition to the dominance of the neoliberal hegemony'.[81] The erosion of traditions of critique in favour of 'positive' – and often positivist – research in the humanities and social sciences accompanies the steady neoliberalization of higher education. This is a point worth emphasizing in the US context today, where the current administration's assault on higher education amplifies and extends the neoliberal aversion to critique to any science that threatens the fantasies of the oligarchs. Rather than disproving the solutionist hypothesis, these developments reveal the irreducible unity of solution and critique. When we accept that actual solutions to the ecological crisis will be built on unflinching critique of the very systems that now exert stricter ideological control over research agendas, we prefigure a truly liberatory science grounded in the histories and ontologies of the dispossessed.

Resistance gives us the opportunity to – as Brown put it in the quote above – 'strengthen our political hand'. Judith Butler argues that critique dissolves epistemological certainties that foreclose the 'possibility of thinking otherwise'[82] while also unveiling the relations of domination a text may work to obscure. It can be both the tactic of 'sabotage and disruption'[83] associated with Foucault and an 'emancipatory practice oriented to the definitive resolution of the social problem or ill'.[84] If critique *explains* social dynamics, it can also aid in finding 'hitherto unrealised possibilities for change in the way life is currently organised'.[85] As Butler shows in their appraisal of Foucault, through undermining received epistemic certainties critique enables a process of 'desubjugation and self-making'[86] that clears the path to subjectivities and imaginaries that stand a chance of addressing the problems we face.[87]

## Conclusion: Building on critical foundations

If we recognize the importance of critique, we can proceed to build an ecointersectional approach that does not take the goodness of 'green' discourses for granted. An approach to the world's problems rooted in critical social theory is something that CDS shares with political ecology, which arose from dissatisfaction with positivistic approaches to ecosystems that excluded human socio-economic life.[88] Nancy Peluso, for example, has over decades drawn attention to the violent gendered and racialized effects of discourses of extraction and conservation in Southeast Asia.[89] At the same time, political ecologists have tended to be chiefly concerned not with fine-grained textual analysis, but with detailed accounts of political economy, the materiality of accumulation and dispossession, subjection, and injury. The strength of this work is its consistent connection between the symbolic and the material. This project can only be strengthened by developing the toolkit of its analytical approach to text, even as the phenomena being analysed shift and evolve. The strong and steady neoliberalization of environmentalism, where 'Nature' is stripped of its relationality, reduced to that which is not human, and produced as just another commodity form for a global elite to consume as tourists, shoppers or social media prosumers,[90] is a phenomenon that will undoubtedly intensify as the consequences of climate breakdown and biodiversity loss become harder and harder to ignore.

Ecofeminists have launched important and intersectional critiques of elements of environmental discourse such as Deep Ecology: its whiteness, maleness, middle-classness and disconnection from the labour of social reproduction.[91] Ecofeminists read nature as a site of explicitly political contestation where hierarchies of the human are negotiated, for example, in its form as a wild 'Other' that must be tamed, or as a measure of morality, where that which is 'unnatural' is abhorrent and subject to disciplinary power.[92] This vein of analysis has also informed queer ecology, which moves from the conflation of natural/unnatural and straight/queer, 'a strongly evolutionary narrative that pits the perverse, the polluted and the degenerate against the fit, the healthy, and the natural'.[93] In both ecofeminist and queer ecological critique, the role of the wilderness as a place where (white) men prove their masculinity, whether in subduing, destroying or protecting it, or communing with it in the manner of a Romantic poet, is subjected to thoroughgoing scrutiny.[94] Gender, race and sexuality are intimately wound up in our understandings of the natural, which can serve a regulatory function, legitimating the 'encroaching politics of normalized subjectivities'.[95] This critique has drawn attention to how the anxieties of dominant groups to the coupling of racial or sexual Others – or 'queer acts against nature'[96] – structure the imaginary space of the nation as vulnerable to both waves of sexual dissolution and the ever-present threat of the climate migrant. Biological and evolutionary discourses within dominant social scripts produce hierarchies, often subtly resting on eugenic notions of reproductive fitness that construct queers, people with disabilities and people of colour as 'enemies of the species'.[97] Ecofeminists and queer ecologists have thus mounted substantive intersectional critique of heteronormativity, whiteness and class domination in environmental discourse, not least in the production of a racist, sexualized and gendered 'wilderness'.

Bringing the analytical strands of critical Indigenous studies, political ecology, ecofeminism, queer ecology, critical environmental communication, and CDS into a singular approach is bound to be a tendentious exercise, largely because there are so many robust traditions and conflicting theoretical precepts to choose from. In the next chapter, I thus propose a partial, and clearly partisan, approach towards this end.

2

# Ecointersectional discourse analysis

Taking ecointersectionality seriously means developing 'genealogical sensibilities' that help disentangle the historical exploitation of humans and environments.[1] An ecointersectional approach refuses a strict nature–culture division between social and ecological relations of domination and exploitation and is alert to their having shared origins. If we can see this clearly, then the idea that freedom, equality and peace on earth would ever emerge for *Homo sapiens* under the current mode of environmental devastation becomes unthinkable. The oppression of humans ruins ecosystems just as the ruination of ecosystems oppresses humans. Our analytical approach thus needs to take very seriously both the entwining of the social in the ecological and the historical in the present. In this chapter, I argue that a 'post-foundational' theory of discourse is well-suited to this task if supplemented by the fine-grained textual analysis tools of the CDS tradition. I then propose five key dimensions of analysis that enable ecointersectional critique. I will argue that understanding discourses as imaginaries (composed of subject roles, social practices sedimented into institutions, imagined connections between specific space-times and particular affective stances) facilitates a critical and genealogical reading of environmental discourse. Six general-purpose sets of textual analysis concepts are then outlined that cut across these dimensions. Before I lay out the specific approach I am advocating in this book, however, I lay out an argument for the necessity of such a move. Though CDS theorists are the intellectual lodestars of many of my proposals in this chapter, I argue that ecointersectionality demands that we move away from the rationalism, mono-epistemicism and textualism associated with this tradition.

## Decolonizing, queering and turning to affect

The CDS tradition is strongly associated with liberatory aims.[2] One of its central tenets is that systems of representation do not merely 'reflect or represent social entities and relations, they construct or "constitute" them'.[3] In the analytical approach developed by Norman Fairclough, the fine-grained lexico-grammatical *description* of texts is combined with their *interpretation* as produced by and productive of discourses, which in turn may be *explained* as enmeshed in broader socio-cultural processes.[4] Discourse and structure are understood as in a dialectical, co-constructive relationship where neither supervenes on the other.[5] As CDS has become a theoretically broad church, it calls for some familiarity with the specific social theory invoked, which is 'part and parcel'[6] of its analytical methodology. CDS is furthermore an openly political tradition that, rather than 'affecting a spurious neutrality',[7] seeks to align itself with oppressed groups and against dominating ones.[8]

The logical links between text, discourse and structure in CDS have however been critiqued as relying on an overly rationalist model of politics. As Felicitas Macgilchrist argues, the idea of a universal and objective rationality underpins the 'hidden violences of a secular, liberal humanism'[9] associated with the coloniality of knowledge. The 'critical' in critical discourse studies is associated with the Frankfurt School, which tended to trace a Western conception of moral and political progress as constitutive of modernity. The feminist philosopher Amy Allen argues that though later Frankfurt School figures had moved on from naïve conceptions of progress, a vestigial notion of a specific 'developmental, unidirectional, and cumulative moral-political learning process'[10] aligned with Enlightenment thinking still structures their work. The reason for this, as the decolonial sociologist Gurminder Bhambra argues, is not only that critical theorists failed to engage in any serious way with colonial history and its racial structures, but that the theoretical presuppositions that shaped their categories of critique were themselves expressions of an unacknowledged colonial modernity unable to think outside of its own conditions of possibility. While early Frankfurt School thinkers inherited from Marxism a tendency to reduce coloniality to a class relation, later figures such as Habermas and Honneth dispensed with the class dialectic altogether without 'consideration of the issues of coloniality that were missing, or displaced, in its earlier versions'.[11] The way out of this problem in Allen's account is the 'radical openness and open-endedness that make problematizing genealogy ideally suited for the kind of internal decolonization that critical theory sorely needs'.[12]

The need for a 'problematizing genealogy' has also been expressed by critical discourse analysts drawing attention to the limits of CDS. Yunana Ahmed thus identifies a 'void' at the heart of CDS: the 'coloniality of power has been entirely left out'.[13] In Sylvia Wynter's account, the coloniality of power names the overrepresentation of a particular conception of *Man* conceived in the West that comes to dominate definitions of the human, but also of human freedom, rationality, social harmony and progress.[14] In order to break with the assumed universality of these categories, Ahmed builds on Macgilchrist's useful tripartite model of how coloniality suffuses texts, including academic ones. Macgilchrist identifies (1) 'colonial residue' in texts that preserve denotations and connotations forged in the context of unequal power relations (such as civilized/uncivilized, advanced/backward, etc.); (2) the centring of a Western perspective (or 'eye') on the world (i.e. ocularcentrism); and (3) 'mono-epistemicism' which holds that 'only one way of being and knowing is valuable (i.e. modernity, development, progress)'.[15] A genealogical approach to CDS is an antidote to the coloniality of power if it can identify these colonial 'residues', challenging Western ocularcentrism and mono-epistemicism.

Another critique of CDS is its textualism. Though CDS has its roots in linguistic analysis, it has become increasingly multimodal. This shift has, however, not rid the tradition entirely of a dogmatism which, Crispin Thurlow argues, 'leaves us struggling to read between the lines, to understand the gaps and the traces, the unspoken and the unspeakable' and is 'inevitably caught up with reproducing an inherently flawed neatness around/about language'.[16] Taking inspiration from the 'trouble-making' tradition of queer theory, Thurlow argues for queering CDS such that it can take into account the 'embodied, affective, and altogether sensory' world of contemporary ideology.[17] This does not mean denying the importance of language, only recognizing its limits.[18] Experiences of embodiment, emotion, and even euphoria, though we try to make sense of them in discourse, are not always easily reducible to a specific code. What is called for in queering CDS is thus a more robust engagement with bodies and affect, and with scholarship as a kind of performance that apprehends the 'transitory, ephemeral, and intangible or invisible' instead of always preferring the 'enduring, the tangible, and the visible'. While there are many possible ways to respond to the critiques of Macgilchrist, Yunana, and Thurlow, in what follows I will propose that post-foundational discourse theory (PDT) offers a robust theorization of socio-political discourses that is (potentially) poly-epistemic, affective and queer.

## Why post-foundational discourse theory?

Much like ecointersectionality, PDT arose from a dissatisfaction with single-issue analysis.[19] Introducing their argument in *Hegemony and Socialist Strategy* in 1985, the political theorists Ernesto Laclau and Chantal Mouffe situate their challenge to class reductionist orthodoxy as motivated by 'the rise of the new feminism, the protest movements of ethnic, national and sexual minorities,

[and] the anti-institutional ecology struggles waged by marginalized layers of the population'.[20] They were by no means the first to gesture towards combining class, anti-racist, feminist and ecologist critique, and I am not arguing here that PDT as it has subsequently developed is the only or even the best way to think about these intersections. My more modest claim is that it is a *productive* way to approach ecointersectionality. This productivity can be attributed to four aspects of PDT. First, it is a poly-epistemic approach to discourse which refuses granting the analyst privileged access to the 'truth', which both undermines the coloniality of power/knowledge and enables a queer subversion of regimes of the 'normal'. Secondly, it has inherited from the Marxist tradition a compelling and subtle account of material-symbolic interactions that helps make sense of how the 'real' has discursive effects. Thirdly, it theorizes affect as fundamental to the emergence of the subject in relation to the 'logics' that structure a discursive regime, bringing bodies and feelings into discourse analysis as a way out of the rationalism that characterizes some strains of CDS. And finally, as it accounts for processes of the discursive constitution and dissolution of social orders, it explains how social change can happen while enabling genealogical critique.[21] I discuss each of these dynamics in more detail below.

## A poly-epistemic approach

By definition, post-foundationalism entails scepticism about claims to be able to find a stable, fixed, unmoving anchor for meaning, located in the realm of the 'real' and unmediated by discourse. While, as per the structuralist insight, meanings are produced through a web of differences between signs, post-foundationalism proceeds from the post-structuralist premise that there is no transcendental signifier – God, Reality, Nature, Reason – to give any part of the web a *necessary* as opposed to a *contingent* character. Post-foundationalism thus rejects the necessity while asserting the contingency of grand metanarratives such as capitalism, religion, science or any version of truth, which are understood instead as sites of 'permanent political contestation'.[22]

Whereas for strands of CDS such as CDA there are discourses whose claims to truth are non-ideological, for PDT this position is seen as 'naïve'[23] as there is nowhere to stand that is outside of ideology. What appears on the surface to be an accurate reflection of reality and to be stable over time may be shaken to the foundations by 'dislocations' (discussed below) that reveal the contingency of all representation. If stable, the subjects of a discourse make claims about what is true/false and good/evil in ways that are discursively consistent, and these judgements are thus relative to their discursive contexts.

As Macgilchrist argues, this does not mean that post-foundational analysts must reject the idea of reasoned judgement in weighing claims, only that they must recognize the contingency of their own ethical and epistemic commitments.[24] Because all subjects are embedded in their own discursive contexts, rationality is not an uncontested and universal concept but a 'particular, political, contingent'[25] one – and thus still useful. A parallel may be drawn to Sylvia Wynter's account of epistemic relativity, which names the 'truth of solidarity' as an alternative to (colonial) 'objective' Truth. Powerful and marginalized groups alike have their own 'truths of solidarity'. As the pragmatist philosopher Richard Rorty, who developed the concept, famously observes, 'there is no answer to the question "Why not be cruel?" – no noncircular theoretical backup for the belief that cruelty is horrible'.[26] The fact that we justify our cruelty or lack thereof in 'particular, political, contingent' ways does not mean that our reasoning is any less powerful or important. In fact, rather than some kind of moral or epistemological nihilism, it forces us to reflect on our own complicity in the reproduction of cruelty and falsity in ways that appealing to an extra-discursive necessity simply does not.[27]

## The material and the symbolic

PDT adapts from Marxism the insight that people's subjectivity 'coincides with their production, both with *what* they produce and *how* they produce'.[28] As distinct from more reductionist strains of materialism, however, the symbolic is understood as always already enmeshed with 'reality' with neither completely determining the other. To understand the significance of this, physical events provide a useful illustration. Laclau and Mouffe clarified that events such as 'the falling of a brick' or an 'earthquake' exist independently of human will, but that 'whether their specificity as objects is constructed in terms of "natural phenomena" or "expressions of the wrath of God"'[29] depends on how the discursive field is structured. In this way, their very constitution as objects has discursive conditions. Discourse itself is a material practice, meaning that it is composed not merely of ideas but 'embodied in institutions and apparatuses'.[30] The emergence of objects in discourse lends it its wholeness and systematicity, allowing judgements about what is 'objectively' there to be shared between people.

Put another way, our shared sense experience of material reality does not *on its own* reliably yield the objects required by human social organization. Discourses are thus attempts to 'fix' the meanings of phenomena, to make them make sense, to afford them the presumption of solidity and coherence required by social life.[31] The elementary discursive practice through which this attempt is made is termed 'articulation'. Articulation can be thought of as a metaphor of both mechanics and speech: to articulate is to utter, but to articulate is also to be joined together in a way that allows for a range of motion, in the same way that a trailer or a limb may be 'articulated'.[32] The identity of any element of a discourse could be otherwise if articulated differently. The practice of articulation brings discursive elements[33] into new relations 'such that their identity is modified as a result of the articulating process'.[34] Articulation '[pierces] the entire material density'[35] of the world in that it organizes the world into objects. These moments of apparent fixity are facilitated by discursive 'nodal points' which represent the central value(s) or organizing principle(s) of a given discursive formation that lend it its unique character.[36] A 'successful' nodal point is one that facilitates the uptake and thus reproduction of a particular discourse by a number of people. In order to do this effectively, nodal points tend to be more surface than substance, empty signifiers onto which people may project their aspirations.[37]

Practices of articulation follow logics of equivalence or difference. In PDT 'chains of equivalence'[38] connect the elements articulated as internal to a particular discourse, the 'surfaces' of which are determined by their difference from other elements. The articulations of a discourse and the success of its nodal points can, however, be thrown into disarray in at least two ways. At a symbolic level, ongoing struggle between social actors to fix meaning results in competing articulations differentially taken up by others. But the discursive field is also materially constrained. Changing material conditions, events of various magnitudes and other experiences may render discourse systems incommensurate – in other words, unable to yield their discursive objects persuasively or convincingly.[39] These dislocations give rise 'not only to negative consequences but also to new possibilities of historical action'.[40] The implication of this, as Laclau points out, is that dislocation is the 'source of freedom'.[41] Because our own agency is constrained by prevailing discursively produced subject positions and their related objects,[42] dislocation presents the opportunity for innovation, for re-articulation and thus for social change.[43] As a recent proposal for the development of PDT within CDS observes, this discursive instability 'serves as a condition of possibility for the evolution and transformation of practices, [and] also opens up a vantage point for rethinking ideological critique'.[44]

## Affect and subjectivity

When social actors participate in a discourse it is because they have in some way recognized themselves as its subjects. In PDT, this moment of recognition relies on the 'passionate affinity'[45] of the subject. As Yannis Stavrakakis explains in his discussion of the psychoanalytic premises of PDT, this success is not reducible to the apparent logical coherence (or 'closure') of a discourse.[46] While it is important that a discourse appears to explain social life to its subjects, affect enters into the picture in at least two ways: first in that the nodal point itself offers the promise of 'imaginarised *jouissance*'; and secondly, in that the process of *identification itself* has enjoyment as one of its 'symptoms'.[47] He thus argues that 'critique of an ideological system of meaning cannot be effective if it remains at a purely deconstructive level; it requires a mapping of the fantasies supporting this system and an encircling of its symptomatic function'.[48] The readiness of social subjects to accept a particular representation of social life as truthful is thus partially explained by their enjoyment of identifying with it, which works to 'cover over'[49] the constitutive lack of any unchanging, positive identity 'fixed' by a transcendent extra-discursive force.

Mapping fantasies, and 'encircling' the symptomatic function of a discourse, requires analysis of the discursive traces of affect. It is important to note here that this account does not pit reason against emotion but rather affirms the facilitative and indeed constitutive role of emotion in making reason possible. Fantasmatic 'logics' are thus interwoven with social and political ones.[50] In PDT 'rational' is not synonymous with 'scientific' nor 'emotional' with 'ideological'. As we have already seen, and in contrast to CDA, ideology in PDT is not theorized as false consciousness or as an obscuring veil, but as an inescapable condition of human social life. Because the identities of discursive elements are always contingent and can never be permanently fixed, reason alone must break down if not alloyed with affect. The possibility of making meaningful observations about the world – in other words, of 'making sense' of it – depends on individual agents taking up positions as subjects within a discourse, which, in turn, requires the affective 'covering over' of the non-necessary, contingent nature of any such identity.[51] People thus come to believe (in Tomas Marttila's formulation) that 'discourses do not actively constitute but objectively represent their authentic social beingness'.[52]

## Genealogical sensibilities

Given its alertness to contingency and dislocation, PDT reads discursive regimes not as static formations but as moments of ongoing struggles to fix the meaning of social life. These struggles leave traces which can be read off the articulatory practices of subjects. As Marttila argues, the routine practices associated with specific subject roles accumulate through processes of *discursive sedimentation* as institutions, material artefacts and infrastructure.[53] Sedimented discursive elements and relations have a certain heaviness, a weight built up over successive articulations and through generations of subjects, which gives them the appearance of solidity, inevitability and belonging to the natural order of things. Just as with any articulation, however, this apparent necessity is an illusion that covers over their constitutive contingency.

The metaphor of sedimentation might suggest archaeological rather than genealogical critique. The former is often understood as requiring time-slice (synchronic) analysis of the episteme, of what is known in a particular sedimentary 'layer', while the latter is constructed as demanding a diachronic account of emergence or formation – the origins of subjects, institutions and regimes of truth. Nancy Tuana cites Ladelle McWhorter's definition of genealogy:

> Genealogies begin with the birth of the being to be accounted for. They note the supporting conditions at the time – those that brought forth the new creature and nurtured it – and then move back in history through the ever-increasing confluences of contingencies that engendered those supporting conditions.[54]

To develop a 'genealogical sensibility' for ecointersectional analysis is thus not merely to point to the sedimentation of the routine practices of subjects in institutions or artefacts but to retrieve the conditions of struggle that occasioned their formation in the first place. Because there is no history of a human 'subject' that is not also the history of practices of identification,[55] and because these practices leave textual traces, historical discursive struggles may be analytically reconstituted through attention to claims to truth, and dissent, but also to absences in the archive, which index what is repressed or excluded. The analytical object of PDT is, therefore, not what people 'really mean' when they engage in discursive activity, but rather the historical emergence of enunciative conditions linked to social, political and fantasmatic logics that scaffold the possibility of social life. This point is gestured to by the late sociolinguist Jan Blommaert, who wrote that every moment of meaningfulness consists of the 'semiotic simultaneity' of elements handed down through long histories of power struggle, domination, subjugation and subjectivation rooted in social structures.[56] What textual analysis does is contextualize these moments as semiotic acts in ongoing discursive struggles.

## CDS with post-foundational characteristics

I have argued that PDT offers a useful starting point for ecointersectional discourse analysis in that it resists mono-empiricism without lapsing into idealism, makes analytical space for affect, and enables genealogical critique.

But it is not without its challenges. Its psychoanalytic and high post-structuralist jargon can be off-putting, for starters. A more significant and long-standing problem is its so-called methodological deficit.[57] PDT offers very little guidance on how to deal with granular textual analysis,[58] surveying the world from a god's-eye view far away from the textual fray. Indeed, PDT's founding ambitions were macro-level theorization of 'the political' – i.e. the structure of hegemonic orders – and not detailed analysis of everyday text and talk.[59]

A methodological elaboration of PDT is, however, well underway, thanks largely to its engagement with CDS.[60] Leading proponents of PDT have argued that operationalizing the approach's discursive theorization of the 'social' would have to go 'beyond guidelines for textual analysis'.[61] Simply put, textual analysis must be connected to the theorization of politics offered in PDT. In the next section, I show how ecointersectional discourse analysis can contribute to this project, in the spirit of both critical linguistics' connection of linguistic theory with critical social theory and queer methodological 'scavenging'.[62] I thus offer a toolkit for analysis that borrows from the CDS tradition while making the theoretical links back to the premises of PDT.

## Ecointersectional discourse analysis in five dimensions

Below, I offer definitions of five possible dimensions of ecointersectional discourse analysis, with examples of each. The encompassing concept in my proposal is that of the 'imaginary'. Imaginaries are discourse regimes that organize the world into objects and the subjects that perceive them. They are (very broadly speaking) not so different from Stibbe's 'stories we live by'. The second dimension focuses on the subject positions that affirm the imaginary. As affect is decisive in securing this affirmation – or 'passionate affinity' – with the imaginary, it is analysed as the third dimension. The repetition of discursive acts within an imaginary has a sedimentary effect, observable in the institutions of an imaginary: its routine practices, typical objects and other structural dynamics, which are analysed in the fourth dimension. The fifth dimension picks out the discursive relations that locate elements of the imaginary in time and space. Following the Bakhtinian practice of other CDS scholars, I will introduce this dimension as the 'chronotope' and locate it within an analytic of scale. A diagrammatic representation of how I understand these dimensions as interacting with one another is offered in Figure 2.

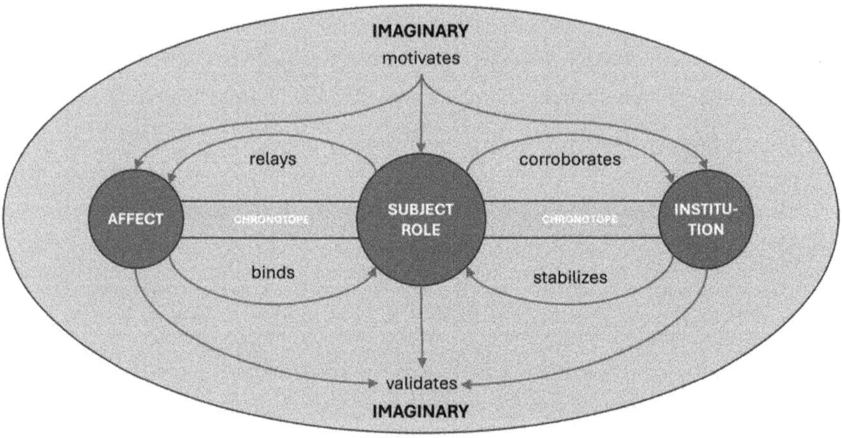


**Figure 2** Diagrammatic representation of the five dimensions of ecointersectional discourse analysis.

## Imaginaries

An imaginary is a specific kind of discourse. Fairclough argued that discourses 'not only represent the world as it is (or rather is seen to be)' but that 'they are also projective, imaginaries, representing possible worlds which are different from the actual world, and tied in to projects to change the world in particular directions'.[63] Projects to change the world by representing it in a particular way are intrinsically *political* acts. When such projects are tied to the social vision of a specific social group, in PDT they are referred to as 'myths'. Myths interweave political logics with fantasmatic and social ones that together make a particular articulation of how the world works, or a particular social vision, appear robust, persuasive and attractive. To the extent that this articulation succeeds, it may extend its discursive range over one or more social groups to become an 'imaginary' – and no longer mere myth. Given the macro-political remit of PDT and its debt to Gramscian analysis, this distinction between myth and imaginary is between the social vision of a single social group or bloc and one whose acceptance has broadened to others. When myths become imaginaries, their *literal* connections to the self-understanding and social demands of one bloc come *metaphorized* as representing those of many, purporting to provide an ordering principle for society.[64] Metaphorization goes hand in hand with sedimentation and hegemonization, in that the connection of the imaginary to a specific set of social demands is obscured, and it becomes routine, mundane and unexceptional, determining the 'horizon of intelligibility'[65] beyond which the world no longer makes sense. The imaginary is thus not an object, but the 'condition of possibility

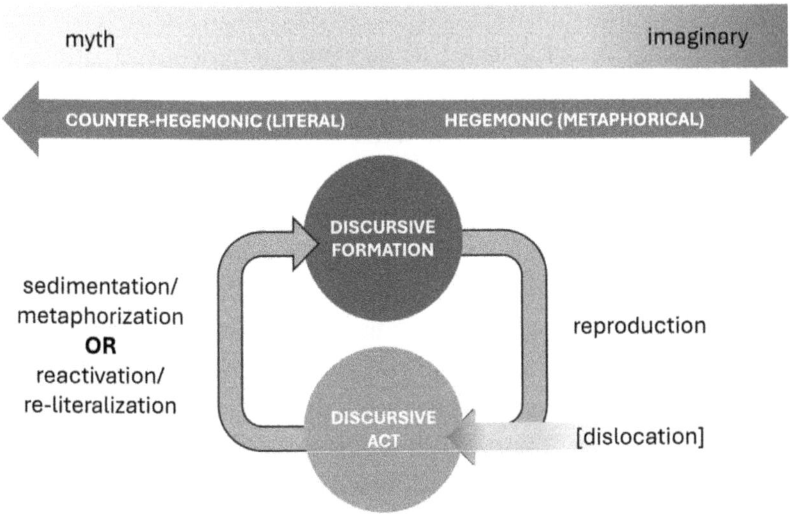


**Figure 3** From myth to imaginary in post-foundational discourse theory.

for the emergence of any object'.[66] Social actors who transgress this horizon are considered irrational, insane or absurd. However, as any discursive structure is prone to dislocation, the foundations of an imaginary may be *re-literalized* through drawing attention to their origins in the myth of a particular group. These kinds of discursive actions are considered counter-hegemonic in that they draw attention to the specificity of the social demands an imaginary satisfies. It is important to restate that in a poly-epistemic model, the metaphorical and the literal do not map onto 'true' and 'false'. This principle is shared with CDS: metaphor is not hostile to truth, but fundamental to making sense of the world.[67]

These distinctions (myth/imaginary; metaphorization/re-literalization) are more helpfully thought of as picking out directions or tendencies than as distinct entities. The useful question for discourse analysts interested in everyday text and talk is therefore not whether a particular discursive act belongs to a myth or to an imaginary, but rather whether it works to sediment/metaphorize a set of social demands or reactivate/re-literalize them through drawing attention to their narrowness and contingency. Reactivations may be set off by dislocations such as natural disasters, changes in economic conditions, new technologies, novel articulations, or other interdiscursive events. Among the many ways that this distinction is already accounted for within CDS, we might say that the former direction picks out processes of naturalization and hegemonization, while the latter picks out denaturalization, counter-hegemonic discursive practice, or even 'queering'.[68] These directions are illustrated in Figure 3.

A further clarification that facilitates analysis is discursive 'interpenetration'.[69] We should not expect to find pure and discrete 'imaginaries' distinct from 'myths' in mundane communication. Nor should we expect to encounter texts that reliably and comprehensively reconstitute a singular imaginary (though this is not impossible). Imaginaries 'often overlap with no clear boundary of where one imaginary stops, and another starts'.[70] As is well understood within CDS, intertextuality and interdiscursivity are the norm. What we are instead likely to find are a variety of myths and imaginaries in combination, with some discursive acts clearly reproduced by an existing formation and confirming it through sedimentation, and others the result of moments of dislocation where the formation is challenged or reactivated.

While textual analysis grapples first and foremost with discursive acts, extrapolating to formations (myths or imaginaries) requires identifying the logics of articulation (social, political or fantasmatic) structured around core values (nodal points). As already noted, nodal points that successfully metaphorize social demands are vague, often to the point of emptiness. They are surfaces onto which multiple social groups may project their aspirations and demands. We should clarify that the analyst is not called upon to identify one, unique nameable nodal point that can be pointed at with a stick. A classic PDT-informed study of environmental discourse clarifies this point. Yannis Stavrakakis analysed the platforms of 'green' political parties in the late 1990s, arguing that they had achieved a novel articulation of existing elements around a 'family' of nodal points including 'green', 'nature' and 'ecocentricity'.[71] Given that these nodal points are given heterogeneous and often contradictory meanings in party platforms, he argues that 'green' is a two-faced sign in that 'on the one hand it is presented as a consideration of supreme meaning and, on the other, it is revealed as an empty name without substantive content'.[72] The signs at the heart of these political projects come to be 'invested with every possible value, they become total embodiments of our lacking fullness, objectivity, positivity, harmony'.[73] The signifiers 'green' and 'nature' and 'ecocentric' are equivalent and therefore interchangeable, standing in for the incommensurable desires of broad swathes of society.[74]

The concept of the imaginary has been used quite extensively in scholarship that takes a critical approach to environmental discourse. Ruth Machen, Simin Davoudi and Elizabeth Brooks have thus recently noted a 'turn to imaginaries'[75] in scholarship on climate change. They conceive of the imaginary as a terrain of discursive struggle, where moving towards a liberatory ecological future requires a feat of the imagination that informs concrete political action, while noting that 'materiality might be accounted for in the production, circulation, or contestation of imaginaries in more flexible and emergent ways'.[76] As I have argued above, this is a task to which PDT's theorization of dislocation is especially well-suited. In another example of ongoing scholarship on ecological imaginaries, Kim-Ly Thompson and Natalie Ban show how imaginaries that centre Indigenous sovereignty and resurgence contradict and undermine the Western and imperial imaginaries of mainstream conservation.[77]

In operationalizing the imaginary in analysis, theoretical emphasis tends to be placed on the concept's political (Gramscian) or psychoanalytic (Lacanian) lineage.[78] What the psychoanalytic line of thinking offers is an account of how, within imaginaries, *fantasy* enables subject formation. Fantasmatic logics are thus essential for the success of the imaginary. Jelle Behagel and Ayşem Mert note that in psychoanalytic theory fantasies express mechanisms of splitting (between the self and the world) and of projection, 'attributing inner feeling of anxiety and joy to outer objects'.[79] Because the fantasy of a split between 'nature' and 'society' is a logic that structures so many environmental, imaginaries, they argue that careful analysis of 'naturecultures' is an important new direction for CDS. Analysis of fantasies furthermore offers 'a more purposeful engagement with political ecology, science and technology studies, feminist, queer, and post-human ontologies, within and beyond environmental social science'.[80] What they are pointing to, in other words, is the rich potential for ecointersectional analysis opened up through taking the role of fantasies within imaginaries seriously.

## Subjects

The wealth of concepts and tools for analysing subjectivity within the CDS tradition can be mined in a PDT-informed approach to ecointersectional analysis. The imaginary *motivates* subject roles, institutions and affective stances, which in turn *validate* the imaginary (assuming that discursive acts are sedimentary and thus moving in a hegemonic direction – see Figure 4). Critical discourse analysts have often noted how the 'concept of "subject" is felicitously ambiguous between passivity (being subjected) and activity (as in "the subject of history")'.[81] They note that it is in relation to differences between the discursive and material resources available to discursive participants and thus to prevailing power relations that subjects act and are acted upon. Both CDS and PDT reject the strong version of Althusserian interpellation, which holds that accepting the role of addressee when hailed determines subjectivity entirely.[82] In PDT, as we have seen, moments of dislocation open up possibilities for resistance, while for scholars such as Fairclough, though the formation of the subject is constrained by available discourses, the 'activity of combining them in ways that meet the ever-changing demands and contradictions of real social situations is a creative one'.[83]

The textual traces of subject roles are many and varied. Blommaert argues that discourse analysis is always also analysis of voice: of who is speaking, and from which perspective they speak.[84] How fragments of discourse are 'represented, re-spoken, or re-written sheds light on the emergence of new orders of discourse, struggles over normativity, attempts at control, and resistance against regimes of power'.[85] Fairclough argues that texts orient both 'retrospectively' to what previous speakers have said, written or done and 'prospectively to the anticipated utterances of the next speakers'.[86] The multi-vocal character of discourse – what discourse analysts call intertextuality, or (after Mikhail Bakhtin) 'heteroglossia' – supplements the question of who is speaking with the question of which other voices are oriented to, and which stances assumed. The voices of a text can be thought of as in ongoing dialogue with each other. Though historically underutilized within the CDS tradition,[87] the conceptual tools of linguistic anthropology are also useful for analysis of this *dialogic* emergence of the subject through stance and voice. Speakers voice others (in direct or indirect speech) or hybridize their voices through 'troping'.[88] Troping may be parodic or confer the authority associated with a specific social domain and its typical personae on the speaker.[89] Voicing is intimately wound up in assigning social value to different 'types' of people: shoring up or confirming the authority of some while undermining or denying the authority of others. When a person performs a particular style of speaking (or some other socially meaningful action) this style takes on its character in that it may be distinguished from the styles of others in value hierarchies. Styles may be idiosyncratic or highly localized to specific communities of practice; they may also become broadly recognized as connected to a specific type of person, in which case they are defined as 'registers'.[90] Register in linguistic anthropology names the way that language 'makes people up',[91] distinguishing the *characterological figure* of the 'Surfer Dude' from that of the 'Posh Brit' (in Penelope Eckert's oft-cited contrast).[92] Whereas individual stylistic innovation pulls meanings centrifugally outwards and in unexpected directions, registers act centripetally to form through inward accretion the stock figures of a particular culture that may thereafter be troped on.[93] In a similar fashion to imaginaries, styles and registers in text and talk will not necessarily be found as discrete and non-overlapping entities, but as the stereotypes and hybrid figures that participants in discourse enact to contest or affirm dominant ideologies, performing subject roles in ways that emerge dialogically through interaction with other voices.[94]

Fairclough distinguishes three 'orders' of discourse: discourse per se (ways of representing), genres (ways of acting) and styles (ways of being).[95] Roughly speaking, these orders map onto imaginaries (discourse), institutions (as sedimented social practices and thus 'ways of acting') and subject roles (styles or, as I have just argued, registers). I have addressed imaginaries above and will return to institutions below. For now it is important to keep both the agentive and the passive senses of the 'subject' in mind when thinking about styles. We do not just style ourselves, we style others in our interactions with and representations of them, and the range of available stylizations is limited by established meaning potentials. We may choose to represent others or suppress their presence entirely; we may represent them as active agents or beneficiaries; up close and personal or at arm's length; by name or by profession; as unique individuals or as tokens of some type.[96] The social semiotician Theo van Leeuwen has built a number of influential taxonomies for classifying how social actors are allocated roles in discourse across modes. Whether people are nominated ('Harold') or categorized ('a nurse'), personalized ('you would … ') or impersonalized ('one would … ') does not, as one would expect, have any fixed meaning, but instead constructs power-laden social relations along hierarchies of value that vary with context.[97] The ideological effects of how subjects are constructed must be argued for in situ. Subjects are represented as kinds of actors, for example, and under racial capitalism whether they are represented as engaging in practices with primarily material or primarily cognitive outcomes is a relevant dynamic of subjugation.[98] Visual representations of faraway people can work to create social distance, lack of eye contact can suggest submission, and depiction in a crowd of similar people suggests a collective rather than an individualized identity.[99] Even how people are represented as moving through space can convey meaning, when men, for example, are 'represented as oriented towards action, seeking to occupy the maximum amount of space, and standing erect, in no need of support'.[100]

Analysis of styles/registers as 'ways of being' and the discursive construction of subject roles may be complemented by sociologist Patricia Hill Collins's theorization of the *controlling image*. Collins shows how the mammy, the matriarch and the jezebel – stock characterological figures – are woven into racist subjugation of Black women as *objects*.[101] She cites bell hooks:

> As subjects, people have the right to define their own reality, establish their own identities, name their history. As objects, one's reality is defined by others, one's identity is created by others, one's history is named only in ways that define one's relationship to those who are subject.[102]

In the racial structure of the United States, Collins argues, Black women's ability to oppose these controlling images is constrained by their differential access to the powers of full citizenship that white and wealthy Americans enjoy.[103] In a similar fashion, Fairclough defines power in discourse as residing in 'powerful participants *controlling and constraining the contributions of non-powerful participants*'.[104] The capacity to effect social change and alter the course of events depends on differentially available resources, as does being able to secure desired outcomes through the agency of more powerful actors.[105] Controlling the representation of Black women thus robs them of authority and influence, and is a key site for the discursive reproduction of inequality. At the same time as Blackness and womanhood are controlled in this manner, whiteness and maleness tend to work through 'ocularcentrism'[106] which positions the perceiving white subject as outside of racialization, and in a non-relation to other identities. This allows white perspectives to masquerade as views from nowhere, aligning them with impartiality and universality, and the 'objective' process of turning subjects into objects.

## Affect

CDS has until very recently not engaged significantly with affect.[107] Emotions[108] are misrepresented as if they were problems hindering the otherwise smooth operation of public reason, and scholars have tended to treat them as mysterious forces that exist 'beyond talk, words and texts, beyond epistemic regimes, and beyond conscious representation and cognition'.[109] These presuppositions clearly fail to account for the manifold ways that affect shapes daily interactions and enduring ideologies, informs decision-making and indeed is enmeshed in our relations with each other and the world. The discursive psychologist Margaret Wetherell builds on work that shows how conventional cultural terms mentally organize what might start out as inchoate sensation, thus 'completing' affect in discourse, to argue that the appropriate unit of analysis is the *affective practice* rather than the bodily sensation itself.[110] In a similar vein, linguistic anthropologist Scott Kiesling proposes that affect be understood as in a dialectical relationship with the stances taken by participants in interaction, in that particular stances may provoke emotions which in turn affect the stances of participants.[111] *Stance* is theorized by socio-cultural linguists as having a triangular structure, with two corners representing two subjects that align (or disalign) with each other through their evaluation of an object, represented by the third corner, where such evaluation works simultaneously to position the subjects.[112] Any communicative act that involves evaluation, whether epistemic ('grass is green'), deontic ('you should go') or affective ('I love zebras!'), is also an opportunity for other participants in discourse to dis/align with that stance. Kiesling argues that as discursive practices, stances are learned and repeated by specific groups such that they become relatively stable structures within a speech community that index facets of identity such as gender.[113] The masculine 'affect of ease'[114] he identifies among fraternity men in the United States thus functions as a shared understanding of how being laidback and cool is linked to the social status of being a 'dude'. This imagined figure as performed through stance stimulates 'authentic' affect in participants, who in aligning themselves with the collective, set the affective tone for interaction.[115] Analysis of stance-taking in discourse is thus a potentially powerful way to identify the textual traces that link affect to subject positions, and a helpful tool to approach Sara Ahmed's observation that emotions 'produce the effect of surfaces and boundaries of bodies'.[116]

Theorizations of affect by Ahmed, Wetherell[117] and Kiesling complement the fantasmatic model of subject formation adopted within PDT to the extent that we accept that subjects derive an enjoyment from identification, which papers over the cracks of the fundamental contingency of all identities. Subject positions are desired at an ontological level – we want to *become* these people – and emotions are 'felt as constituting the desired and identified-with characterological types of people'.[118] Being a certain type of person is equated with feeling a certain way, a feeling which may or may not always be realized, but which is aspired to. To be discursively potentiated, this feeling needs to be feeling *about* something (and thus a stance): about the self, other people, institutions and the organizing logics of the imaginary within which subjects are formed. In this way, subjects work to *relay* affect to others through taking stances, and these affects in turn work to *bind* subject roles to actors and actors to each other. Affect similarly works through these subjects to *corroborate* specific institutions. Subjects are in turn *stabilized* by institutions, a topic to which I now turn.

## Institutions

Institutions are constellations of practices, subject roles and artefacts materialized through repetition through a process of discursive *sedimentation*.[119] Sedimentation lends discursive formations the appearance of solidity and stability, of being self-evident features of our objective reality. It works to stabilize subject roles and the meanings of discursive elements within an imaginary through lending them the appearance of necessity. Institutions are typically approached in CDS through analysis of *genre.* Textual genres are understood as sustaining structural relations between institutions such as schools, universities, the media or the family, in their specific socio-historical contexts.[120] Genre emerges from complex interplays between types of activities, how these activities are structured or staged, which social relations they constitute, the technologies they use, the norms of interaction (such as turn-taking, dialogue or topic setting) they enact, the assumptions and ideologies that inform them, and the narrative structures that underlie them.[121] Genres codify the rituals of a society – its roadside political posters, job interviews, prison architecture, focus group discussions, podcast formats – and make this discursive material available for recontextualization and reference across other scales and contexts.[122] Institutional genres (and thus institutions) are, therefore, in a mutually constructive role with their subjects – the politician, the job-seeker, the prisoner, the researcher or the podcast bro. Institutions work to name, categorize and classify subjects, associating them with the performance of particular functions.[123] While institutions work to stabilize subjects, the sedimentary discursive acts of their associated subjects *corroborate* institutions.

The shaping of institutions from discursive materials is an effect of prevailing power relations. In other words, the sedimentation of a particular articulation as institutional genre relies on repeatedly controlling and constraining the contributions of less powerful participants. As Andrea Mayr shows, institutions enable subordination, and critical discursive research thus tends to focus on three broad areas: how subordinates challenge the institutionalized form of their oppression, how subordinates perpetuate their own oppression and how dominant groups construct and reproduce their own positions of power.[124] Every utterance potentially enacts an orientation to some institution, as Blommaert argues.[125] Analysis of the epistemic, deontic or affective stances of subjects can thus reveal whether attention is being drawn to the contingency of sedimented discourses (through re-literalization) or whether the metaphorical foundations of the imaginary are reaffirmed (and thus sedimented) in text and talk. The commitment of ecointersectional discourse analysis to a genealogical sensibility necessitates attention not merely to institutions in their current forms, but to the social struggles that occasioned their historical sedimentation. While subordination may be one of the effects of contemporary institutions, critical analysis reveals the contingency of their historical emergence, and to their potential dislocation and re-literalization in discourse.

## Chronotopes

The fifth dimension of ecointersectional discourse analysis picks out a subset of discursive objects fundamental to environmental imaginaries: time-space and how action is narrativized within it, or what the Russian philosopher and literary theorist Mikhail Bakhtin refers to as a *chronotope*. Bakhtin argues that in literary texts 'temporal and spatial determinations are inseparable from one another, and always colored by emotions and values'.[126] Bakhtin shows how chronotopes shape character and genre in artistic works, representing the indissoluble unity of subjects and their range of motion within temporal frames and moral universes, arguing that the 'image of man is always intrinsically chronotopic'.[127] In the theorization of discourse I have presented here, where social visions are understood as fundamentally metaphorical, these literary concepts can be put to good use.[128] Regimes that promise a 'new order', for example, typically adumbrate their ideology in both spatial and temporal terms, often populating them with a 'new man' who is the prime mover within them. For Laclau, the formation of a myth and its transformation into an imaginary require the appearance of 'fullness,' which is secured through discursive 'spatialization'.[129] As we have already seen, the spatial metaphor of a 'horizon' is key to understanding how the imaginary is theorized as a limit to intelligibility. Time, as the 'fourth dimension of space',[130] is furthermore inseparable from this metaphor and its commitment to the temporariness of discursive stability in an ontology characterized by flux and contingency. The chronotope thus picks out how the spatial and temporal dynamics of an imaginary give meaning to both subjects (as characters) and institutions (as genres) in the context of specific affective and normative values.

The chronotope also usefully provides an analytical staging point for conversations between geography and discourse theory that are essential for ecointersectional analysis. Space is theorized by human geographers not as a fixed reality, but as a fluid process open to agentive influence, always in flux and contested, and in a dialectical relationship with embodied experiences of it.[131] Critical accounts of race and space consistently point to how racialization relies on the surveillance and control of bodies in space. Combined with other vectors of social difference such as class, ability, gender, faith or sexuality, we can analyse how some people are able to move around freely and at their own pace while others are slowed down, hurried up, corralled in an unchosen direction or forced to stay put.[132] Mobility across scales and boundaries, and the extensivity of such movement (its distance and direction) work to underwrite 'private property and the construction and orderly maintenance of segmented social space, from gated communities to redlined districts, from nature "preserve" (including, for example, all-white golf courses) to office towers (white by day, brown and black by night)'.[133] In this 'grid epistemology'[134] of social differentiation, power lies in being the (e.g. white, male, straight, abled-bodied) subject that surveils without itself being surveilled, the eye/I at the centre of the map whose particularity is made to appear universal and eternal.

Space is an abiding concern of Indigenous and decolonial scholarship under the rubric of 'critical place inquiry'.[135] Indigenous ontologies and #LandBack decolonial activism emphasize the defining cultural and eschatological connections between land and people, critically drawing attention to how the commodification of places under colonial racial capitalism coincides with the trampling underfoot of lifeways connecting the human and more-than-human worlds. Human connection to place is thus a key element of a variety of social and political struggles over spatial definitions and usage.[136] These relations to place connect with memory practices and imagined futures, justifying or resisting social and material developmental projects.[137] It is in unpicking how the pasts, presents and futures of place are enmeshed and made politically salient in discourse that the chronotope can pack its analytical punch.

In terms of the reproduction of racial regimes, the geographer Andrew Baldwin has shown how the future is constructed as a 'terrain upon or through which white racism will get resolved'.[138] Fundamental to this critique is the insight that the racial hierarchy is renovated in relation to a future that is invoked in the present, in an imagined tomorrow that works to give meaning to contemporary identities and actions. Nationalisms of the left and the right, as well as liberal overseas development assistance discourses, invoke the spectre of racial futures in order to structure their imaginaries, shaping geographies of place, nature, mobility and space.[139] In their introduction to a special issue of the journal *Geoforum* entitled 'Political Ecologies of the Future'[140] Jevgeniy Bluwstein and Salvatore Paolo de Rosa observe that while the envisioning of just futures linked to radical transformations of society is an important part of decolonial environmental justice, these projects are up against the fact that the ideological elaborations of capitalist, colonialist, racist and patriarchal systems dominate how the future may be imagined. While the past is remembered, retrieved or memorialized in the present in ways that are intrinsically political, so too are invocations of the future unavoidably imbricated with struggles over the power to represent, and thus to determine what will happen next.

The textual traces of the chronotope are numerous. It may be thematized explicitly through references to location (in time or space) and extent (as in duration or distance). The analytical question is to identify the chronotopes of here/there/now/yesterday/tomorrow and then to investigate how they are connected.[141] The chronotope may feature in a text through practices of comparing features of spatiotemporal location or extent, which the linguistic anthropologist Susan Gal refers to as 'scaling'.[142] Scaling 'implies positioning and, hence, point of view: a perspective from which scales (modes of comparison) are constructed and from which aspects of the world are evaluated with respect to them'.[143] Analytical questions here include who is observed, who is the observer and from whose perspective are comparisons made? Identifying the observer brings us back to questions of voice. The voices in a text are indexical of a chronotope in that people speak *from* specific times and places in ways constrained by their range of action and the demands of narrative.[144] Often these times and places can be identified through linguistic variation – the variety of a language being spoken, which can point to geography and/or historical period. They can also be identified through the invocation of the specific attributes that constitute a space-time onto which 'senses of belonging, property rights, and authority can be projected'.[145]

## Cross-cutting analytical tools: 'No sign is innocent'

While the relations between signifier and signified may be arbitrary in structuralist linguistics, the use of a particular sign in a particular context is not. Gunther Kress argues that since the sign positions its user in a socio-political system of interests, using one resource instead of another is inherently ideological, and thus 'no sign is innocent'.[146] All aspects of texts (whether written, photographed, sculpted, filmed, built, landscaped, etc.) are open to critical readings. Critical reading practices rely on an understanding of how sign systems enable and constrain meaning-making. This requires some familiarity with the elements of the system (or *lexicon*), its rules of construction (or *grammar*), and its interaction with its spatial and socio-historical *context*. Though it is neither possible nor necessary to present every critical reading tool available to CDS in this section, I will briefly mention six useful sets of concepts that can serve as a basis for textual analysis.

*Action and agency.* A good starting point for analysis of texts across semiotic modes that represent social action are the questions: Who is the agent of the action, and who is affected (i.e. **agent** vs **patient**)? This question relates not merely to the subject positions of an imaginary but to institutions and social processes which may be turned into agents through **nominalization**. Instead of spelling out, for example, that a corporation laid off their workforce in order to maximize profits, an economist may invoke 'globalization' as the prime mover. Specific responsible humans are occluded from the scene, and the machinations of capital generalized as outside of history, inevitable, eternal, universal and imposing rules of action on humanity.[147] Nominalization typically involves the loss of semantic detail.[148] In a similar fashion, **passivization** allows for **agent deletion**, as does the choice of **transitivity pattern**.[149] In the active voice, a journalist might write, 'The soldier killed the child during a military operation'. If the journalist writes, 'The child was killed during a military operation' the specific agent of the action has disappeared. If an intransitive verb is chosen instead, as in 'The child died during a military operation' even more detail is lost, and the situation cloaked in mystery. A photographer's decision to include the soldier in the frame, or a filmmaker's choice of cuts, has similar effects.

*Force.* The **epistemic** (related to truth) and **deontic** (related to obligation) force of a textual fragment may be deconstructed through analysis of **mood** and **modality**. As Louise Green argues, in making statements such as 'if temperatures were to rise by two degrees … ' humanity is increasingly living in the subjunctive mood, which relates to that which is uncertain but might be, thus signalling the unpredictability of climate change and its future effects.[150] Grammatical moods such as the declarative (A is B), interrogative (Is A B?), or imperative (A, fetch B!) furthermore provide clues as to who is in control of the communicative situation.[151] Modality meanwhile relates to the level of certainty a speaker commits to in knowledge exchange (epistemic modality: is, could be, maybe, etc.) and activity exchange (deontic modality: must, ought to, should, etc.).[152] Under the rubric of deontic modality we may further distinguish between **offer**, where a speaker commits to act on an obligation, and **demand** where the speaker asserts the existence of a strong obligation. These concepts have been fruitfully applied to the analysis of images, where (dependent on context) direct eye contact may constitute a 'demand', for example, and a downcast gaze an 'offer'.[153]

*Intertextuality.* The multi-vocal and heteroglossic nature of discourse discussed above requires analysis of voicing and citation practices. The author of a particular discursive fragment may invoke other texts – for example, through using vocabulary associated with specific religious, scientific or literary works – that establish what can be assumed or left implicit in the new text. Analysis of this **intertextuality** reveals 'presuppositions, logical implications or entailments, and implicatures'.[154] The citation of a text is always also a **recontextualization** of that text. What the text assumes to be the case arguably has high modality, as it takes for granted the epistemic or deontic force of the text in its new context.

Different types of presuppositions may be identified within a text: **existential assumptions** (about what exists), **propositional assumptions** (about what is, can or will be) and **value assumptions** (about what is good or desirable).[155]

*Persuasion and legitimation.* The fourth analytical subset unpicks how social actors seek to persuade others through **legitimation.** Theo van Leeuwen identifies four main types of legitimation: **authorization, moral evaluation, rationalization** and **mythopoesis.**[156] Authorization includes appeals to tradition, custom, law or the institutional authority of individuals; moral evaluation includes appeal to value systems; rationalization refers to legitimation that references the utility of the institutionalized action and the validity of such knowledge; while mythopoesis involves legitimation through narrative.[157] I have marked the critical role played by affect in all of these strategies through choosing to foreground **persuasion**: ultimately the rhetorical power of a particular statement may be attributed to pathos as much as to its apparent validity (logos) and the authority of the speaker (ethos).[158]

*Indexicality and iconicity.* 'X **indexes** Y' typically means that X points to Y without resembling it. To the extent that an index resembles what it is pointing to, it becomes **iconic** of its referent, and no longer indexical. Roland Barthes's classic distinction between **denotation** (the generally accepted meaning of a sign) and **connotation** (the cultural associations of the sign) is similar but not identical. Except for onomatopoeic words, the vocabulary of a language tends to be indexical rather than iconic, and both the denotations and connotations of words are similarly indexical. Images, on the other hand, tend to represent their objects as icons, while their connotations rely on questions of context that create **indexical ties** between the image and its socio-cultural environment. Analysts in the CDS tradition have theorized how to read the indices of non-verbal texts in a systematic fashion.[159] Colours, for example, may be analysed through their properties of differentiation, saturation, purity, modulation, value and hue, which create context-specific indexical ties that when read in relation to each other can be thought of as governed by a kind of 'grammar'.[160] Architecture may similarly be read as a text, as may the built environment more broadly.[161] Objects can be approached through analysis of the material that is used, its texture and shape, any writing thereon and the typography used, in addition to the iconography employed.[162] The meanings of these elements tend to fall somewhere between denotation and connotation and between index and icon, and analysis will thus often draw on references to how the meaning potentials of an attribute are realized in other contexts, in order to make the case for indexicality/iconicity in the specific case of the text being analysed.

*Discursive strategies*. The final set relates to how discursive fragments can be thought of as serving specific purposes in their contexts and integrates concepts from the other five subsets. In thinking of discourse as purposive, it is important to note that CDS does not claim to be able to get 'into the head' of social actors and makes no claims about their intentions. **Discursive strategies** may be read off the text directly, without the attribution of conscious intent. Ruth Wodak and John Richardson identify six discursive strategies: **reference** strategies which name discursive elements; **predication** strategies which assign qualities to these elements; **process** strategies which represent events or actions; **argumentation** strategies which represent opinions and reasoning; **perspectivization** strategies which involve voicing the self or others; and **intensification** or **mitigation** strategies which regulate predicated qualities.[163] An argumentative strategy that we will see often in environmental discourse is the **topos.** In argumentation theory, as Ruth Wodak argues, topoi are the 'parts of argumentation which belong to the obligatory, either explicit or inferable premises [...] or "conclusion rules" which connect the argument or arguments with the conclusion, the claim'.[164] They are the presuppositions of a discourse and therefore have particularly high epistemic modality. Whereas in CDA topoi tend to be 'fallacies', in a post-foundational approach the accuracy of a particular premise is neither here nor there. Topoi instead reveal the specific 'logics' that are assumed to be able to explain social life. A different kind of strategy theorized by Michał Krzyżanowski is the **discursive shift**. A discursive shift is an interdiscursive strategy consisting of three steps: first, a discourse is strategically **enacted** within a novel discursive context, this combination is then **perpetuated** through contextualization across diverse social fields, until it is **normalized** as ethical and rational.[165] Krzyżanowski's model is particularly helpful in thinking through the processes by which discourses thought of as extreme or absurd can move from the fringes of social life into the centre, undoubtedly a defining feature of the current political moment.

## Conclusion

With these concepts and tools in hand, we are ready to commence ecointersectional discourse analysis. Before starting, however, I would like to make some remarks on the importance of reflexivity, especially seeing as I identified this as a weakness of ecolinguistics in the previous chapter. In her exploration of feminist critical discourse analysis, Michelle Lazar offers a very useful model of what reflexivity in CDS might look like. She draws attention to how the 'sign-value' of (feminist) critique can be appropriated in reactionary 'post-feminist' consumerist discourses[166] as well as to how scholarly investments in knowledge projects reveal imperialist ambitions. A ubiquitous reality in the neoliberal academy is the 'state of generalized competition'[167] that calls on academics to think of ourselves as individual enterprises measured in terms of our 'impact' – and writing a book detailing a 'new' approach to discourse analysis clearly fits the bill. Both of these critiques could be levelled at my project here.

In terms of the 'sign-value' critique, shallow intersectional critique of environmental discourses could provide cover for 'business as usual'. A recent viral tweet featuring a BBC interview with the president of Guyana, Irfaan Ali, is a good example. In the clip, President Ali berates the interviewer for his belligerent questioning of Guyana's decision to exploit oil reserves given what this would mean for global carbon emissions.[168] The clip has been viewed over 20 million times. Much of the buzz reflected positively on the Guyanan president highlighting the irony of a white journalist from the imperial metropole lecturing the head of state of a formerly colonized and underdeveloped nation on the causes of climate change. However, as the political economist Keston Perry was quick to point out, this reaction is only progressive on the surface. While it is good to call out the hypocrisy and arrogance of the North, this should not distract us from the fact that Exxon, the US government and other powerful transnational interests were going to be allowed to profit from Guyana's decision.[169] Even an apparently radical decolonial critique may thus be quickly co-opted by the hydrocarbon extractivist complex if it focuses on isolated elements of a discursive event, rather than on the fuller systemic picture. In other words, the symbolic 'win' of the president's dressing down of the interviewer should not distract us from the material 'loss' occasioned by the consequences of the exploitation of Guyana's oil reserves in a neocolonial world system. As discourse analysts, we must thus constantly question our own complicity in socio-ecological harm.

The critique of the imperial ambitions of this knowledge project is also apposite. Proposing a novel approach to environmental discourse analysis can generate citations from proponents and detractors alike. When an approach gets attached to a specific acronym, journal or set of scholars, some may promote it as 'the road to salvation for everyone'.[170] Michael Billig critiques these 'processes of marketization' that see critical projects subsumed under catchy initialisms.[171] For her part, Lazar observes that feminist practitioners had been doing feminist work for decades without any need for an 'FCDA' label to get behind.[172] In my view, a closer engagement by practitioners of CDS or PDT with fields such as political ecology does not fundamentally change or rename an approach, as much as enrich it. In proposing that CDS should be ecointersectional, I do not claim to be founding a new approach, but rather to be identifying strategies for textual analysis that are already widely practised and bringing them into conversation with each other, a task to which I now turn.

3

# Green capitalism and the conscientious consumer

This chapter turns to the cluster of discourses organized around the idea that capitalism can save us from ecological catastrophe, through technological innovation and through meeting the increasingly enlightened needs of eco-conscious consumers. Environmental issues are given prominence in the marketing of big brands whose businesses rely on constant growth and resource extraction. By shifting the work of environmental protection onto their customers, who must now make the 'responsible choice' about which product to buy, whether to recycle or whether to 'compensate' for the climate impact of their purchase, often by paying a premium for it, corporations create the illusion of proactively addressing ecological problems while continuing to extract surplus value, capturing the initiative from alternatives that might require top-down statist, internationalist or bottom-up solidaristic solutions to environmental crises. Billionaires present themselves as the ecomodern solutions to environmental problems, while radicals are constructed as standing in the way of innovation. Key to the success of this imaginary has been the articulation of the subject position of the 'conscientious consumer' who affirms the wisdom of markets and the primacy of individual choice. Institutional life is organized around these subjects in ways that corroborate their sense of self to the same extent as they work to give licence to the institutional forms of late capitalism. The sedimented discourses spun around these privileged positions come to have their own climate, which consists of all of the threads of daily life in 'eco-conscious' late modernity, where spaces of luxury consumption have proved especially important in affirming the sense of a proactive, conscientious and responsible pattern of consumption. I start this chapter by exploring one such 'climate'.

## Climates of conscientious consumption

Like many urban passages in Hong Kong, getting from the Park Royal Hotel in the New Territories to Sha Tin station is possible through seemingly limitless combinations of arcades, skybridges, corridors, staircases, and escalators that connect hotels, malls, offices, apartments, restaurants, sports facilities, transport hubs and public spaces. I move from one gleaming, fragrant labyrinth to another, along subway carriages and through revolving doors, without having to face the humidity for more than a minute or two, in a perfectly air-conditioned tube all the way from Sha Tin to the new international airport on its specially constructed island. From Hong Kong I fly in comfort to Johannesburg, and after a few minutes on a platform I am whisked away in the plush capsule of the Gautrain to Rosebank, where another sparkling shopping mall awaits, with its high-end restaurants and luxury hotels. Later that southern summer I pass from luxury cabin to luxury cabin through the chilled duty-free bonanzas of Dubai into the heart of the Canadian midwinter, into mile after mile of the maze-like heated indoor malls of Toronto, which connect the cities' shops, businesses, homes, schools and universities, with little need to venture outside into the cold. I do this all without breaking a sweat or feeling a chill, enveloped in a climate designed especially for citizens of the world, for people like me.

A perfectly controlled climate is not the only thing I sense in the air. Wafting through nearly every space I have passed through from Hong Kong to Johannesburg and Dubai to Toronto, I recognize two dominant notes: the rich, dark smell of coffee with a hint of pastry, and a mix of fruity, spicy, intoxicating fragrances from soaps, perfumes, creams and other beauty products. How many Starbucks have I walked past? Ten? Twenty? And how many Body Shops? At any one of them I could have stopped for a coffee or a pastry, a facial scrub or some exquisitely fruity shower gel. If I miss this one I could always go to the next one; there will always be another opportunity to shop. The fragrances coming from Starbucks and The Body Shop remind me of the earth and its fruits, their ripeness, the appetising combination of savoury and sweet, and I start to sense that I am, after all, an organic being, even though I spend so much of my time in contact with the sterile steel and glass surfaces of shopping malls, airport terminals and conference centres. Starbucks and The Body Shop are always there to connect me back to Creation. While getting a cappuccino, I learn that my purchase is contributing to a 'resource positive' initiative – where the multinational chain has a plan in place to 'give more than we take from the planet'.[1] When I eventually pop in to the beauty store to replenish my green tea and lemon moisturiser, I am heartened to learn that I can use an aluminium refillable dispenser for certain products, which is part of this brand's commitment to 'empower' me 'to embrace the circular economy and help eliminate waste'.[2] Not only am I now clean and caffeinated, I'm also reassured of my harmonious relation to the natural world. I am giving more than I take, I am eliminating waste and making responsible decisions about what I will and will not spend my money on.

When I'm ready to check out of my hotel, the Fairmont Royal York, connected by underpass to Toronto's Union Station and miles of shopping mall beyond, I go down to the reception to drop off my keycard. I flip it over in my hands, looking at it carefully for the first time. One side, which contains only close-up detail of the 'F' of the flamboyant calligraphy of the Fairmont logo, is redolent of old-time elegance and sophistication. The Fairmont group has its origins in the nineteenth century, and the Royal York hotel on Front Street in central Toronto first opened its doors in 1929.[3] The extravagance of the logo connects to the aristocratic pretensions of the 'Royal York' moniker, the stolidity of its architecture and a heroic colonial Canada, indexed in the internal branding of the hotel on framed posters and station clocks as a railway hotel connected to Union Station across the road, and to the historic role of the Grand Trunk Railroad in opening up the interior to white settlement. The rightward-sloping calligraphic font suggests an artefact hand-crafted in a time before computers and neon lights.[4] The fact that the detail of the 'F' was chosen speaks furthermore to its aesthetic value. It is something to be admired up-close, its harmonious curves and delicate lines the work of a skilled artist.

Keycard locks have been around since the 1970s and tend to be made of plastic. This keycard however is made from lightweight blonde wood with open grain. In product design, wood is often associated with luxury products hand-made by artisans for an exclusive clientele, as in the packaging of expensive cigars or whisky.[5] But this wood is light and unstained, suggesting both natural simplicity and the kind of unvarnished brightness associated with Scandinavian interiors. This is not some dreary, dark Victorian maple or oak: it is cheerful and modern. The mixing of old and new is confirmed by the light-brown etching in the keycard, clearly burned by laser, leaving traces of a technological advance not common until fairly recently. In running my fingers over its surface, I feel warm natural fibres and etched lines. This is an artefact with a past and a future – something perhaps I should keep as a souvenir?

This course of action is however advised against by the text etched on the other side:

> THE GRANDEST GIFT OF ALL IS OUR PLANET
>
> Help us reduce our environmental impact by returning your key card.
> Made with FSC certified wood.

Because the card has the look and feel of a luxury item, my conscientiousness in returning it is all the more impressive. How diligent I am! How committed to doing my duty, responding as implored to in the imperative mood: 'Help us'! And here I am, helping!

I am well aware that I would not be charged if I kept it. A large part of the utility of keycards is that, if lost, no lock needs to be changed or key cut; all it takes is a new RFID code to keep the hotel door secure, and the cards themselves cost very little.[6] But I am clearly a person who, having recognized the value and appreciated the aesthetics of the keycard, is now willing to surrender it at the front desk. This small act of reducing waste, ensuring that the keycard can be reprogrammed for use by another hotel guest, is given great significance by the all-caps headline on the card: 'THE GRANDEST GIFT OF ALL IS OUR PLANET'. Owing to the simultaneity of these messages, a lexical tie is created between my action in returning the card and the act of giving a gift. And not just any gift: the 'GRANDEST' one. This choice of superlative echoes the calligraphic font of the Fairmont logo: not the simpler and straightforwardly Anglo-Saxon 'best' or even 'greatest' but the flamboyant, vaguely Romance and patently old-fashioned 'GRANDEST'. I am interpellated as if I had old money to spend, thinking of the grandest gifts I could give, with the world at my feet in more ways than one.

As Crispin Thurlow writes, the 'semiotic landscapes of super-elite mobility [...] do not just get into one's head; they also get under one's skin'.[7] Spaces of elite consumption *feel* a certain way. I am cosseted and cradled by the rich aromas, comfortable textures, and velvety sounds of hotels, malls and aircraft cabins. Everything is accommodated to my pace and tailored to my preferences. All the spaces I move through are smooth[8]; nothing gets in my way. The world seems organized around my desires. And when everything is designed to satisfy me, to lure me in, promising euphoria and self-actualization, the very kernel of my being, the space in which I have the strongest sense that 'I' still exist, is in the act of *choice*, assembling for myself a way of being, a style of life, that reflects my identity. For I am not merely following instructions: I am a discerning person, someone with taste, who won't accept just anything! The very architecture of the spaces I move through encourages me to flow with them, following my desire to realize myself through my choices.[9] Here in the man-made climate-controlled birthing canal of capitalist consumption I feel communion with nature, but on my own terms: that it is I who am in control.

The space I move through without impediment – thanks to the structural conditions of able-bodiedness and how I am racialized, classed and gendered – has an inside and an outside, and much of the work of keeping its ecosystem of signs correctly articulated must be done by preventing the intrusion of semiotic matter out of place across an invisible membrane. The moral feedback I get on my behaviour – that I am a good person making wise choices that are in harmony with the world – requires the background assumption of the ecologically unsound practices of others. People on the other side of the climatic membrane are simply less empowered to participate in the circular economy – and they take more from the planet than they give back! I think of those faraway people, living in filth, where the dominant notes are not coffee and fruit but excrement and coal. These contrasting environments, as Nancy Tuana argues,[10] are 'racial climates' where death-dealing differences quicken and thrive. Geographers of race and class have for decades studied how space-management practices and technologies assert harmony with nature as a means to establish the 'character' of a place that aligns with social divisions.[11] The spaces I travel through thus connect my patterns of consumption to a positive in-group image that may be explicitly or implicitly positioned higher up the hierarchy of the human than a racial, ethnic, national or class out-group.

We move through ecosystems of signs, some 'green' and some not, materialized in our everyday environments. We wear them on our bodies as signs of class belonging; we live in them; we eat them; some of us even drive them to work. When the Twitter account of erstwhile US president Joe Biden shared a picture of him behind the wheel of a massively oversized SUV announcing that '[o]n my watch, the great American road trip is going to be fully electrified'[12] as the stars-and-stripes waves in the background, what is affirmed is the integrity of the nation and its character, which will continue to be defined by its appetite for more speed, more space, more resources, more fun. It is clearly unimaginable that there could ever be an 'American' way of life not premised on thousands of miles of interstate highways asphalted over stolen land. These roads too are in the system of signs, the sediments of discourses of automobility which emerged in the twentieth century as the circulatory system of petrocapitalism.

This articulatory practice, where the shift from the internal combustion engine to battery power is smoothed over as an unthreatening and even patriotic act, pierces 'the entire material density'[13] of the world, affirming the inevitability of capitalism and the inviolability of a whole way of life. But a lifestyle is never inevitable, as the ecocritic Louise Green observes, it is a 'complex negotiation of desire, choice, necessity, attitude, culture, history, and innovation. It is an orientation towards the world and to other people, a conscious and unconscious distribution of what is considered valuable.'[14] What may thus at first seem to be taking Nature's part in fact subordinates the more-than-human to a value it derives solely from the sustenance of a human lifestyle. Put plainly, it is ecological indifference dressed up as environmental concern.

## Beyond 'greenwashing'

The term closest to hand to describe the affirmation of elite consumption and the marketing of SUVs, coffee shops and cosmetics brands as agents of environmental protection is 'greenwashing'. Since it was coined in 1986 to name the clearly disingenuous (though enduring) practice of hotels upbraiding guests to reuse their towels, official definitions of greenwashing have tended to hinge on an intention to misinform or mislead.[15] Because accusing a brand of 'greenwashing' challenges its legitimacy and value, portraying the people behind it as dishonest actors, public relations professionals and strategic communications scholars have devoted a considerable amount of energy to pinning down what does and what does not count as greenwashing, effectively regulating how the term may be used. 'Greenwashing' has thus been decisively captured by industry actors. The advertising industry was recently emboldened to announce that the 'greenwashing era is over'[16] due to the intervention of regulators in the United Kingdom and the European Union. Evidence for claims such as 'net zero' or 'carbon neutral' would henceforth have to be backed up with better evidence, supporting greenwashing's assumption of a legal character.

Coverage of these developments in *The Guardian*, which cites only advertising industry insiders, provides a paradigm case of the 'elite capture'[17] of what was once a term of critique. The president of a prominent EU ad agency is quoted as saying:

> [For] good brands with good intentions and responsible messaging, I expect there will be little change. But for companies that have oversimplified and overstated their sustainability claims, then life is about to get complicated [...] If all brands

> can claim they're green, then you remove the incentive to win consumers based on superior commitments to the environment. This will hopefully make being an environmentally responsible brand even better for business.

The executive's point is that if a brand makes a claim, it should not be allowed to do so too cheaply: companies must bear similar costs for the claims they make or at least do some honest innovation instead of deception if they want to drive these costs down. A similarly industry-supported regulatory initiative in the United States was led by the Securities and Exchange Commission, which sought to strengthen the evidentiary basis for claims made about so-called 'ESG' funds – grouped investments in corporations which supposedly score highly against environmental, social and governance metrics.[18] In the words of one BlackRock ESG fund manager, '[greenwashing] is a risk to investors and detrimental to the asset management industry's credibility'.[19] The potential for 'greenwashing' to play a critical role is thus cancelled out by its reduction to an obstacle in the way of fair competition between fund managers or popular brands. As captured by industry, it works to affirm the wisdom of markets and the fundamental value of consumer choice. Its semantic field is delimited and determined by capital, contained completely within a profit-seeking logic.[20]

Ecointersectional critique targets the roots of ecological crisis in the overlapping world systems of coloniality, racism and capitalism. Instead of individual claims of 'net zero' or 'environmentally friendly' what it aims to deconstruct is the *corroboration* and *legitimation* of the routine representation of these systems as if they were enabling bases for ecologically sound futures. This requires a thoroughgoing critique of green capitalism. In the social visions of ad agencies and fund managers, competition between green brands is possible only if they bear similar costs for environmental responsibility. As we will see, this idea, rooted in neoclassical economics, encapsulates a solution to the latest crisis of capital. To understand what got us here, a brief tour through historic modes of production is useful. The point is that where nature once was cheap, the correction envisioned by capital is that certain forms of it should now cost a bit more, but that otherwise nothing about the status quo needs fundamental change.

## 'Cheap natures' and the crises of capitalism

Capitalist modernity, as Jason W. Moore argues, is built primarily on the 'Four Cheaps': labour power, food, energy and raw materials.[21] If the illusion of profit is to be maintained, these siphoned-off energies and resources of the human and more-than-human world and their complexes of systemic oppression must go unaccounted-for. There have been attempts to develop economic models that include the actual cost of this 'free' energy, which are termed 'externalities'.[22] One report by an environmental NGO calculated that if the world's largest corporations had to pay for the over $7-trillion worth of value they take out of the earth every year not one of them would still be profitable.[23] Environmental services are not the only value that profit devours. As Nancy Fraser argues, we need a more expansive definition of the conditions of possibility of capitalism, which include not only 'nature', but also social reproduction, political power and the wealth of others.[24] On her account, capitalism lurches from crisis to crisis precisely because its hunger for the very sources of value on which it depends is internal to its logic, in a way that is not the case in pre- or post-capitalist societies. As Fraser explains, capitalism's 'cannibalization' of its enabling conditions lies at the root of its periodic developmental crises:

> In the case of its ecological conditions, what is at risk are the natural processes that sustain life and provide the material inputs for social provisioning. In the case of its social-reproduction conditions, what is imperiled are the sociocultural processes that supply the solidary relations, affective dispositions, and value horizons that underpin social cooperation, while also furnishing the appropriately socialized and skilled human beings who constitute 'labor.' In the case of its political conditions, what is compromised are the public powers, both national and transnational, that guarantee property rights, enforce contracts, adjudicate disputes, quell anti-capitalist rebellions, and maintain the money supply. In the case of capital's reliance on expropriated wealth, what is endangered is the system's professed universalism – and hence its legitimacy – and the ability of its dominant classes to rule hegemonically through a mix that includes consent as well as force.[25]

As noted in the introduction, Fraser argues that in the current conjunction the capitalist world-system faces not merely an ecological crisis but a complex of overlapping crises that cut across all four of these enabling conditions. To understand how the world has got to this point, it is useful to read the history of modernity through its relations to the ecological, social-reproductive, political and expropriative conditions of possibility. We may begin with the 'jump-starting'[26] of capitalism-as-we-know-it around the sixteenth century through two parallel processes: the enclosure of the commons in the 'Old' World and the expropriation of land in the 'New'. These moments of acquisition-by-fiat signalled a 'constitutive spatial moment unlocking the epoch-making potential of endless accumulation'[27] where capitalism was turbocharged by the detachment of parcels of 'nature' from complex human social systems and their integration into global markets. Unprecedented access to enclosed land in Europe and to enslaved people's labour and land stolen from Indigenous people legitimated through racial 'science' delimited new ideological horizons of possibility for the mercantile class. In the expanding imperial periphery, agents of capital implemented 'brutal systems of socio-ecological extractivism [ ... that] worked land and labour to the point of exhaustion, making no effort to replenish what they expended'.[28] Western intellectual production of this period provided obliging pretexts for this brutality. In laying the philosophical foundation for liberal property rights, John Locke, for example, stipulated that only land that had been 'improved' by human labour was valuable and that all the rest was 'Waste'; that taking land from people who are living there but not 'improving' it is justified through the inclusion of such people in financial markets; and that unlimited accumulation of land is justifiable so long as it is being opened up to investment.[29] Locke's arguments, which are fundamental to modern liberalism, serve as an ethical license to Europeans to dispossess and exploit Native peoples in their colonies.

In the nineteenth century, the muscle power of livestock and indentured people was supplemented by fossil fuels, which 'appeared to liberate the forces of production from the constraints of land and labour'.[30] The prodigious potential of the fossil fuel age turned coal into a globally traded commodity managed through military dominance, often resulting in war. Industries eschewed hydropower, which was free but connected to rural social systems and tangled up in traditional rights making it harder to control, in favour of coal power in urban areas where the labour could be concentrated through influxes from country to town, and thus more easily exploited.[31] The result was that cities became polluted and rural soils exhausted through intensive cultivation, requiring further expropriations from the colonies to bring home the sugar, tobacco and fertilizer needed to keep workers productive. This 'environmental load displacement'[32] fed the imperial metropoles with the sources of energy it needed to sustain profit margins, while shifting expropriation and a significant amount of violence to the periphery. A similar pattern of peripheral violence sustaining metropolitan prosperity endured after the United States supplanted Europe as global hegemon in the twentieth century, ushering in the automobile age and the ecocidal vortex of petrocapitalism, industrial agriculture, and neocolonial military and economic relations. Over the past fifty years, the oil industry has extracted $3 billion in profits *every single day*.[33] What appeared to be gains for working people in this period – such as the establishing of social security nets in

Western democracies – were in fact extracted from 'intensified private plunder of nature in the Global South'.[34] The triumph of this 'modern/colonial world system'[35] is perhaps most pithily expressed in how the ethical duties generated by centuries of Euro-American exploitation are constructed not as leaving the rich world in debt to the poor, but the reverse. Despite all the wealth that was extracted from them, former colonies and dominions are designated 'highly-indebted poor countries' (HIPCs).

This story of where capitalism gets its energy from suggests periodic and revolutionary change. Power sources did not historically replace each other, however; in fact, supplementing an old power source with a new one tended to increase demand for both, as Jean-Baptiste Fressoz has shown.[36] The economic structures built around each of these ages worked in similar fashions to exploit labour and enclose public goods. Some of these structures are discursive, working to legitimize acts of exploitation and appropriation based on phantasms of rationality, growth and prosperity, as well as 'civilizational' and thus racial assumptions. Locke's arguments, for example, invoked 'Indians' in North America specifically as separated from the 'Civiliz'd part of Mankind',[37] living lives of indolence and squalor that justified their being expropriated of their ancestral lands. That European settlers on stolen territory have a stronger claim to it than its traditional caretakers is a central pillar of what Brenna Bhandar calls 'racial regimes of ownership'[38] encoded in colonial legal systems, many of which persist to this day. The entangling of race and property rights rings down the ages from these colonial beginnings, and is nowhere better encapsulated than by W.E.B. Du Bois in 'The Souls of White Folk':

> 'But what on earth is whiteness that one should so desire it?'s Then always, somehow, some way, silently but clearly, I am given to understand that whiteness is the ownership of the earth forever and ever, Amen![39]

From at least the sixteenth century onwards, class stratification in Europe was mirrored in a global hierarchy of the human that offered the exploited workers of the Old World the psychological and economic consolations of racial superiority, a promise that most would eventually cash in on.[40] The property of being white in colonial North America, for example, was equated with the property of *not* being enslaved, and the ability to own land expropriated from Native Americans, which, as Cheryl Harris famously argues, made whiteness itself a kind of property.[41] This property has significant staying power, in that it survives evolving juridical regimes. By the twentieth century, the presumption that white people were *entitled* to the fruits of historical exploitation and expropriation was enshrined in affirmative action legislation which looked only forward, foreclosing redistributive or reparative action based on historical abuses. Thus what appears to be progressive social change is often old wine in new bottles. As Aileen Moreton-Robinson argues, the racial regime shaped the formation of European settler subjectivity around private property. The 'white possessive' is a kind of property relation grounded in self-possession, a sense of the centrality of one's own control over mind, body and social environment that worked historically to produce Europeans and their desires as the source of rational social ordering.[42] It is worth remembering in relation to this that for Locke, individual private property rights were the basis of a just social order that cleared away the influence of the church and nobility. From the perspective of the mercantile class, it was a liberatory imaginary reliant on commodification as a strategy to upend what was constructed as an outdated and superstitious social order.[43] Liberalism's articulation around private property is presented as if it were liberatory for everybody, when its historical conditions of possibility were in fact war, dispossession and exploitation.

Racial capitalism was thus spawned by primitive accumulation through landgrabs, genocides, and systems of appropriation and exploitation that endure in the present.[44] Each energy-stage of capitalism relied on the exploitation of marginalized and racialized people around the world and the enforcement of sustained access to energy extraction through military dominance. The overarching patterns of this world order have been sustained even as the racial composition of the capitalist class has changed. From the earliest days of the mercantile order, comprador classes of dominated peoples would 'bow to the power and ideology of empire and readily oppress and exploit their own people in order to secure their own position and wealth'.[45] The alliance of racialization with land enclosure has been an enduring one, for example, as is attested to in the complexities of non-Western neocolonialism in Africa that mobilize fictions of race, gender, class and enslaved descent.[46] Racial capitalism is thus a flexible and mobile matrix for the justification of accumulation based on a human hierarchy. It is also, as we shall see, an enduring feature of green capitalism.

## Capitalism goes 'green'

The idea of a 'transition' from petro- to 'green' capitalism has been driven by mounting evidence for the exhaustion of cheap natures, and a serious scientific attempt to calculate the harms being done to ecosystems and the future risks these portend. A fundamental claim of green capitalism is that, if correctly priced, environmental 'services' previously thought of as mere externalities would increase in value, allowing markets to determine the 'value' of mountains, rivers and animals while the 'costs' of carbon emissions, pollution and other harmful effects would also be set through trade. Because commerce grows value and keeps expenses down (so the logic goes) this 'new' capitalism should be let loose on environmental problems, which are so large and complex that only the system of dispersed decision-making offered by markets could properly solve them. No individual stakeholder could possibly know what is best for Nature, and thus the wisest course is to let the market decide.

As Louise Green has evocatively pointed out, however, the 'nature industry' thrives in conditions of scarcity, where commodification and loss 'exist together in one breath', the precise value of any part of an ecosystem being its 'precarious evanescence'.[47] Even the chief economics op-ed editor for the *Financial Times* has recently had to admit that the welfare of future humans is 'irrelevant' to green financial markets, and that 'market forces are not enough to halt climate change'.[48] Many of the most celebrated and well-publicized market-led biodiversity and carbon-reduction initiatives have been shown to increase poverty, displace communities, disrupt lifeways and negatively affect human rights.[49] Where transitions to non-fossil forms of energy are undertaken in order to maximize shareholder returns, as in the vast solar energy plants of South Africa's Northern Cape, '[low] wages and long working hours undermine households' ability to provide for their basic needs and wellbeing, exacerbating the crisis of social reproduction'.[50] The 'green' stage of capitalism is thus unsurprisingly as cannibalistic of its conditions of possibility as any before it, though it pretends to be able to 'give back' to ecosystems. On closer examination, this latter claim, too, cannot be sustained. The idea, for example, that a forest's capacity for carbon sequestration can be transformed into a privately owned commodity retraces the primitive accumulation of enclosure and colonial expropriation, wrenching ecosystems away from the complex socio-ecological relations that have historically sustained human and more-than-human life.[51] Examples abound of carbon sequestration programmes that have replaced biodiverse grasslands with monocultural tree plantations, some of which have not survived.[52] 'Clean energy' in the UK and elsewhere has resulted in logging and then burning old growth forests, on the logic that more carbon will be captured by the plantations that replace them.[53]

The people who benefit most from this wave of carbon enclosures are often bankers who restyle themselves as environmentalists, resignifying profitable financial transactions as philanthropic acts that claim to help the planet but are in fact often ineffective and sometimes harmful.[54] The thoroughgoing valorization of the entrepreneur in late capitalism has engendered a complacency about the role of rich people that Nima Shirazi and Adam Johnson have termed the 'billionaire knows best' trope.[55] The logic of late capitalism attributes to the famous and wealthy the moral and intellectual profile required to lead the world out of its current predicament. In a social order that values more than anything the sovereign individual who can leverage (what appear to be) their own resources and talents in order to amass vast amounts of money and prestige, these figures are given the licence to play the role of priest, seer and feudal lord all rolled into one. The male tech entrepreneur (or what the journalist Carole Cadwalladr calls the 'broligarch'[56]) has an especially prominent role to play. Neoliberal cultures conflate entrepreneurship with scientific acuity, forging an 'ecomodern masculinity'[57] – a heroic gender role periodically represented in film in the form of the 'cowboy scientist'[58] – through which consumer publics can manage anxieties about environmental breakdown. Rich men who appear to be far more intelligent and successful than the rest of humanity are depicted applying their bold thinking and maverick personalities to the climate crisis. Ecomodernism and its institutions – the futuristic-sounding corporations and research institutions with which these men are associated – work to lull the public into complacency about ecological breakdown. As James Brooks and Leah Wingard show, while outright denial of anthropogenic climate change has mostly come to seem a reactionary political position, the embrace of ecomodern thinking promotes 'implicatory denial'[59] that indefinitely delays any kind of systemic change and affirms the justness of the status quo. Ecomodernism's search for 'solutions' is an unreliable source of emancipatory futures for humanity, furthermore, as tech billionaires seem more interested in creating colonies on Mars than on saving Earth[60] and are on record as writing off the bulk of humanity as not intelligent enough to be taken along for the ride.[61] Ecomodernism constitutes a souped-up version of a strict Man/Nature dichotomy that privileges Man, reconfirming, as Tuana argues, 'all the lineages from which hyperseparation is animated – modernity, colonialization, extractivism – and from which ecological indifference and systemic racisms co-emerge'.[62]

The individualism of the green capitalist resonates strongly with the Lockean paradigm of free property owners looking to the state only to protect their property rights, staying as far away from interference in the economy as possible. What unites Locke's historical arguments for the expropriation of the commons and the schemes of green capitalists is that their primary concern is to 'imagine and legitimate particular social orders'[63] and not to 'save' Nature or even use it wisely. In other words, the centring of the liberal capitalist order on property rights prioritizes the perspectives and decisions of the vested interests that already control access to financial markets, making competition rather than emancipatory relations to each other and to the more-than-human world their guiding value.[64] The relation to 'Cheap Natures' is thus not fundamentally changed, only reoriented, and the modern/colonial world system thereby renovated.

This is nowhere more apparent than in the continuities between previous eras of extraction and the current rush for rare earth minerals to build batteries and other green technologies, which has expanded the frontiers of energy extractivism from Middle Eastern petrostates to the Global South. As is by now well-known, the massive amounts of lithium, cobalt, copper and nickel that are needed to make 'sustainable' technology are being mined in ways that cause social and ecological devastation. Nations that seek to reduce the socio-ecological harm of this extraction – such as Bolivia under Evo Morales's *Movement for Socialism* – are subjected to foreign interference, including US-backed coup-d'états and other forms of coercion. As the political scientist Thea Riofrancos has shown, the pressure to supply global markets squares off even ostensibly socialist governments against the communities affected by mining.[65] Development sociologist Max Ajl argues that the 'Green New Deals' of the progressive factions of centre-left political parties in the United States and elsewhere represent an accommodation with capitalism that perpetuates the world-system of extraction for profit, which like all phases of capitalism before it demand 'a robust and nimble military to secure accumulation on a world scale, including through the erasure of Arab, African, Asian, and Indigenous life'.[66] He proposes that the only 'Green New Deal' that would rise to the moment would be one that prioritized climate reparations, land restitution in both the imperial core and the periphery, and internationalist solidarity that centred on environmental justice.

The current state of international relations could hardly be further from this vision. Instead of solidarity between nations, intergovernmental agreements affirm a framework of rules within which powerful nations may compete for economic and military dominance as well as moral authority. Despite bearing the lion's share of responsibility for the ecological crisis, the Global North is able to position itself as planetary saviour. Former imperial powers have harnessed the 'redemptive' power of multilateral institutions such as the UNFCCC, the IMF and the World Bank which work to 'maintain the ecological and political economic order'.[67] Within this order, nations jostle with each other for leadership of the 'green transition' as a path to growth and prosperity in ways that confirm the justness of the world system, while working to increase rather than decrease disparities between nations.[68] A neoliberal approach to environmentalism led by the West is thus normalized.[69]

The irony of this supposed leadership is that it relies (in a similar fashion to previous energy stages of capitalism) on environmental load displacement. Apparently ecologically sound prosperity at the metropolitan centre is enabled by ecological devastation and exploitation at the peripheries, both in the Global South and in 'sacrifice zones' closer to the hearts of empire. As already discussed, the flow of environmental benefits to some populations and harms to others is the hallmark of environmental racism.[70] The imminent possibility of climate catastrophe has led not to solidarity and a reckoning with historical guilt for exploitation and oppression, but a frantic drive to secure the lifestyles of the global elite at whatever cost to racialized and working-class people, resulting in what Kai Heron terms 'eco-apartheid'.[71] Extractivism, 'green' financing and the renewable energy transition have been theorized as forming a single necropolitical nexus[72] that deepens colonial power inequalities between nations. Meanwhile, rich-country diplomats are dispatched to the Global South to upbraid its leaders about their responsibility to deal with climate change. In 2022, US climate envoy John Kerry called on African nations, that account for a negligible proportion of global carbon emissions, to reduce these even further, as 'Mother Nature does not care where those emissions come from'.[73] One instrument of this neocolonialism is debt, where technology transfer and foreign investment targeted at HIPCs are offered in return for concessions written into repayment terms that may be exploited by multinational corporations headquartered in rich countries.[74]

In other cases, the colonialism is even less subtle, with some governments conceding vast swathes of their territory to carbon credit vendors. The comprador class of Liberia, for example, recently attempted to sell around 10 per cent of its territory out from under the feet of the people to UAE-based Blue Carbon. Sometimes carbon credits are conjured from historical colonial landholding patterns, where, for example, white landowners in Northern Kenya work with 'community conservancies' to urge local herders to move their animals across grasslands more swiftly, generating spurious carbon-sequestration potential which is sold for hundreds of millions of dollars on global carbon markets.[75] Because so many of these schemes have resulted in disrupting the lifeways of Indigenous peoples, and in certain cases in their dispossession, Survival International has dubbed them 'blood carbon'.[76] Laurie Parsons names the global system at the heart of this model 'carbon colonialism'. He argues that whereas the world operates like a 'global factory' with profits flowing to owners and costs borne by workers, the fiction of a national carbon footprint allows rich, post-industrial Northern nations to pretend they are not a part of the same polluting world system as the Global South. Rich nations, instead of ceasing their contributions to environmental harm, accumulate resources to mitigate its impact. When the issue of historical culpability is addressed, it is often done disingenuously. The Danish government, for example, recently contributed $13 million to a climate reparations fund to near-universal acclaim. The Climate Action Network was a lone voice in pointing out that about a third of their spending would go to private insurers based in the European Union. Meanwhile, Switzerland is paying for the installation of cooking stoves and lighting in five million households in Ghana, crediting the Ghanaian carbon savings to the Swiss account. Though it would take nearly seven Ghanaians to emit as much carbon in a year as a single Swiss person,[77] it turns out to be within Swiss power to buy up other nations' negligible emissions rather than doing anything about their own, all while appearing as if they are doing good in the world.

Green capitalists can for the most part rely on fawning media coverage. The semiotic landscape of green capitalism is filled with representations of celebrity environmentalists on dramatic landscapes, the right kinds of humans enjoying natural beauty or communing with charismatic animals. What is rarer is any kind of systematic picture of the roots of biodiversity loss and climate crisis.[78] Far from getting to the 'truth' about nature, the social media logics of 'platform capitalism' have taken the baton from legacy media in representing the interests of the dominant forces in society, and lulling consumers into a sense of security through interaction with digital facsimiles of nature, or what Bram Büscher and his colleagues have termed 'Nature 2.0'.[79] In late capitalist modernity, these signs are cultural commodities, and as described above, they permeate climates of 'conscientious' consumption.[80] Because consumption is now oriented around conscientiousness, significant amounts of money are chased by communications entrepreneurs in service of capital. An excellent example of this is the Stockholm-based media start-up 'We Don't Have Time' which claims to be 'the world's largest media platform for climate action'.[81] As their CEO Ingmar Rentzhog explains in his pitch to investors, the sense of urgency promoted in the company's name is alloyed with a focus first on their own profitability and then on the profitability of its clients: 'We're focusing on profitability because if we can't be profitable on doing sustainable business, we can't do business in the long run'.[82] The dream of green capitalism is so nicely encapsulated in this statement: that one never has to choose between a profitable business and ecological harmony. The question never even arises, as it is beyond the horizon of capitalist intelligibility. Profit is constructed as the *only way* out of environmental catastrophe; if your plan for environmental protection is not profitable, then it is also foolish, absurd, impossible. It is difficult to overstate the extent to which this is precisely the same pattern of thinking that caused the environmental crisis in the first place.

## Summarizing the green capitalist imaginary

In the first part of this chapter, I laid out the discursive dynamics of green capitalism through reviewing a fair chunk of the critical literature on this and related phenomena. In this second part of the chapter, I summarize the components of the green capitalist imaginary, before presenting three case studies of how this imaginary is realized in textual form. My account summarized in Table 1, is of necessity a partial one, and I therefore encourage analysts to pursue their own interpretations of how green capitalism is discursively materialized, in the hope that the map I present here provides a useful starting point.

In her excoriating 2022 book *The Value of a Whale: On the Illusions of Green Capitalism*, Adrienne Buller notes how people across power and class differences conceive of environmental problems in ways shaped by market logics and economic calculations.[83] Liberal notions of open markets, free choice and maximizing utility through the pursuit of self-interest are thus unchallenged

**Table 1** Summary of the green capitalist imaginary.

| **GREEN CAPITALIST IMAGINARY** | |
|---|---|
| Enlightened consumers making individual choices within free markets will solve the ecological crisis. Our objective economic and scientific calculations will chart the way to sustainability: an ecomodern future state where people, planet, and profit are always in balance | |
| **SUBJECT ROLES**<br>• The conscientious consumer<br>• The unsustainable Other<br>• The benevolent billionaire<br>• The ecomodern visionary<br>• The lunatic activist mob | **INSTITUTIONS**<br>• Green certification<br>• ESG reporting<br>• Recycling practices and infrastructures<br>• Carbon markets<br>• Sustainable Development Goals |
| **CHRONOTOPES**<br>• 'Future eco-perfect'<br>• Smooth green spaces of consumption<br>• Eco-village heterotopias<br>• Faraway wastelands | **AFFECT**<br>• Hope<br>• Ease<br>• Sadness / Anger<br>• Harmony<br>• Empowerment |

as the organizing values, or nodal points, of green capitalism, even as they are joined by the presence of 'ecological', 'organic' or 'green' values. Indeed, the value of 'choice' works even more effectively to secure consent for the status quo when bolstered by the warm glow of identifying as an environmental steward, the *jouissance* of feeling earthy. A 'sustainable' or 'green' capitalist economy where my judicious choices as a consumer save the planet delimits the horizon of intelligibility beyond which lie absurd alternatives: systems that *just don't work*,[84] unfreedom, scarcity, Communism and ruin. The political character of capitalist realism is obscured through the mono-epistemic universality of neo-classical economics which claims 'scientific objectivity'[85] in its understanding of how the world *really* works and, in a classic case of the environmental fallacy, how it thus *ought* to work. Following 'the science' is thus accepting a transdisciplinary knowledge formation that combines meteorological models with financial instruments, and complex biochemical projections with actuarial risk calculations, so as to create the impression of a wholly dispassionate map of the future that is above the political fray.

## Green capitalist subject roles

The modes of subjectivity hailed into being within the green capitalist imaginary are avatars of economic and scientific rationality who interpret this 'profoundly human crisis through an almost puritanically economic lens'.[86] Investment bankers and other money men are authorized to speak, while activists who dare to disrupt the status quo are targets of racist violence,[87] sexist and ableist hate speech,[88] and called 'idiot criminals', 'barbarians', 'terrorists' and 'lesbians and feminists' who deserve to be raped.[89] Positive media coverage is meanwhile enjoyed by 'benevolent billionaires'[90] whose philanthropic schemes deepen power inequalities, while a privileged role is reserved for the ecomodern saviour, the tech *bro* working in *his* lab to invent solutions that will disrupt polluting and unpleasant industries and usher us into a new reality. In the brave new green world, all of our desires may be met without any impact on the planet, and thoughtless consumption made guilt-free again.

Until that day, the most accessible subject role into which people who are not tech bros, scientists, financiers, economists or billionaires are interpellated – as I sketched at the start of this chapter – is that of the 'conscientious consumer'. In preparation for their roles as conscientious consumers, environmental education produces what Malin Ideland has termed 'eco-certified' children.[91] These emergent subject positions, which she analyses in the Swedish context, hail all young people into a singular national identity built on assumptions of national ethnic sameness that problematizes immigrant children as in need of intervention, getting these 'children of concrete' from their urban milieus outside into the healthy, invigorating forests.[92] Similarly raced and classed fictions are observable in other contexts.[93] These are the children who will grow up to always recycle, to reduce their carbon footprint through flying less, or who change their diet to contribute less to both animal cruelty and agricultural emissions. As has been pointed out by a number of scholars, the idea of the 'carbon footprint' was an industry coup, orchestrated by corporate players including British Petroleum, that effectively transferred responsibility for reducing carbon emissions onto consumers rather than producers.[94] The ideal subject of green capitalism affirms that it is an individual's responsibility to reduce their carbon emissions, as opposed to a collective one to shut down extractive and destructive industries, and that there is an ethical lack in people who continue to make unhealthy, carbon uncompensated, 'unsustainable' and ecologically disharmonious purchasing choices.

## Green capitalist institutions

After eco-conscious shopping, the practice of household recycling is the model practice of the conscientious consumer. It is presented as a simple, feel-good activity that is continuous with everyday consumer behaviours, pushing to the background the often very dirty and not so environmentally friendly things that happen with recycled waste.[95] As Per Ledin and David Machin observe, marketing images of recycling 'place saving the environment at the level of personal choice and consumer behaviour rather than something instituted on a large scale by governments'.[96] Because recycling is such a ubiquitous daily practice it has a special role to play in the stabilization of subject roles: we are to a large extent, as per Marx's insight, that which we spend our time producing, and we produce vast amounts of sorted refuse.[97] Other green capitalist institutional frameworks that have emerged to 'solve' the climate crisis, such as 'green' finance, triple-bottom-line financial reporting, carbon trading and the certification of certain products as 'organic' or 'sustainable' rely equally on the idea that an enlightened customer will choose the less environmentally harmful product if given the chance to do so, and perhaps even pay a premium for it. State subsidies for transitioning to renewable energy, meanwhile, tend to be grounded in logics of 'green growth' and energy sovereignty that offer the promise of long-term gains to corporations and the economy as a whole.

The private right to the extraction of surplus value from the economy is thus left inviolate, imagined as the solution to ecological breakdown rather than its original cause.

## Green capitalist chronotopes

Governments that claim to take environmental decline seriously engage in a repeated 'strategy of deferral' that 'involves constant, confident gestures to an idealised future'[98] as Rosemary-Claire Collard and Jessica Dempsey have argued. This 'future eco-perfect' tense is ubiquitous within the ecomodernist visions invoked by tech bros and benevolent billionaires who, working together with liberal institutions, are understood as ushering us into ecological utopia. Pockets of this future are already available in Foucauldian 'heterotopias' – 'counter-sites, a kind of effectively enacted utopia in which the real sites, all the other real sites that can be found within the culture, are simultaneously represented, contested, and inverted'.[99] These 'heterotopias' include farmers markets, city developments around green spaces, eco-villages and other spatial formations where closeness to nature is produced in ways that retrace racial and class divides.[100] What these spaces share is that they contain the kernels of their own negation, seeking to instantiate social and ecological harmony while relying on faraway scenes of capitalist-colonial extraction that must be suppressed, and societal relations of domination barely visible below the surface. Shopping malls, luxury hotels and methods of transportation offer similar insulation from the stinking, polluting underbelly of capitalist modernity, allowing endless mobility through interconnected smooth spaces that oil the wheels of commerce.

## Green capitalist affect

In times of crisis, capitalism presents itself as the only way out. In everyday text and talk capitalism is equated with hope: hope for the future, hope for 'our' children, hope that everything will work out alright in the end.[101] Far from the kind of critical hope that informs the revolutionary liberation practices of activists, capitalist hope is primarily therapeutic. It has analgesic, soporific and tranquilizing properties. It postpones action and shifts responsibility. As the veteran environmentalist Bill McKibben observes:

> When they say, 'You give me hope,' part of what they're saying is, 'I don't want to feel so bad about myself.' The thing that drives me crazy is when people say, 'Well, it's up to the next generation to deal with these problems, and they're so

> good.' That is a cop-out of major proportions. I accepted a long time ago that a large part of my job was just going to be to be a professional bummer-outer of other people [...] I don't think we owe anyone hope. The job of activists is not to make people feel psychologically improved and more hopeful.[102]

The job of green capitalists, on the other hand, is all about aligning hope for the future with *feeling good* about their purchasing choices. Their 'hope' is that branding participation in systemic social and ecological exploitation as eco-conscious or 'environmentally friendly' will keep consumers docile and unquestioning. While sadness at despoliation and anger at polluting Others might be strategically channelled, for consumption to continue customers must be given a sense of *harmony*, of being enveloped in organic goodness and in relations of mutual *goodwill* with picturesque rural coffee-growers, in a ploy to keep them moving through the smooth spaces built especially for them. To keep them moving they must feel at *ease*: comfortable, able to flow and convinced they are the ones in control. They must feel *empowered* to choose, to shop and to not look back.

## Textual analysis

In this section, I present three cases where elements of the green capitalist imaginary are in evidence.

### Case 1: 'Mindful travel'

In August 2022, I received an e-mail from Europe's largest airline, Lufthansa, under the subject line 'The sustainable way to fly'. At the head of the text is a composite of three images with the superimposed headline 'Mindful travel', the Lufthansa logo and the hashtag #MakeChangeFly. The images are of a calm lake reflecting the high mountains around it; a side profile of a blonde young man with the setting sun glowing through his wet hair; and a smaller inset action scene where a dark-haired young white man climbs out of the water onto a wooden jetty while another dives from it into the water. The gentle smile on the lips of the wet-haired youth suggests contentment, reflected in the perfectly still lake. The relative sizes of the images, where the action is diminished in one small scene while serenity is magnified in two large ones, narrativize an afternoon swim with a beginning, a middle and an end. The lake was calm before the boys started diving, swimming and having fun. They could be brothers or best friends – perhaps on holiday. Now they have spent their energy, and the sun is setting high over the mountains. The lake has returned to a state of calm. The youth whose profile we see up close happily surveys the still water, the exhilaration of boyhood camaraderie and of plunging into the lake still fresh in his memory as the lowering sun turns his hair to gold.

Sunsets are often times of peaceful reflection, and the colour palette of the images is appropriately warm and muted. One can imagine an audience that is probably mostly white, possibly middle-aged and heterosexual, seeing in the play of these boys echoes of their own youth, or perhaps that of their children, and comforted by warm, perhaps nostalgic feelings. Underneath the headline and images, the first two sentences of the e-mail read as follows:

> Travel connects people, countries, and cultures. To ensure that you can continue to travel with peace of mind in the future, we are always looking for innovative solutions and effective measures to create a carbon-neutral aviation industry.

Within the chronotope of the aviation industry as presented here, 'travel' is a nominalized agent performing good deeds in world-space, 'connecting people, countries and cultures'. (Other, less noble reasons for travelling – personal status? colonial invasion? sex trafficking? fleeing the scene of a crime? – are not considered.) We see the world coming together in real time, in harmony, a brotherhood of Man, like the two boys at the lake, thanks to this agent. But there is a problem: our kumbaya moment is threatened by the 'peace of mind' of 'you', the person who should be travelling. Though the source of 'your' disquiet is not named explicitly in this text, it has left its traces in how 'we' (the personalized, intimate voice of Lufthansa) are tirelessly searching for the perfect fusion of scientific and commercial rationality ('innovative solutions and effective measures') that will create in the perfect eco-future tense a 'carbon-neutral aviation industry'. This should come as a great relief to you, the consumer, as it is your peace of mind that stands between the world and its harmonious, connected future. Now that you know we've found the solution, it is once again time to let 'travel' do its magic.

Subtly, powerfully and with admirable economy this communication creates a narrative structure where progress towards global connectivity is disrupted by an unseen force, which disruption is then repaired by the good works of a protagonist, who in the final act restores the initial equilibrium.[103] Much like the lake, whose surface is once again calm after being splashed about in, we can return to quiet contemplation and happy thoughts. The role of conscientious traveller is valorized in this drama: you want to be part of the connectivity that brings the world together, but you also want peace of mind. What Lufthansa is doing is *empowering you* to act ethically in this complex world. With their special fusion of science and commerce they have laid a powerful tool at your feet, and you now just need to pick it up (as the rest of the text goes on to detail): a 'Green fare' where 'carbon-neutral flying is already included in the price' that allows customers to 'contribute to making the future of travel more sustainable'. *The ball is in your court.* If you are ethical enough to have your peace of mind disrupted by this unnamed force, then you will also take seriously your responsibility to pay extra for a 'Green' fare.

What must not be brought to mind in this narrative of thwarted connectivity are the many effects of travel on human and more-than-human life, on the intricate webs of climate, labour, hydrocarbons and the other parts of supply chains stretching around the world that make flight possible. When calculated per passenger and tonne of freight, aviation covers about half the distance that rail does in a year, though with *eleven times* the carbon emissions.[104] Meanwhile, only about 4 per cent of the world's population fly internationally in any given year, with the 'frequent flyer' top 1 per cent accounting for *half* of emissions.[105] 'Connecting' the world's 'people, countries and cultures' is thus a course of action available only to the select few, but at a cost to the planet that they do not have to bear equally with all humanity. As relatively well-to-do people, this sector of the global population will be protected (initially at least) from the worst effects of global warming, while the global poor face disease, starvation and homelessness. But if our analysis only takes us so far, we will not yet have seen through to the literal interests beneath the metaphors of connection and placid lakes. The coup de grâce of this text is that it is selling a product – the 'Green fare' – that (though this is left implicit) will come at a premium, meaning that the conscientious consumer is being asked to pay extra for 'sustainable aviation fuel (20 %) and climate protection projects (80%)' that make flights 'carbon neutral'. These precise percentages, which on the one hand (via a topos of numbers) speak to scientific certainty, hint at a more extensive spreadsheet of calculations where the costs of these initiatives are borne not by every passenger, but only by those who *choose* to pay extra. It is this *ethical* and *mindful* choice of the consumer, cloaked in childhood nostalgia, that keeps Lufthansa profitable, allowing dividends to be paid to investors and shareholders. Profit is the literal meaning of this text that metaphors of connection obscure, as it remains sacrosanct even as dark disruptive forces are hinted at. Lufthansa's continued profitability and growth are metaphorized as a childhood holiday suffused with positive affect so that the costs of (small and perhaps insignificant) emissions reductions may be passed on to its consumers, with no serious admission of corporate responsibility for climatic impacts.

## Case 2: Against 'wishful thinking'

The following extracts are taken from an essay published in September 2022 in the 'Ideas' section of *The Atlantic*, a centrist liberal periodical that covers mostly US politics, culture and technology. The author is Alec Stapp, who co-founded the Institute for Progress, a business-friendly think tank based in Washington, DC. Stapp interweaves compelling accounts of progress in renewable energy projects with instances of 'progressive' interference in their realization. For the purposes of this analysis, I will focus on how the problems that stand in the way of 'progress' are represented.

### What Many Progressives Misunderstand About Fighting Climate Change

Wishful thinking hampers the clean-energy revolution
By Alec Stapp

Since the 1960s, fighting for the environment has frequently meant fighting against corporations. To curb pollution, activists have worked to thwart new oil drilling, coal-fired power plants, fracking for natural gas, and fuel pipelines. But today, Americans face a climate challenge that can't be solved by just saying no again and again.

Decarbonizing the economy will require an unprecedented amount of new energy investment. […] And the same laws that environmental groups leveraged in the past to block or delay fossil-fuel projects are now being exploited by NIMBYs in ways that, however well intended, will slow the country's transition to green energy. […]

[…] Jamie Henn, an environmental activist and the director of Fossil Free Media, recently told *Rolling Stone*, "Look, I want to get carbon out of the atmosphere, but this is such an opportunity to remake our society. But if we just perpetuate the same harms in a clean-energy economy, and it's a world of Exxons and Elon Musks – oh, man, what a nightmare." […]

Such comments raise a question: What is the real goal here – stopping climate change or abolishing capitalism? Taking climate change seriously

> requires an all-hands-on-deck attitude and a recognition that technological solutions (yes, often built and deployed by private firms) can deliver real progress on decarbonization before the proletariat has seized the means of production. A massive infusion of private investment, made not for charity but in the anticipation of future profits, is precisely what's needed [...]
>
> The idea that top-down decision makers can choreograph precisely how the clean-energy revolution will proceed runs deep in progressive circles. [...] In the face of uncertainty about the best path to decarbonization, policy makers should think like a venture capitalist – placing lots of bets in the expectation that some technologies will fail but the investment portfolio will succeed as a whole. [...]
>
> Yet permitting reform requires loosening regulations and laws that many environmentalists hold dear. The National Environmental Policy Act requires reviews that give enormous power to anyone who wants to block or delay a proposed energy project, either out of genuine social concern or for self-interested reasons. [...]
>
> Yet we cannot succeed in the fight against global warming without giving many alternatives to the status quo an opportunity to evolve and prove themselves. In reality, the false solution to climate change isn't geoengineering or nuclear energy – it's the belief that we can decarbonize the economy only by upending our economic system, categorically rejecting certain technologies, and spurning private investment.

In this text, the role of ecomodern logic in structuring the green capitalist imaginary is especially clear. This is established through chains of equivalence, where 'decarbonizing' relies on 'technologies' that require 'investment' from 'private firms' anticipating 'future profits' which are thus imagined as the precondition for 'stopping climate change'. The specific kind of ecomodern subjectivity connected to this logic is made explicit in the insistence that policymakers should think 'like a venture capitalist' when facing uncertainty, placing a number of bets on different technologies in the knowledge that the 'portfolio will succeed as a whole'. The replacement of the political by the economic is the hallmark of neoliberal subjectivity, as a number of scholars have shown.[106] It is further fuelled by the centrist fantasy of technocracy, where politics is (in Chantal Mouffe's account) not a 'partisan confrontation but the neutral management of public affairs'.[107] This vision of social life invites all social collectivities to see their own interests as aligned with those of the neatly interlocking institutional realms of for-profit corporations, scientific and technological advancement and finance, which together will deliver a brighter future. In the face of uncertainty about what will be the most effective technology, the rational choice is to spread bets, safe in the knowledge that the market will deliver some solutions (outlined in the excised parts of the essay) that outperform others.

The primary aim of this text is the construction of figures of personhood that antagonize the attainment of this future eco-perfect world: the 'Progressives' introduced in the headline who 'Misunderstand' the fight against climate change. As a nominalization, 'progressive' simplifies the complexity of left-wing political views into the essence of a character: they do not merely believe in social progress, they are a specific *type* of person. In US politics the 'progressive' is the mirror image of the 'conservative' – which usefully makes room in Democrat discourse for a happy medium that is not dogmatically partisan, that is neither socialist nor right wing, but that merely follows the 'evidence' of science and the market, achieving perfect neutrality. 'Progressive' also indexes members of the Congressional Progressive Caucus, a group of Democratic lawmakers chaired at its founding in the 1990s by Bernie Sanders (who is one of the people cited in this text as a 'progressive') and now includes members of the 'Squad' such as Alexandria Ocasio-Cortez, Rashida Tlaib and Ilhan Omar.

While these Congressional 'progressives' come from different racial and class backgrounds, there are some clues in the text that suggest Stapp is targeting middle-class white progressives specifically in his critique. In starting his argument in the 1960s, he invokes countercultural environmentalism in the wake of Rachel Carson's *Silent Spring*, a movement that from the start was dominated by white hippies. The reference to NIMBYs in the second paragraph, on the other hand, indexes a privileged, probably white, homeowning subject.[108] Jamie Henn's exclamation 'oh, man, what a nightmare!' in contemplating the proliferation of Elon Musks gives voice to a slightly spaced-out 'dude' – a figure of personhood that tends to be coded as white. Moreover, when Stapp writes that we can make progress on decarbonization 'before the proletariat has seized the means of production' the abrupt change of register into Marxist terminology indicates a change in voice and characterological figure, signalling that he is troping on this voice parodically. These semiotic choices assemble a specific style of person: a person who talks about the proletariat and says 'oh, man!', who is both a hippie and a privileged NIMBY, and who 'misunderstands' important things about science and finance. Importantly, this person is given 'enormous power' by the government, which they often use for selfish or short-sighted reasons. Because Stapp has made them seem like privileged white 'radicals' he has constructed a controlling image that it is viable to poke fun at.

The image of the irrational hippie is illustrated in detail throughout the essay. In the subhead, this person is said to engage in 'wishful thinking'. They say 'no again and again'. Their activism may be 'well intended' but it slows down the energy transition (i.e. they cannot think through the consequences of their actions). They have the idea that top-down decisions can 'choreograph' the market – because these people like the performing arts, clearly, and think of everything in terms of dance. They are organized in 'circles' (as hippies tend to be), they cling to the regulations they 'hold dear' and 'spurn' private investors as if they were lovers. Stapp consistently draws from the lexical fields of Flower Power and the childish, irrational and solipsistic excesses of 1960s counterculture. The 'progressive' is a person who has refused to grow up, who is lost in a haze of fantasies about the way the world ought to be, and like a child throwing a tantrum, will only say 'no again and again'. There is just no reasoning with these people: they are overgrown babies, probably high, who have surrendered to their emotions and cannot think through the impacts of their own actions. And yet they are so powerful!

The ideological nub of this essay lies in Stapp's assertion that taking 'climate change seriously requires an all-hands-on-deck attitude'. Whereas corporations may have been deserving of critique in the past, the implication is that at this historical juncture doing so is a dereliction of duty, a kind of mutiny against enlightenment. We are in a crisis, and crisis means we must *cooperate with the ship's officers*. This is no time to question, only to do. Stapp thus establishes high deontic modality: we are obliged to conform. In this chronotope of ageing flower children at odds with modernity, 'progressive' unreason and emotion are threats to humanity. They do not propose a coherent alternative, and have no social vision, as they are too lost in their own dreams and wishful thinking. They may thus be safely ignored, kept below decks or – if they prove too much of a nuisance – forced to walk the plank.

## Case 3: An 'elegant' idea

On 3 October 2024, *The New York Times* published a news story in its Climate section by Manuela Andreoni under the headline 'An "Elegant" Idea Could Pay Billions to Protect Trees'. The subhead explains that 'Brazil is proposing a fund that would pay countries to protect tropical forests that are crucial to curbing climate change. It would generate returns, too'.

Andreoni's story introduces the Tropical Forests Forever Facility (TFFF), a scheme that would pay $4 per hectare of protected forest per annum in a bid to enable governments in 'developing' countries such as Brazil to put an end to deforestation. At an anticipated $125 billion, the TFFF would be the 'world's biggest pot of money to help fight climate change and biodiversity loss'. Though the story mentions reasons to be sceptical that the scheme will work, most of the people quoted in the text think it is worth trying. An investment banker states that it is 'crazy in a kind of interesting way'; a US State Department official says that it is 'elegant' and 'seems to have some political momentum'.

According to the story, the TFFF would require no donations, only 'investments'. The first $25 billion would come from governments and big philanthropies in the form of interest-bearing loans. The rest would come from private investors who would be paid returns slightly higher than those of Treasury bonds. In essence, the TFFF would work as a bank, investing in a 'diversified portfolio' generating returns that go first to its investors, with the 'excess' returns divided up between seventy nations with tropical forests. Countries with high rates of deforestation would be ejected from the scheme, while those with lower rates would be docked $400 from their payout for every hectare lost in a given year. According to the numbers in the article, the annual payout to protected forests would be in the region of $1.8 billion (or a 1.4 per cent yield on $125 billion) compared to the over 4.5 per cent rate of return offered to private investors. The TFFF thus works on the assumption that a large enough gap exists between what investors will accept as a decent return on their money and what effective portfolio managers can actually do with it in order to generate the cash that will curb deforestation. While there are many parts of this scheme that warrant critique, interesting perspectives on the TFFF are offered by ministers for the environment in the governments of Brazil and Malaysia:

> If the fund was already operating, [Brazil] would have received some $600 million this year from its existing forests, minus what it lost. That's almost double the yearly budget of the country's environment ministry.
>
> 'We have long talked about the advantages of conservation, but haven't been able to, let's say, translate that into concrete things people can feel,' said Marina Silva, Brazil's environment minister.
>
> But with the fund disbursing real cash for standing trees, she added, politicians who now fight environmental protection policies at every turn may start 'counting each hectare of preserved forest and thanking each Indigenous person inside their state.'
>
> Governments of key forest countries, such as Colombia and Malaysia, have voiced their support for the proposal and its design.
>
> Finding the resources to rival the economic power ravaging forests, said Nik Nazmi, Malaysia's environment minister, 'will be crucial for us.'

At least two things stand out in Brazilian minister Marina Silva's rationale for supporting the TFFF. The first is her claim that for 'people' the existence of the forests is not a 'concrete' advantage of conservation that they 'feel'. Given that she mentions Indigenous people soon afterwards, these cannot be the 'people' she refers to, as it is a fair assumption to make that they do in fact 'feel' the advantages of not (yet) having their home destroyed. The 'people' she refers to neither live in the forest nor own it. Their demands to 'feel' advantages from its conservation are naturalized in her statement as existing within the appropriative colonial logic of the state, which is constructed as existing to facilitate the exploitation of territory by its citizens in ways that place them at an 'advantage' with respect to other nations. She thus opts not to challenge, but to affirm the logic of global market competitiveness through Indigenous dispossession. The second thing to note is the future state of affairs she imagines being brought about by the TFFF: that 'people' who were champing at the bit to dispossess Indigenous people of their forest home would now 'thank' them for preserving the return-generating hectares. Indeed, she accounts for the conservation of the Amazon in ways uncannily similar to the Lockean argument for Indigenous dispossession. In Locke's account, Native Americans stood outside of civilization leading lives of indolence and squalor as they neither 'improved' their land nor attached monetary value to it. This 'waste' of land was what made its appropriation by white settlers who would make it productive the *ethical* basis for the liberal-colonial social order. In a green capitalist spin on Locke, da Silva rehearses a logic where the reason Indigenous people ought not be dispossessed of their land is that it is *no longer waste*: it has been shown to have economic value in its 'unimproved' state, and should therefore be left as it is, as it is in the financial interests of the Brazilian settler-colonial state. The Amazon has thus been enclosed by proxy, as the state extracts rent from forest and the lifeways of its Indigenes in ways that pass on financial rewards to settlers.

The Malaysian minister's statement meanwhile is tinged with bathos. The 'resources' they hope will 'rival the economic power ravaging forests' are in fact the infinitesimally small proceeds from the normal operations of the very economic power they say they need to fight. At 0.0018 per cent of world GDP (which is well over $100 trillion) the proceeds from the TFFF that are envisaged as allowing developing countries' environmental ministries to 'rival' this 'economic power' are placed in more sobering perspective. Private investors are getting about three times the returns that the forests are. Furthermore, as the TFFF model is built on investment for profit in the world economy as it is, warts and all, the returns it is ploughing into forests themselves consist of the surplus value extracted from Fraser's four enabling conditions of capitalism: (other) natural resources, social reproduction, political power and the wealth of others. The forest fund is thus no less a snake eating its own tail than any other financial instrument under capitalism.

Though the story gives most of the credit for the TFFF to Garo Batmanian, director of the Brazilian Forest Service, it concludes with a legitimation strategy that works to authorize the fund through reference to the mainstream and Euro-American institutions of neoclassical economics and the World Bank:

> The general idea of an investment fund that rewarded standing trees was first imagined some 15 years ago by the economist Kenneth Lay when he was treasurer of the World Bank. Since then, it has been making rounds in the development finance and conservation circles.
>
> While the project has changed since Brazil took it on last year, it continues to rest on a fairly simple idea, Mr. Lay said. The fund would give developing countries something they have tirelessly fought for: the kind of cheap money to which only wealthy nations have access.
>
> Many developing nations simply wouldn't be able to borrow money at a low enough interest rate to fund forest protection on this scale.
>
> When he talks about the idea, Mr. Lay often gets asked the same question: Couldn't this work to fund many other of the world's most difficult causes, such as the distribution of vaccines in the developing world or curbing plastic pollution?
>
> 'You're right,' he always responds. 'But let's start with the trees.'

Lay's mention of 'cheap money' references the facility with which large industrialized nations are able to create endless lines of credit for themselves and pay little or no interest, while HIPCs have to choose between much higher interest rates or taking loans conditional on neocolonial and neoliberal policy reforms. This is a complex topic in its own right but, as discussed above, what is important to note is how the image of countries 'tirelessly' fighting for cheap money is disconnected from histories of colonial domination and extraction. The lack of a genealogical approach to developing world debt constructs poor countries as lacking resourcefulness and rich ones as deserving and benevolent. Being able to secure the money to embark on projects that provide social goods such as education, health and welfare was one of the key negotiating chips during the Cold War, and much of the infrastructure of the World Bank and

International Monetary Fund was established in order to integrate the poor world into global markets in ways that favoured Euro-American capitalist interests.[109] What is on full display here in Lay's remarks is the defence of the world economic system and of neoclassical economics *at any cost*, where the existence of drastic inequalities and life outcomes is seen as a normal and entirely acceptable feature of the world, so much so that it is only worth skimming a couple of percentage points off planetary return rates to address these matters of life or death. The cavalier and dismissive riposte – 'let's start with the trees' – bespeaks the ethical poverty that lies at the heart of the green capitalist imaginary.

A final point on the logic of this imaginary is that the adjective chosen to describe the TFFF in the headline as an 'elegant' idea pithily expresses the affect of ease and harmony that suffuses the spaces through which the conscientious consumer moves. The fund managers behind the TFFF will doubtless be excited about a new product they can sell to clients that appeals to their sense of proving themselves ethically through consuming and profiting. That it is easy, 'elegant' and smooth is a key element of how it must be marketed. As I have noted from the outset of this chapter, the climate of green capitalism dictates that we must carry on shopping, we must follow our desires, like sharks in water we cannot stop moving through the smooth spaces designed especially for us. When we live in hope that capitalism will save us, feeling complacent about our own ethical participation is precisely the kind of attitude the conscientious consumer is valued for.

## Conclusion: Is there no alternative?

That great harbinger of late capitalist modernity, Margaret Thatcher, is often associated with TINA ideology, the idea that 'there is no alternative' to liberal market capitalism. And yet if there really was no alternative, the constant need to legitimate capitalism, to renovate its foundations, to valorize its avatars and envelop its subjects in positive affect, wouldn't be worth all that effort. Masses of people know very well that capitalism transforms the 'live green earth [...] into dead gold bricks, with luxury items for the few and toxic slag heaps for the many'.[110] As the effects of the ecological crisis are increasingly felt throughout the world, capitalist ideologues are forced into defensive positions, having to justify themselves, having to repair the damage to the edifice of business-as-usual caused by wildfires, hurricanes, floods and droughts, and called on to articulate just why it is that the future should belong to them.

There are, in fact, alternatives. One vision that is gaining in popularity and force is offered by the politics of 'degrowth' – which is sometimes framed as the discovery of a 'greener' Marx.[111] To 'degrow' the world economy and thus keep human economic activity within planetary boundaries, while delivering social justice, developed nations must significantly curb and even reverse their economic growth, while allowing other nations to focus on extending social care and building infrastructure. Put simply, the idea is that growth be redistributed from rich to poor[112] and that instead of serving as the base for individual wealth accumulation, technology should serve the interests of the people. Degrowth thus potentially forms part of an anti-colonial political project that has as its aim reparations and restitution for the wealth historically extracted from colonized and enslaved people.[113] Degrowth envisions planning the economy at a global scale, but without the rapacious extractivism associated with historical socialist regimes.[114]

Ecomodernist Marxists have however argued that degrowth would never be a viable popular strategy as workers' unions of industrialized nations would not accept a 'less is more' ideology.[115] This critique has been met with the rejoinder that to locate the revolutionary force of history as solely in the factories of the developed world, and not in global coalitions of marginalized people, repeats the race-insensitive analysis of historical Marxism.[116] Scholars and activists have started calling for a People's Green New Deal,[117] for a Red Deal[118] and for 'an anti-imperialist eco-communism that understands how relations of dependency and uneven ecological exchange devastate ecologies and exploit workers in both core and periphery'.[119] Though there are tensions and contradictions, these are all critical projects that seek to historicize the present conjuncture and forge a way out of that does not merely repeat the mistakes of the past. What the green capitalist imaginary demands of us is that we see it and it alone as the realm of solutions, and anything else as 'wishful thinking'. These are the ideological blinkers that we must urgently remove.

# 4

# 'Barbarian hordes': Overpopulation and the climate migrant

The spectre of the climate migrant haunts environmentalist texts. Occasionally this figure is the focus of an empathetic concern for the suffering of others; more typically it is invoked as a source of fear for what will happen if millions or even billions of people worldwide are displaced by environmental breakdown and start heading for safer, richer places. In this chapter, I consider how the controlling figure of the climate migrant is articulated within broader discourses of overpopulation. Metropolitan calculations about the 'carrying capacity' of specific environments motivate a zero-sum politics, where any gain for an extra-national or ethnic other is seen as a loss for the 'indigenous' population. These discourses tend to fixate on the threat of uncontrolled population growth as largely a 'Third World' problem, working to shift blame for the planetary crisis to poor and often racialized people, against all evidence to the contrary. Demography becomes a question of Western security, and the militarization of borders part and parcel of climate 'resilience'. Anxiety about the low fertility of 'white' populations in comparison to the rest of the world's 'dark' populations energizes this imaginary, as well as the ecological nationalisms we will discuss in Chapter 6. Overpopulation discourse is however ubiquitous in Western texts across the political spectrum and should not be thought of as the exclusive preserve of the right.

## The jungle and the garden

In his 2022 address to the European Diplomatic Academy, the EU's foreign affairs chief Josep Borrell contrasted the continent to the rest of the world in colourful terms:

> Europe is a garden; we have built a garden. Everything works. It is the best combination of political freedom, economic prosperity and social cohesion that

> humankind has been able to build [...] The rest of the world [...] is not exactly a garden [...] Most of the rest of the world is a jungle, and the jungle could invade the garden [...] But [the gardeners] will not protect the garden by [...] building walls. A nice small garden surrounded by high walls in order to prevent the jungle coming in is not going to be a solution because the [...] jungle has a strong growth capacity, and the wall will never be high enough in order to protect the garden. The gardeners have to go to the jungle. The Europeans have to be much more engaged with the rest of the world, otherwise the rest of the world will invade us by different ways [and] means.[1]

This powerful image of an encroaching fecundity, an excess of life with its 'strong growth capacity' spilling over onto the neatly manicured lawns, geometrical hedgerows and roses of Europe sparked an online firestorm. On social media the speech was condemned for its 'condescending, out-of-touch and racist undertones [reflecting] the Western sphere's superiority complex over the Global South'.[2] In response, Borrell published a defence on his official blog under the title 'On metaphors and geo-politics'.[3] He begins his reply pointing out that the garden/jungle distinction is not his invention, but the subject of a perennial debate: 'should the international order be based on principles accepted by all, regardless of the strength of its actors, or should it be based on the will of the strongest, which is commonly called "the law of the jungle"?'.[4] It is not the rest of the world in general, but Russia's invasion of the Ukraine that he claims justifies the garden/jungle metaphor:

> [...] War is back in Europe and around the globe we see a transformation of geopolitics. We face a world of power politics with the weaponisation of interdependence and more examples of countries using force, intimidation and blackmail to get their way. The growth of this lawless world and disorder is what I meant when talking about the 'jungle'. My reference to 'jungle' has no racist, cultural or geographical connotation. Indeed and unfortunately, the 'jungle' is everywhere, including today in Ukraine. We must take this trend seriously and that was my message to the students. Certainly, we should not retreat behind the false security of walls and isolationism.[5]

Borrell thus adopts a specific discursive strategy: instead of predicating 'garden' status on countries of the Global South he implies in naming Ukraine that there is a 'white' country (Russia) that is jungle-like, using 'force, intimidation and blackmail to get their way'. This suggests that Russia is the exception, 'Indeed and unfortunately', to a general, non-white rule to which he cannot think of a single counter-example.

Metaphors are seldom innocent; they inculcate 'mental models that serve the interests of specific, usually dominant, social groups'.[6] The metaphorical predication strategy of referring to specific territories as jungles has an interesting history. The notion of the 'law of the jungle' enters the Western idiom via one of the British Empire's most celebrated praise-singers, Rudyard Kipling, whose *Jungle Book* is set in India and codifies a 'Jungle Law' inspired by a Darwinian 'survival of the fittest' ideology[7] which comes into Borrell's discourse as 'the will of the strongest'. 'Jungle' – a loanword from Hindi and Urdu[8] – carries with it connotations of a broad range of warmer equatorial and southern climes. The historical racial Othering of the denizens of these 'jungles' firmed up an imperial sense of European whiteness. It is important to note that scientific racism was built on environmental determinism, the idea that climate in particular was responsible for the development of the different characters of races.[9] In Montesquieu's *The Spirit of the Laws* (1748), for example, civilizational values are determined by how easy it is to feed oneself off the earth:

> The barrenness of the earth renders men industrious, sober, inured to hardship, courageous, and fit for war; they are obliged to procure by labor what the earth refuses to bestow spontaneously. The fertility of a country gives ease, effeminacy, and a certain fondness for the preservation of life.[10]

For Montesquieu this is clearly a value-laden distinction in that life is not there to be preserved for its own sake but channelled into manly war-like exploits that conquer new territory and defend the homeland. But another sense of 'the preservation of life' is encapsulated in lines from Lord Byron's satire *Don Juan* (1819): 'What men call gallantry, and gods adultery, / Is much more common where the climate's sultry'.[11] The association of the warmer regions of the 'South' with sexual license and indolence and the cooler climes of the 'North' with industry, abstention and reason is perennially productive of racial difference. Even within Europe, a hardworking and efficient north constructs itself in contradistinction to a sleepy and sensuous south. In a fractally recursive manner this idea also works within nations, dividing, for example, the industrialized light-skinned more 'European' north of Italy from the perennially corrupt, darker-skinned and more 'African' *Mezzogiorno*.[12] As was discussed in Chapter 3, one of John Locke's arguments for the dispossession of Native Americans was that they were indolent: they had too much land and were doing nothing to 'improve' it.[13] The construction of racial others as lazy and licentious is a foundational precept in the liberal conception of property that animates the institutions of racial liberalism.[14] The characterological figure of the undeserving native malingers in the white 'cultural unconscious' which Frantz Fanon defined as a 'constellation of postulates, a series of propositions that slowly and subtly – with the help of books, newspapers, schools and their texts, advertisements, films, radio – work their way into one's mind and shape one's view of the world of the group to which one belongs'.[15] As Fanon details at length in *Black Skin, White Masks*, the association of dark-skinned people with warm climates, sexual decadence, savagery and lawlessness has a long history in Western cultural production. This history is furthermore very much still visible in a definition of the 'law of the jungle' in a popular online dictionary: 'a system or mode of action in which the strongest survive, presumably as animals in nature or as human beings whose activity is not regulated by the laws or ethics of civilization'.[16]

Thus, while it is certainly possible to protest, as Borrell does, that one is 'totally opposed to any form of contempt, or racism towards anyone'[17] it is another matter entirely to attempt to evacuate by fiat any 'racist, cultural or geographical connotation' from the garden/jungle metaphor. One cannot merely wish away the chauvinism of an idiom coined by Kipling, which *by definition* draws boundaries between people who are 'civilized' and those who are not. Russian aggression in the Ukraine was furthermore, despite Borrell's claims to the contrary, not the primary context for his original remarks to the Academy. His address had focused mainly on what he termed 'not developed' countries:

> The big difference between developed and not developed [countries] is not economy, it's institutions. [...] I cannot go to emerging countries and build institutions for them. They have to be built by them. Otherwise, it will be a kind of neo-colonialism. The big difference between us and [most] of the rest of the world is we have institutions. [...] The world needs Europe. My experience of traveling around the world is that people look at us as a beacon. Why [do] so many people want to come to Europe? Are there flows of illegal or irregular migrants going to Russia? Not many. No, they come into Europe – but for good reasons, for good reasons. So keep the garden, be good gardeners, but your duty will be not to take care of the garden itself, but the jungle outside.[18]

The 'jungle' according to Borrell is thus identical with countries onto which he has predicated specific qualities: lacking institutions, 'not developed ... emerging' and likely to object to 'neo-colonialism'. While Russia is indeed an 'emerging' economy, its exclusion from this set is entailed in Borrell's construction of the country not as sending migrants, but as a potential receiver of them. Russia is also not a formerly colonized nation that could plausibly claim 'neo-' colonialism. And given that he raised earlier in his speech how the challenges of war, climate change and poverty were creating a 'dramatic situation' in the Horn of Africa, one of the world's migration hotspots, the 'strong growth capacity' of the jungle is clearly that it is growing as a threat, its fertility spawning not economic success and independent institutions but populations, populations which will become 'flows of illegal or irregular' migrants from the Global South into Europe. This 'flow' is a liquid metaphor associated with negative evaluations of migration.[19] Receiving countries in the West are often constructed as inundated – 'flooded' or 'swamped' – by migrants.[20] Both the 'jungle' and the 'flow' thus invoke feelings of overwhelm, of encroaching otherness, of a once orderly place flooded and overgrown with weeds.

In addition to Borrell's reproduction of racist distinctions, we might draw attention to the ecocidal dimensions of an ideology that values gardens over jungles. The central claim of this book is that if we don't analyse the intersections of biodiversity loss, climate change, cisheteropatriarchy, racism and colonialism, we will have failed in 'articulating the full dimensions'[21] of any one of them. Buried beneath Borrell's 'garden' lie histories of oppression, genocide and environmental devastation. To get to the garden required cleared woodland, industrial waste, enclosed commons, murder, war, slavery and dispossession. These are the spoils that nourish the roots of today's European prosperity. The suppression of the conditions of possibility for this wealth mirrors the harm that must be done to ecosystems for gardens to thrive, as well as to the bodies of the cheap labourers who clear the land. Gardens are artificial creations that impose human designs onto ecosystems, and have a lasting association with coloniality.[22] The root of colonialism, after all, is the Latin *colere* – to cultivate.[23] Borrell acknowledges part of this history when he states that 'the beauty of the European experience is you and me overcoming this heritage of the past and offering to the world the recipe for peaceful coexistence, cooperation, integration and development'.[24] In other words, now that European harmony has been achieved, the fact that it was won through the commission of countless unspeakable crimes may simply be wished away, and its societal norms presented as a 'recipe' for the rest of the world to follow.

A final point on Josep Borrell is that his pronouncements reflect the liberal, humanitarian face of the European project. He is an opponent of 'Fortress Europe' and, as we saw, he warns against the 'false security of walls'. There are many politicians to his right who would advocate for higher fences and even more violent borders.[25] Borrell articulates a case for overseas development assistance and friendly humanitarian intervention in the interests of safeguarding the prosperity of Europe. His discourse may be no less racist in its effects, but its violence is (at least partially) obscured. This violence, as I argue in the next section, subsists in its construction of some populations as inherently problematic and in need of management, and others as self-managing and continent, under the cloak of the supposed universality of 'overpopulation'.

## A plague on the earth

One of the world's most prominent environmental heroes is the British 'national treasure [and] global superstar'[26] Sir David Attenborough. Along with other famed naturalists such as Jane Goodall, Attenborough is a leading proponent of overpopulation discourse, in part through his patronage of the charity Population Matters. In a widely reported 2013 interview, he made the case as follows:

> We are a plague on the Earth. It's coming home to roost over the next 50 years or so. It's not just climate change; it's sheer space, places to grow food for this enormous horde. Either we limit our population growth or the natural world will do it for us, and the natural world is doing it for us right now [...] We keep putting on programmes about famine in Ethiopia; that's what's happening. Too many people there. They can't support themselves – and it's not an inhuman thing to say. It's the case. Until humanity manages to sort itself out and get a coordinated view about the planet it's going to get worse and worse.[27]

One of the interesting things in this passage is how Attenborough's usage of pronouns shifts. Pronouns are *deictic*: to understand what they refer to requires some contextual knowledge. So how does deixis work in this text?

Attenborough begins by talking about a universal 'we' being 'a plague on the Earth' – in which he and his global audience are ostensibly included. The headline of the *Telegraph* article reporting on his comments – 'David Attenborough – Humans are plague on earth' – picks up on the broad claim that if you are human, you're a part of the problem. This supposed universality faces some contextual difficulties however in addressing the statement 'Either we limit our population growth or the natural world will do it for us' to a general audience. As is well-known, not only is the UK, like all high-income countries, in the throes of demographic decline, with fertility well below the 'replacement level' of 2.1 children per woman, but it has been so for many decades.[28] The low fecundity of Britons was in fact a key concern of social improvers, eugenicists and planners for much of the twentieth century.[29] So 'our' population growth clearly cannot refer to any citizen of a rich country. Demographers in fact describe a 'demographically divided world'[30] where the populations of rich countries shrink even as poor countries continue to grow. Familiarity with this division would thus arguably prevent the deictic 'we' from including most of his audience and attach instead to the denizens of the developing world.

It is in contexts such as these that the 'natural world will do it' – i.e. limit population – 'for us'. He claims that this is what is currently happening in Ethiopia. Attenborough must have been referring to some generically African place, as there was no famine in Ethiopia in 2013. The mention of Ethiopia however shifts his pronoun use from talking about 'we' and 'us' to talking about 'them': 'Too many people there. They can't support themselves'. This is an inaccurate picture of the infamous Ethiopian famines of the mid-1980s which, like the ongoing Tigrayan crisis, have their roots in the combination of drought and civil war, with human rights abuses that include the destruction of otherwise fertile farmland by militias.[31] What 'we' are now doing is 'putting on programmes' about the famine: 'we' are spectators of these faraway famines, observing how the natural world is controlling population for 'us' and, because we don't like these images, we prefer to see population controlled through contraception than death and disease. This shift in deixis from a universal 'we' to a specific 'them' somewhere vaguely African also dislodges 'our' inclusion in humanity, which is now relegated to the third person: until 'humanity manages to sort itself out' there will only be problems. Humanity is no longer in the familiar, near context of the first person plural, but an amorphous mass in the faraway third person, giving us a vividly racialized picture of just who this 'humanity' is that needs to have its growth 'limited'.

Attenborough is elsewhere on record as claiming that sending bags of flour to countries ravaged by famine is 'barmy'.[32] This line of argumentation is a familiar one, associated with supposedly 'rational' approaches to historical famines. During the mid-nineteenth-century Great Famine in Ireland, one million people died and two million became migrants. While the potato blight was one cause of hunger, it was the ideology of classical economics which sealed the deal, motivating the British colonial metropole to adopt a policy of non-intervention on the logic that it would clear the land of 'surplus population'.[33] Famine was seen as a 'natural corrective' to the excessive fertility and potato-dependence of the Irish, and food exports continued to flow from Ireland to England throughout the Famine, under military escort.[34] It should come as no surprise that the Irish were both colonized and racialized at the time.

Proponents of overpopulation discourse style themselves as brave truth-tellers, unafraid to face uncomfortable facts and articulate unpopular positions. Attenborough has to affirm that it is 'not an inhuman thing to say' and elsewhere he has claimed that speaking out about this must deal with 'the last sensitivity – and the most tricky of all – is the fact, when you talk about world population, the areas we're talking about are Africa and Asia, you know'.[35] A similar call for overcoming sensitivity to face the truth of overpopulation was made by the British primatologist Jane Goodall at the World Economic Forum in 2020, where she stated that 'we cannot, we cannot hide away from human population growth because you know it underlies so many of the other problems'.[36] While some people want to 'hide away' or pander to 'sensitivity' these figures present themselves as unafraid to face reality. Speaking with scientific authority, they claim what Alana Lentin has called 'Not Racism™' – a discursive strategy to control the definition of racism from a structurally white perspective.[37]

The Black British artist and environmental justice activist Susuana Amoah writes that 'I have become used to white environmentalists telling me, with an air of bravery and confidence, that they think that families like mine who have lived highly sustainable lifestyles for generations and are currently dealing with climate crisis situations that they did not create, should refrain from having large families'.[38] Amoah's point, backed up by extensive scholarship, is that given the environmental impact of Northern consumption, both historically and presently, population declines in the South will have a negligible effect on climate change and biodiversity loss. There is clearly something other than a simple concern for harmonious ecological relationships, underlying this fixation on the fertility of Asian and African populations.

## Overwhelm anxiety and 'white affect'

On a page with a jumble of images of huddled crowds of migrants and piles of blankets and discarded clothes, the right-wing British tabloid the *Daily Express* published an opinion piece in 2015 by erstwhile leader of the United Kingdom Independence Party (UKIP) Nigel Farage, under the headline 'It's official … Britain's population surge is TOTALLY out of control'.[39] While the photographs are mostly of masses of dark-skinned men, in his argument, he also mentions the threat of their women: 'As many as 1.7 million births in the UK between now and 2039 are projected to be to foreign-born mothers'.[40] In the context of declining white birthrates, his message was clear: the surge in population was going to be coming from 'foreign-born' people and not from 'native' Britons. While the mass of men seems threatening and violent, the wombs of the women are constructed (in a similar fashion to that described in a British National Party pamphlet) as 'a baby-bomb poised to breed the British people out of existence'.[41] Health and social services would be overrun by outsiders. The following June, the UK voted to leave the European Union – in part, to '[take] back control of our borders'.[42]

While Farage does not mention climate change, the climate migrant features in dominant discourses in ways that echo UKIP's political commitments. As mentioned above, the liquid metaphor of a 'surge' is a standard element of anti-immigrant rhetoric. Andrew Baldwin argues that thinking about the ecological crisis in terms of migration 'generates a particular racial orientation to the phenomenon of climate change'[43] – an instantiation of 'white affect'. While the generation of this affect has racial effects, it seldom explicitly thematizes race. It relies on the figure of the climate migrant, a 'threatening, monstrous figure from the future'[44] who provokes fear, but also desire 'to preserve the apparent normalcy of an imagined social order that the monstrous, future-conditional climate change migrant threatens to overwhelm'.[45] The 'ungovernable excess' of the Other – the sense of being flooded, overwhelmed, swamped – is thus linked to an 'anxiety of loss'[46] which works to motivate the preservation and thus endurance of whiteness as a social norm and thus of the power and privilege of white people.

As environmental arts scholar Andil Gosine points out, much environmentalist discourse evinces an 'anxious preoccupation with the sexualities of nonwhite peoples, which undermines their rights and denigrates the fullness of their humanity, while affirming the authority of whites to colonize and govern the world'.[47] These anxieties are made discursively visible in government policies, the pronouncements of NGOs, the media, in official climate change communication, but also in popular culture. In their analysis of apocalyptic tropes in Hollywood films, Mabel Gergan, Sara Smith and Pavithra Vasudevan note the irony of imagining how to survive the end of the world on a continent where Indigenous worlds had already been ended by settler colonialism.[48] Hollywood routinely exculpates white Americans of their complicity in environmental disaster in order to centre instead 'a universal human frailty that ends with triumph, a clear moral, and a clean slate'.[49] Many of these films stage the trope of 'humans versus the horde'[50] – where the horde often consists of undead zombies, close-to-human but not-quite-human, a cultural space in which fantasies of race and post-race are routinely explored.[51] These cultural anxieties about being overwhelmed by racial Others are also in evidence in technologies of surveillance in white institutional spaces that determine who 'fits' in them, and who is a 'space invader'.[52] Being able to tell the difference between a zombie and a real person during the apocalypse is after all a matter of life and having your brains eaten.

Cinematic worlds mirror the spatial management of racial difference, where institutions such as health and social care, policing, and education control access to social power and privilege. National policy logics also play their part, as Susanne Schultz has shown in the case of Germany, which uses demographic markers signalling racial populations projected into the future to determine immigration policy, combining an in-country biopolitics with extra-national necropolitics that regulates which migrants are desirable as labour, and which not.[53] Baldwin's important observation that white affect is as much about *desire* for the preservation of a social order as it is about *fear* of its being overrun is thus a crucial insight into the spaces where migration and environmental catastrophe meet. In performing the subject role of a future-oriented eco-conscious citizen, white environmentalists can thus make space for unnamed and unrecognized racial anxieties that are contingent 'not solely on the events of an imperial past, but on some form of future Other as well'.[54]

## Malthus, Ehrlich, Hardin, Kaplan: From population to securitization

Anxiety that 'civilization' will be swamped by overly fecund subaltern populations has a storied scholarly profile. The English clergyman Thomas Malthus's *An Essay on the Principle of Population*, which was published and revised in a number of editions between 1798 and 1826, argued that the 'power of population is indefinitely greater than the power in the earth to produce subsistence for man'[55] – to wit, that population can expand exponentially, while agricultural production increases in a linear fashion. What may seem on the surface like a bland mathematical observation was in fact embroiled in a profoundly political project. In the late eighteenth century, the French Revolution polarized British politics, with a small camp of radicals seeing it as exactly the societal upheaval the UK needed while dominant Whigs considered revolution a catastrophe to be avoided at all costs. Malthus's work was resolutely in the latter camp, and in fact arose from a critique of any form of social 'levelling'.[56] The *Essay* launches a no-holds-barred conservative attack on the possibility of social improvement. Malthus wrote (in sections excised from later editions) that poor people *ought* to be left to the 'punishment of nature' – i.e. starvation and famine – lest they overrun 'nature's mighty feast' and bring society to ruin.[57] In a slightly more humanitarian humour, he also argued that since efforts to better the conditions of poor people would lead to population explosions, and 'since the rich have sufficient economic means at their disposal to procure food even in conditions of dearth, the suffering brought on by overpopulation will be inflicted on the poor'.[58] The *Essay* was claimed by politicians of the day as incontrovertible scientific proof of the impossibility of a social order where everybody could have plenty to eat, and thus of the undesirability of revolutionary change.

What is notable about Malthus's argument is the idea that feeding the poor only exacerbates *their* suffering, increasing the number of people who will eventually have to die. A redistribution of plenty, informed by principles of justice, is not considered. He thus presents himself as having the long-term interests of the suffering masses at heart while at the same time refusing to critique the interconnectedness of social classes and the exploitation of the lower down by the higher up. Instead of the rich being forced to share the feast – which could only end in ruin anyway, as mathematically 'proven' by Malthus – the poor should practice 'moral restraint'. In other words, they should simply not procreate. This pattern, where controlling the fecundity of subaltern populations is imagined first as in their interests, and secondly as one of the root causes of their poverty in the first place, has been repeated numerous times in a variety of contexts in racial capitalist modernity.

The key contemporary popularizers of overpopulation thinking are Paul and Anne H. Ehrlich, whose 1968 book *The Population Bomb* predicted the mass starvation of hundreds of millions of people around the world in the 1970s.[59] Their dramatic claims, which have been repeatedly proven wrong, turned the Ehrlichs (and especially Paul) into intellectual celebrities who are lauded to this day by environmentalists for working to 'alert humanity to the dangers of ecological carelessness and arrogance'.[60] Their argument follows a similar pattern to Malthus's, though in signature American style they do not cite him. The debt to Malthus is however acknowledged in another 1968 publication, ecologist Garrett Hardin's 'The Tragedy of the Commons' in the academic journal *Science*.[61] Despite the fact that its author was a white nationalist who openly held racist and eugenicist views,[62] and the unflinching anti-humanism of its argument, this hugely influential essay, beloved of economists, management gurus and environmentalists alike, continues to be reproduced in contemporary environmental anthologies.[63] In it, Hardin argues for the benefits of 'children of improvident parents [starving] to death' and against the idea that human beings have the right to procreate, or to put it in the blunt terms he uses: 'Freedom to breed will bring ruin to all.'[64] The supposed 'tragedy' of the commons named in Hardin's title is that universal individual self-interest supposedly incentivizes people to take more than their fair share from that which is collectively owned.

They thus tend to out-'breed' the 'carrying capacity'[65] of land, and the only solution is to force them to stop. In the 1970s, he developed the notion of 'carrying capacity' into an allegory of territories as lifeboats, where rich countries with stable populations are under an ethical obligation *not* to admit the 'excess' populations of poor countries into their 'lifeboats' lest everybody sink.[66]

Though there is much to critique in the ethical visions of Malthus, the Ehrlichs and Hardin, their work does not pass muster even at the empirical level. The sleight of hand that attributes poverty and hunger to the sector of the population that in fact consumes the least – both totally and per capita – combined with extensive evidence that subsistence does not in fact increase in a linear fashion that will never keep pace with geometric population growth has invalidated the mathematical presuppositions of Malthusianism.[67] When people go hungry it is not because of the limited 'carrying capacity' of the land but differential access to it and the distribution of its fruits, and thus on socio-political factors. The Irish and the Ethiopian examples are again relevant: colonial domination in the former and civil war in the latter combined with environmental factors to produce famine. As for Hardin, actually existing commons tend not to be scenes of 'tragedy' as predicted, but models of social co-operation enabled by community institutions.[68] It is the destruction of these institutions by colonial and capitalist enclosures that is in fact associated with environmental degradation. So while Malthus and Hardin presented their arguments as dispassionate analyses of universal truths, what they were in fact doing was projecting their class and racial ideologies onto reality, conveniently finding their prejudices reflected back at them.

Building the legitimacy of overpopulation discourse in the North occasioned extensive 'population propaganda' that supported the rapid rollout of services linked to economic incentives in the South, creating a globally racialized population control regime that persists in the present.[69] The insidious logic at the heart of overpopulation discourse rests on the truism that those never born cannot go hungry, an eliminationist fantasy. This logic works in the subjunctive mood; if these people did not exist, they would not now be making demands on us. That their suffering is caused by war regimes of extraction, global inequality, environmental degradation, or any other factor is simply too troubling and complicated to contemplate: it would all be so much easier if they simply weren't there. Population thus becomes the primary lens through which socio-political problems are seen, a default position that claims self-evidence.

While much discursive energy is spent on imagining how much better the world would be if poor people just did not exist, there is also significant investment in constructing the ones who are here as a near and present danger.

A central dynamic of contemporary overpopulation thinking is thus the idea that 'surplus' people are a cause of conflicts that threaten to engulf the planet in chaos. A key moment in setting the scene for this discourse was the publication in February 1994 of an essay in *The Atlantic* by US journalist and public intellectual Robert D. Kaplan. Entitled 'The Coming Anarchy: How Scarcity, Crime, Overpopulation, Tribalism, and Disease Are Rapidly Destroying the Social Fabric of Our Planet',[70] Kaplan argued that

> it is time to understand 'the environment' for what it is: the national-security issue of the early twenty-first century. The political and strategic impact of surging populations, spreading disease, deforestation and soil erosion, water depletion, air pollution, and possibly, rising sea levels in critical overcrowded regions like the Nile Delta and Bangladesh – developments that will prompt mass migrations and, in turn, incite group conflicts – will be the core foreign-policy challenge from which most others will ultimately emanate, arousing the public and uniting assorted interests left over from the Cold War.

Kaplan's text was cited in a speech by President Bill Clinton soon afterwards and has had a considerable impact on US foreign policy thinking.[71] While his argument draws on academic work that describes the complexity of causal factors underlying conflict while arguing for some independent causal role of environmental scarcity in driving institutional dysfunction and social fragmentation in poor countries,[72] the vision of the future that Kaplan presented in *The Atlantic* is decidedly bleaker and less subtle. Kaplan names Malthus as the 'prophet' of scenes of 'tribal' conflict in West Africa. As Nancy Peluso and Michael Watts observed soon afterwards, Kaplan 'conjured up a surreal picture of an African continent in the throes of an apocalyptic crisis: overpopulated, undernourished, and driven to barbaric acts of violence by irrational spirit power […] once again the "Dark Continent"'.[73] The geographer Simon Dalby argues that Kaplan's simplistic geographic determinism treats societies as 'failing' due to purely internal factors, disconnected from historical and present interdependencies with the rich world. Issues of redress or reparation for global disparities rooted in a shared global political economy are thus foreclosed, while inequality is framed as demanding a primarily military answer. This elision of history is facilitated by presenting overpopulation as cold and incontrovertible fact, or 'nature unchecked' which as Dalby argues

> operates to support the Malthusian fear of poverty-stricken mobs – or, in Kaplan's terms, young homeless and rootless men forming criminal gangs – as a threat to political order. Economics becomes nature, nature in the form of

> political chaos becomes a threat: the provision of security from such threats thus becomes a policy priority. In this way 'nature unchecked' can thus be read directly as a security threat to the political order of postmodernity.[74]

The picture of a world of ubiquitous war driven by climatic changes can trace its genealogy to the kind of geographic determinism espoused by Montesquieu, where Southern excess is imagined as inundating the frugal North.[75] Though this simplistic picture lacks evidentiary basis and is rejected by most critical scholars,[76] it is repeatedly used as the basis for decisions by 'policy and defence actors, most prominently the US military'.[77] This invocation of 'environmental security' as a priority has had a significant effect on the militarization of development.[78] It follows the logic of the 'bunker' – a special kind of fortress that can withstand drastic environmental changes – intrinsic to the neoliberal concept of climate 'resilience'.[79] When a world of generalized competition meets with environmental catastrophe, the rational neoliberal subject eschews cooperation or mutual aid in favour of self-preservation. These policy positions are constructed not as inhumane, nor even as political, but as rational administrative solutions entirely above the fray of emotion or doubt.[80]

In climate security discourse, the poor are raced and gendered. As Nancy Tuana shows, the climate migrant is a controlling image that works biopolitically to separate 'hard-working citizens' from free-riding immigrants, and 'more civilized from less civilized'.[81] These images fixate on the fecundity of dark-skinned figures, whose prodigious women and hypersexualized men are imagined as reproductive invaders who will overwhelm the orderly Euro-American world of democracy, welfare and the rule of law. While this is clearly an instance of 'reproductive racism',[82] as with Josep Borrell there is ideological room for a spectrum of positions on what the policy response should be: from the relatively liberal diplomatic creed that the 'gardeners have to go to the jungle', to the militarized fortification of borders.

Environmental security when animated by overpopulation discourse is allied to a racial project that creates a sense of urgency through images of 'barbarian hordes' at the gates of 'civilization'. As the political ecologists Robert Fletcher, Jan Breitling and Valerie Puleo point out, this dramatic staging of overpopulation exceeds any rational assessment of the evidence, making it a *scapegoat* for a repressed reality, 'a fantasmatic construction serving to conceal the gap between the symbolic order of international development and its persistent failure in actual practice'.[83] While in policy discourse it may be the failure of development that needs to be concealed, it is arguably the interconnectedness of the world system more generally that is being repressed. Precisely because it is the flow of surplus value to the rich, who take a disproportionately large share of the world's energy, resources and labour power, scapegoating those who consume the least as being populations 'out of control' who will soon, in the near future, overwhelm our manicured garden, protects well-to-do psyches from facing up to their primary culpability for the ecological crisis.

What is thus hidden by the overpopulation scapegoat, in the words of the UN's population division director John Wilmoth, is that 'the problem that is spiraling out of control is consumption'[84] – not population. As discussed in the previous chapter, the energy consumption of the richest 1 per cent would secure a decent standard of living for the poorest 1.7 *billion*.[85] And when it comes to representations of overpopulation in Africa, as economist Yarri Kamara points out, the continent's population density is 'a tenth of India's, a fifth of the UK's, and a third of [China's]'.[86] When placed in the appropriate historical context of colonization and the slave trade, Africa's *underpopulation* can be seen as an effect of relations of domination, and not the cause for alarm generated in the fever dreams of rich-world neo-Malthusians.[87]

## Between barbarian hordes and cornucopia

While overpopulation thinking informs policy positions from liberal humanitarianism to ecofascism in the rich world, it is not the only show in town. Cornucopianism – an ideological tendency within ecomodernism that predicts that technological improvements will support growing populations, and that growing populations will support endless economic growth – is a standard capitalist riposte to population doomsayers within their ranks. Danish statistician and political scientist Bjørn Lomborg is one of cornucopianism's leading lights, and he notes that population growth is predominantly an urban phenomenon, arguing that cities not only provide people with better living conditions but are 'power centers which provide greater economic growth'.[88] Working off similar datasets to people who fret about overpopulation, cornucopianists point to the historical record of an ever-larger global population leading (ostensibly) ever-healthier and longer lives, attributing these improvements to capitalist hegemony and innovation through competition, the triumph of science and the spread of liberal democratic values. Cornucopianists do not deny that planetary limits to raw materials exist, only that these limits curtail economic growth, which they predict will continue apace due to human innovation and pioneering new frontiers of value, such as converting plastics to food, or mining asteroids.[89]

Struggles between overpopulationism and cornucopianism are thus entirely internal to neoliberal late capitalism, lacking an emancipatory politics beyond the racist-capitalist-colonialist horizon. In this imaginary, liberation is equated with humans as units of utility working smoothly within a system of value extraction, living longer lives under a mono-epistemic, horizontally integrated and deterritorialized world with no safeguards against vertical exploitation, injustice, or ongoing cultural and ecological annihilation. Even ostensibly left-leaning cornucopianism retraces a capitalist horizon, while reducing the non-human world to mere means for human thriving.[90] As the philosopher Émile Torres argues, the 'extreme optimism' of such models not only reads the historical record selectively while discounting the chance of any cataclysmic interruption to growth, it also allows for absolute increases in human suffering and death as long as there appear to be relative declines that can be chalked up to progress against ever-bigger numbers.[91] Because multidimensional definitions of justice are not easily integrated into population statistics, and because the fairness of the system as a whole is only ever compared to invented pre-colonial or post-capitalist dystopias, cornucopianism is not an answer to overpopulation discourse that meets the requirements of an ecointersectional approach.

A similar tension is noticeable in migration discourse more broadly, where the standard liberal response to fears of being overwhelmed by migrants is the assertion of the value that these 'surplus populations' create in their new homes. People are thus only seen as valuable if they plug gaps in social care or other enabling conditions of capitalist value extraction.[92] Global historical injustice and the ongoing interconnectedness and neo-coloniality of the world system are ignored, while the acceptance of the 'new' population is made conditional on their cultural assimilation, and regimes of reproductive surveillance that construct them as a potential demographic threat.

## Summarizing the overpopulation imaginary

The first part of this chapter has offered insights into the genealogy and trajectory of overpopulation thinking. The second will coalesce these elements into an overpopulation imaginary that might serve as an aid to ecointersectional discourse analysis. In the following section, I will present four cases of how this imaginary is textualized. As with all the other chapters, the way that I

**Table 2** Summary of the overpopulation imaginary.

| **OVERPOPULATION IMAGINARY** | |
|---|---|
| Cold, hard, demographic facts shown on charts and irrefutable logic prove that overpopulation is the primary social problem. Nature will curtail these populations in situ, but if they are allowed to move they will do as waves that overwhelm orderly civilizations. | |
| **SUBJECT ROLES**<br>• Brave truth-tellers<br>• Objective demographers<br>• The monstrous climate refugee<br>• The barbarian horde<br>• The progressive and 'feminist' reproductive health advocate | **INSTITUTIONS**<br>• Family planning NGOs<br>• Overseas Development Assistance<br>• Structural Adjustment Programmes<br>• Militarized borders<br>• Immigration control regimes |
| **CHRONOTOPES**<br>• Garden/Jungle/Wasteland<br>• Surges, waves, inundations<br>• Lifeboats and Bunkers<br>• 'Waiting for the Barbarians' | **AFFECT**<br>• Bravery in the face of critique<br>• 'Not Racism™'<br>• Fear of overwhelm: 'white affect'<br>• Pity for those who starve |

represent the imaginary (in Table 2) is proposed as a useful starting point and not as the final word.

On its surface, the overpopulation imaginary is cold hard mathematics. It is structured by economic and demographic logics visible in charts, formulae and infographics encoding portents of looming crises. The 'carrying capacity' of a territory incontrovertibly limits its population. The originary Malthusian assertion of arithmetic (linear) growth in subsistence not keeping pace with geometric (exponential) growth in population is vividly diagrammatic; it provides a sense of inexorable momentum in a particular direction. Riding on these impervious formulae are the problems we derive from them: parcels of land that are divided up until hillsides become a patchwork of meagre plots, soil that grows measurably less fertile every year and kilolitres per hectare of water that isn't there. This is an instance of the 'topos' of numbers The reduction of people to populations, of complex social processes to demography, and of agriculture to agronomy is an abstraction worthy of deconstruction in itself in that it represents a choice to engage in universals instead of particulars, and algebraic calculations instead of, for example, ethical ones. Numbers have a certain solidity to them; they carry associations of expertise and rationality. It is not irrelevant that a common idiom in English tells us that 'the numbers don't lie'. As Michael Billig shows, numbers can take on a 'semi-magical' character however when entrained in political rhetorics.[93] So while mathematical rhetoric implies certainty, precision, objectivity and having a handle on things, it is no more or less 'ideological' than any discursive articulation.

The truistic logic of 'population primacy' – that non-existent people can't starve or otherwise make demands on rich-world charity, and that population is thus upstream of all other social problems – lends the imaginary an air of irrefutability. Developmental and security policies are thus legitimated as 'post-political', administrative, technocratic and purely rational. They push colonial history, territorial or climatic particularity, and questions of justice into the background. The world is imagined as parcelled out into discrete nation-states that succeed or fail on internal factors alone, metaphorized as 'lifeboats' floating around in an unforgiving sea, 'nature', which will either limit their population through famine and plague or send them scrambling to get into other lifeboats. This latter possibility makes room for a specific dynamic of the larger overpopulation imaginary where surplus people are imagined not at home, but abroad, sparking the racial terror at the heart of climate security: that a global lumpenproletariat of dark-skinned refugees will overwhelm the West. The lifeboat analogy invokes zero-sum calculations, where every Other who makes it to a different lifeboat pushes somebody from the in-group out, and every gain for the forces of lawlessness and racial degradation is a loss for the 'best combination of political freedom, economic prosperity and social cohesion that humankind has been able to build'.[94]. The necropolitics of climate security is thus in tension with the biopolitics of other forms of 'reproductive racism' that seek to regulate the flow of racial Others into the nation for the purposes of doing the 'shit work'[95] of the imperial metropole in ways that keep them in positions of permanent second-class citizenship.

## Overpopulation subject roles

When on the move, the subject of overpopulation is the 'climate migrant', that 'monstrous' figure from the future who serves as the controlling image of environmental security discourse. Jungle people spilling over into the garden are framed as barbarians who will overrun orderly societies, sack the metropolitan centres of civilization and leach off their hardworking citizenry. When at home, the potential migrant is Attenborough's 'enormous horde': an undifferentiated mass of humanity living cheek-by-jowl in a jumbled mess. Photographs show the 'press of population'[96]: buses moving through crowds, people washing clothes in lagoons, or farmers eking out a living on small patches of land. Whether the population is urban or in some 'jungle' it is packed, crowded, with little space for individualization.[97] The specific hopes or projects of social actors cannot come into focus when they are represented as amalgamated into a stereotype of a person, a prodigiously fertile 'imago'[98] whose 'breeding' is out of control. The imago is gendered in a binary fashion: men are oppressive and hypersexual; women are the hyperfecund victims of their menfolk. Women demand contraception which patriarchs withhold it from them. This mass of people is thus both to be pitied and feared, in that their present or imminent suffering demands a moral response from an audience at the same time as signalling desperation, and the potential threats of a starving peasantry becoming a marauding horde.

This audience and their icons, such as David Attenborough or Jane Goodall, meanwhile, are the purely rational, 'Not Racist™' subjects of late capitalist modernity, who only have the best interests of faraway populations at heart, and see population control as meeting humanitarian, ecological and even feminist ends. They are defenders of the global order, Borrell's 'gardeners' who are precise in their measurements and predictions, quantifying the land and its capacity in comparison to the number of people that it will need to support. The demographic and economic rationality of their positions may make them vulnerable to critique, of neocolonial meddling in societies far away from their own. They are however not afraid of critique, as they are brave truth-tellers who communicate difficult truths.

## Overpopulation institutions

The key institutional types of overpopulation discourse are development and security. In the former, what the Organisation for Economic Co-operation and Development (OECD) calls 'overseas development assistance' (ODA) is a primary arena for a constellation of state-funded, intergovernmental, and non-governmental development actors to implement sexual and reproductive health and rights programming and influence national policies on contraception and family planning. When ODA is targeted at population, the priorities of healthcare provision may shift towards measures designed to ensure smaller family size. These programmes require the surveillance of populations through censuses and other population registration mechanisms, and the availability of this data through tools such as the World Bank's DataBank or the US Central Intelligence Agency's World FactBook. They also include extensive policy incentives for population management and control. What the subjects of overpopulation discourse corroborate is the 'orderliness' of the West and its legitimacy as the keeper of global order. As Josep Borrell put it, the 'big difference' between the developed world and the developing is 'we have institutions'.

If ODA is 'speaking softly' (in Teddy Roosevelt's aphorism) then the second type of institutional type involves carrying a 'big stick'. The militarization of borders and the securitization of development are key contemporary trends related to the controlling figure of migrants more generally, and the climate migrant in environmental discourse. Both categories of institutions serve to stabilize the subject roles of population activists and the controlling image of the 'horde' and appeal to both humanitarian and foreign policy objectives.

## Overpopulation chronotopes

At the heart of the overpopulation imaginary are vivid place images. The first of these is the fecund 'jungle' with its 'strong growth capacity' and 'irrational spirit power', its 'sultry' climate producing sexed, raced and gendered overabundance. The movement of the 'jungle' into a second place-image, the 'garden' – which is imagined as neatly regulated, clean and orderly – is precipitated by a combination of prodigious procreation and climate change which create the third place, the wasteland, and its filthy cities strewn with rubbish, overcrowded farmlands, flooded river deltas or featureless deserts. This narrative structure of abundance turning to ruin, and of orderliness being overwhelmed by entropy, can be summarized from the perspective of the rich world (with a nod to J.M. Coetzee) as 'waiting for the barbarians'.

The garden is a 'beacon' of freedom and prosperity, which makes it an example to emulate but also an attractive destination for 'flows of irregular or illegal migrants'. These 'flows' are often imagined as waves or floods that will overwhelm, swamp or inundate the rich world. Much like a tsunami, the wave is imagined as having starting with some kind of seismic event, with a 'boom', 'explosion' or 'baby bomb'. The moment of inundation is typically imminent: the crisis will come soon; it looms threateningly over the garden. As is common in hegemonic discourses, this dramatic mise-en-scène lacks historical depth. Borrell, for example, presents Europe as having 'overcome' the 'heritage of the past'. They are a beacon because they have wiped their slate clean; and who can even remember what happened? It is certainly not relevant to this chronotope, which is oriented exclusively to the present excess of faraway Others and their imminent arrival at the gates of civilization. When the barbarians arrive, they will bring disorder, violence and war. To prevent this from happening, a responsible and rational pre-emptive move is to send our diplomats and troops to keep them at home, while also ensuring that our borders are as strong as they can be, with high walls and razor wire and men with guns.

## Overpopulation affect

The dominant affect of overpopulation discourse is overwhelm, a panic at being swallowed, suffocated, consumed by the overabundance, the 'strong growth capacity' of the Other. These fears are rationalized through the topos of numbers to seem entirely rational, rooted in a dispassionate appraisal of the facts about a territory and its population. What stands in the way of the triumph of rationality are other people ('woke' social justice activists?) who want to 'hide' from reality and have given in to 'sensitivities' (race? gender? eugenics?). To face down these sensitive simpering saps requires bravery and fortitude in the face of critique, a kind of self-righteous vigour that is energized by the conviction that preventing overpopulation is the truly humanitarian and caring thing to do. Overpopulation activists are working to empower women to take control of their reproductive futures, and the warm glow of good deeds done thus suffuses their project.

# Textual analysis

In this section, I present an analysis of four cases that exemplify how the overpopulation imaginary is textualized.

## Case 1: 'A crisis, not an achievement'

A number of op-eds reacted to the news in late 2022 that total human population had surpassed 8 billion. In March 2023, Harvard historian of science Naomi Oreskes published a short opinion piece in *Scientific American* entitled 'Eight Billion People in the World Is a Crisis, Not an Achievement: More People Will Not Solve the Problem of Too Many People'. Oreskes asserts that the 'major driver of plant and animal loss is habitat destruction caused primarily by the encroachment of a swelling human population'.[99] To back up her claim, she cites a report that points to how 'agriculture, mining, industrial infrastructure and urban areas' are eating into natural areas. This smooth logical leap, which exculpates those who profit from commercial agriculture, mining and industry while laying the blame 'primarily' at the feet of the poor, is a clear instance of what I have called 'population primacy'. This telling presupposition is combined with the usual liquid metaphors of overwhelm, of a 'destructive surge'. Her article is however primarily a critique of cornucopianism, which she identifies in a *Washington Post* editorial that excitedly talks about millions of new people bringing 'their new ideas and fresh energy' to spur innovation and solve problems. Oreskes rightly dismisses cornucopianism as a fable that ignores the decisive role of massive public investment, and not the free market, in major historical technological breakthroughs. But she goes one step further in the penultimate paragraph:

> It's both counterfactual and illogical to imagine that more people will solve the problem of too many people. Most population growth is occurring in poor countries, where most people lack educational opportunities that might enable them to develop the kinds of ideas and skills they would need to apply their "fresh energy." And, as the biodiversity agreement makes clear, the issue isn't just living standards as measured by per capita income. It's also quality of life, which is threatened by widespread degradation and destruction of nature.

In rejecting the cornucopian logic of free markets spurring innovation, what is also lost in Oreskes's argument is the sense that people in poor countries can be a part of innovation and progress. In this version of events, there is only one way out of ecological crisis, and that is through rich-world state-driven institutional ecomodernism, through schools and universities and well-funded research agencies. Supposedly surplus people are constructed as outside of the kind of civilizational conditions of possibility for showing leadership, developing ideas or skills, or bringing 'fresh energy' to the climate crisis. Not only is this global lumpenproletariat blamed for being the 'primary' cause of habitat destruction, they are also constructed not as people who *have* problems but (in W.E.B. Du Bois's terms) as people who *are* problems. And as so many of them are not going to enjoy a decent quality of life, the conclusion we should draw from this is clear: that the world would be better place if these particular people had never existed at all.

## Case 2: Affirming the heterodoxy

In another opinion piece written to mark the 8-billionth human being, former environment editor at *The Guardian* John Vidal argues that '[i]t's time to ditch the generations-long argument between those who blame overpopulation and those who worry about consumption.'[100] The headline chosen for his piece is 'It Should Not Be Controversial to Say a Population of 8 Billion Will Have a Grave

Impact on the Climate'. Vidal assumes the pose of the brave truth-teller unafraid to share his heterodox opinion with readers. He understands that in the past there may have been 'authoritarian regimes and far-right extremists' who have implemented population control inspired by eugenics while bemoaning that 'genuine concern today in rich countries is often met with accusations of racism or eco-fascism'.

He lays out his argument as follows:

> Until now, the orthodox western intellectual argument has been that the number of people does not matter as much as how people use resources. Consumption and inequality are the problem, not population size. True, the wealthiest 10% consume about 20 times more energy overall than the bottom 10%. So of course the rich must change their behaviour. But making climate breakdown all about consumption has become an excuse for countries to do nowhere near enough to reduce their populations. [...] The hard fact is that in an age of climate breakdown, human numbers matter. And the ecological impact of another 2–3 billion humans will be immense.

The topos of numbers is mobilized by Vidal in an interesting and not very coherent way. If, as he writes, the wealthiest 10 per cent of humans (or 800 million people) consume twenty times as much as the poorest 800 million, then, for example, rich people halving their consumption would free up enough energy for *8 billion extra people* in the poorest decile.[101] That's simple maths, but Vidal does not seem to have done it. If energy consumption is 'nature's mighty feast' (in Malthus's terms) then some people are gorging themselves and others subsisting on scraps. So why *must* policy focus on the latter, when the worst impacts are coming from the former? Why is he so worried that governments are doing 'nowhere near enough to reduce their populations'? The next number he mentions is supposed to clinch the argument: the 'hard fact' that 'the ecological impact of another 2–3 billion humans will be immense'. The sheer volume of this latter number is presented as self-evident, as the decisive fact that proves the 'orthodox western intellectual argument' wrong, and brave heterodox truth-tellers like Vidal right. But the impact of the extra 2–3 billion people joining the poorest decile, on Vidal's own example of energy usage, could be compensated for by a mere 15 per cent reduction in rich-world consumption. Should rich people consume a fraction less, or poor countries be forced to 'control' their populations? Perhaps because he has guessed that his numbers are a bit wonky, Vidal goes on to make another argument for focusing on population:

> This is also about women's rights too [*sic*]. By ignoring population, the needs of women and girls are being sidelined by governments that are too obsessed with consumption to notice how vital education and family planning are in tackling the climate emergency.
>
> [...]
>
> According to the UN population fund (UNFPA), 257 million women have an unmet need for proper contraception, half of all global pregnancies are unplanned, and nearly a quarter of all women do not have enough agency to refuse sex.

This apparent alignment with a feminist politics of bodily autonomy and sexual agency is however quickly revealed as merely instrumental.[102] For if these women and their children are not having such a big impact on climate after all, would their rights be any less important? Moreover, the construction of poor women primarily as victims of local patriarchies obscures their agency while eliding the broader dynamics of the world system of capitalist extraction and domination, the real drivers of ecological devastation. When faraway ethnic strongmen and 'their' fertile womenfolk are blamed for harming the climate, environmental racism and cannibal capitalism are exonerated, while patriarchal assumptions about women's agency are subtly reinforced. This is a strategic choice that pushes colonial histories and present profiteering into the background as non-problems, and the intimate lives of dark-skinned men and women to the fore. Andil Gosine's 'anxious preoccupation with the sexualities of nonwhite people' is thus once again on full display.

## Case 3: 'Climate justice gets harder as world population passes 8 billion'

A syndicated Reuters story published by the *Daily Maverick*, a prominent South African online news company, asserted in its November 2022 headline that 'climate justice' gets 'harder' with an increased population.[103] Though this framing of the story (around 'justice') implies that it thematizes rich-world obligations to other countries to compensate for their historical profits from ecological devastation, the article focuses exclusively on problematizing sub-Saharan African growth, resource scarcity and urbanization, with no historical depth and no mention of any 'justice'-related themes. For the *Daily Maverick* 'climate justice' is thus a signifier that attaches to poor people and their problems without necessitating any discussion of the historical debt of the rich to the poor, a discursive move that empties the concept of its radical potential and valorizes any rich-world intervention in climate breakdown as altruistic and humanitarian rather than reciprocal or reparative.[104]

The article reproduces the discourse of environmental security, reporting that 'experts said' that rapid 'population growth combined with climate change is likely to cause mass migration and conflict in coming decades'. The *Daily Maverick* allows its readers to comment on stories, and one of the commenters picked up on this theme:

> The bio diversity/wild life of our planet earth will disappear for ever. Wars will be prevalent. Our quality of life on earth will reach new lows.
>
> We cannot wait until the population growth, naturally, diminishes to 1% and all people are more educated and concerned over the future of our planet. We have got to act before we get to 2080/2100.
>
> We must actively start educating all people about the horrors of overpopulating our beloved planet and must introduce humane methods of curbing rampant population growth in all population groups throughout in sub- saharan Arica and throughout the world
>
> WE HAVE GOT TO ACT NOW.

Analysis of the deixis of 'we' suggests that two categories of subject are being invoked. 'We' is the agent of mental processes (recognizing urgency/crisis, understanding that 'we cannot wait' and that 'WE HAVE GOT TO ACT NOW'), emotional processes (the planet is 'beloved' by us) and the material processes of educating 'all people' and introducing 'humane methods of curbing population growth'. The patients of this latter category are 'all people' and 'all population groups'. These groups are presented as lacking 'education', and as being responsible for 'rampant population growth'. It is thus clear that 'we' is not included in 'all people' or 'all population groups' – as 'we' is predicated with being educated, being concerned, loving the planet and being able to 'humanely' curb the population. This division arguably tinges the 'Our' of 'Our quality of life will reach new lows'. While the second subject category is the patient of 'educating' and 'curbing', its agency thus lurks behind the comment as the threat of the effects that 'rampant' growth, environmental devastation and war will have on 'our' quality of life.

'Population group' is a term whose definition was firmed up in the early days of apartheid as the preferred bureaucratic name for 'race'.[105] One of the insights of critical whiteness studies is that whites tend to think of themselves not as racial subjects, but as 'normal' humans unmarked by race.[106] Race is something that other people have, and a signifier of their difference, their strangeness. While this certainly does not hold universally and across contexts,[107] in this comment 'population group' attaches to other specifically racial controlling images of rampant growth, wars, lack of education and lack of care for the planet. The writer names sub-Saharan Africa before gesturing vaguely to the rest of the world, and the piece they were commenting on focuses almost exclusively on the continent where 'most of the 2.4 billion people to be added before the global population peaks will be born'.[108] This comment in conversation with the article thus divides humanity into racial populations with those who educate, understand, and care valorized as actors and those whose profligate population expansion harms biodiversity, causes war and wrecks 'our' quality of life.

## Case 4: 'What women seem to be wanting … '

In the wake of Tory cuts to the ODA budget in 2020, in August 2021, *The Guardian* published an article reporting on an NGO-led call for climate funding to be used to pay for reproductive health services, under the deadline '"Use your £11bn climate fund to pay for family planning," UK told'.[109] Though there are many parts of this text worth analysing, I will focus here on how the desires of African women are voiced by metropolitan actors.

> [Subhead] More than 60 NGOs call for spending rule change, saying people on frontline of climate crisis want greater access to reproductive healthcare
>
> […]
>
> Bethan Cobley, a director at MSI Reproductive Choices [ … said] 'We're hearing loud and clear from communities and women and our clients who are most affected by the climate crisis that what they really want is access to reproductive healthcare, so that they can make choices about when or whether they have children.'
>
> […]
>
> David Johnson, chief executive of the Margaret Pyke Trust, said people in developing countries knew what they needed.

> 'From increased risk of unintended pregnancy to dropping out of school, the communities we work with tell us they are already seeing how climate change impacts them, and that they see the connections between their health and the health of their local environment,' he said.
>
> [...]
>
> Prof Susannah Mayhew, of the London School of Hygiene & Tropical Medicine, said the connection, though 'not intuitive' to people in the west, was obvious to those in affected communities, where 'the thing you need to underpin [sustainable living] is the ability to control your own fertility as an individual woman'.
>
> [...]
>
> 'What women seem to be wanting is to have better spaced and healthier children so that they can be economically active and plan for good-quality futures for their children that don't lead to increasing pressures on farmland and bad livelihood practices, which comes [*sic*] with lack of education, so that they're able to live locally in a more sustainable way through having less pressure on land.'

Instead of the activation/passivation patterns noted in other texts, these NGO and scholarly figures are studiously consistent in representing the communities they serve as playing an active role in designing the interventions they need most. This agency is however nested within indirect speech, a form of reporting that means that subaltern populations are being spoken *for*, rather than themselves speaking.[110] The desires of African women are thus being interpreted by a global elite who are empowered to represent what they 'seem to be wanting'.

These representations differ in interesting ways. The first speaker, Cobley, says that clients 'who are most affected by the climate crisis' (a predication strategy) 'really want [...] access to reproductive healthcare' (a perspectivization strategy). The syntactic proximity of perspective and predicate implies but does not commit to a causal link between the two. The reader is left to infer that they want these services *because* they are affected by climate change. The claim of the second speaker, Johnson, is similarly evocative but noncommittal. Family planning is topicalized through fronting the sentence with 'From increased risk of unintended pregnancy to dropping out of school' while the rest of the sentence ('the communities we work with tell us they are already seeing how climate change impacts them') fails to spell out how these phenomena are related. Does climate change cause unintended pregnancy, and if so, how? Similarly to Cobley, just putting these two topics in the same sentence, in some vague relation or conjunction, seems to go far enough for inclusion in the story by *The Guardian*.

The third speaker, Mayhew, comes closest to committing to a causal account. She does this in two places. First, she claims that 'the thing you need to underpin [sustainable living] is the ability to control your own fertility as an individual woman'. The deixis of 'you' in this case works to universalize Mayhew's claim: any 'individual' woman wishing to live sustainably would need to be able to control her fertility. The discourse of sustainability is enmeshed in a normative and moral project; one *ought* to live sustainably.[111] This deontic modality dictates that women by extension *ought* to control their fertility, and that this is the *responsible* thing to do in the context of the climate crisis. Fertility is furthermore constructed as 'the thing' that underpins sustainability, pushing other enablers of sustainable living to the fringes and thus asserting the primacy of population as an environmental problem.

She then specifies the causal chain: better spaced and healthier children allow women to be economically active and to invest in their children's futures in ways 'that don't lead to increasing pressures on farmland and bad livelihood practices, which comes with lack of education, so that they're able to live locally in a more sustainable way through having less pressure on land'. Land degradation is presented as overdetermined by population pressure and the 'bad livelihood practices' of the uneducated peasantry on whose behalf she speaks. The threat that people who are no longer 'able to live locally' will migrate, sparking conflict and strife, points to a devastated future chronotope of waste and disease. What readers are asked to accept is that African communities are not very good at subsistence agriculture while at the same time being highly aware that their own ignorance is caused by too many children. The impact of enclosures, conflicts, and corporate and state greed on rural livelihoods is not even considered. Much as was the case was in the days of the *mission civilisatrice,* the only path out of ignorance and want is clearly Western altruism. The connection of overpopulation logic to a 'feminist' politics of reproductive rights recalls Oyèrónké Oyěwùmí's critique of white, Western feminists as thinking of themselves as taking up 'The White Woman's Burden'.[112]

## Conclusion

Reflecting on how the responsibilities of subjects in capitalist societies are constructed, Christian Chun points to an imbalance of both culpability and blame:

> Those in power now have created the very conditions, whether it is war, poverty, and/or climate change, which have forced people to flee these conditions. Those in power then blame those fleeing, attempting to victimize them yet again. We try to teach and remind our children, our students, our neighbors, our fellow citizens to take responsibility for their actions, yet those in power almost never take responsibility for their actions in the dire conditions they have created.[113]

Overpopulation discourse as it arises in connection to the environment represents a failure to correctly analyse the 'roots' of environmental crisis and the suffering of subaltern populations. It should go without saying that access to family planning services is an important plank of women's empowerment. Political ecologist Lisa Tilley and Max Ajl have dubbed as 'ecologically unequal exchange' (EUE) the global system of extraction, domination and reproductive injustice that sees the resources of an impoverished 'South' flowing to the rich world.[114] When interventions targeting marginalized women fail to address histories of unjust EUE in constructing their beneficiaries as environmental threats, this injustice is merely perpetuated. Much like historical eugenics, the current brand of 'populationism' is animated by anxieties about the sex lives of racial Others and, unfortunately, in evidence across the political spectrum in the West.

# 5

# Colonial conservation: Green violence at the wild frontier

This chapter presents an imaginary that structures how most mainstream environmental NGOs as well as many Western nations and corporations tend to think about biodiversity conservation, thus shaping the popular concept of what protecting the earth should look like. This dominant set of ideas has variously been named 'fortress conservation' (picking out its reliance on high fences and discourses of defence), 'militarized conservation' (pointing to the fact that wildlife rangers are armed to the teeth and frequently exchange fire with human 'intruders'), 'mainstream conservation' (picking out its hegemonic status as the dominant mode of biodiversity protection) and 'neoliberal conservation' (diagnosing its imbrication in late capitalist modes of accumulation and value production).[1] I have opted to describe it here as 'colonial conservation' both as this term encompasses the violence at play on the conservation frontier and because it aptly gestures to the kind of genealogical reading of the present required by an ecointersectional approach. The models of 'success' in conservation presented in mainstream environmental and media texts tend to rely heavily on Romantic visions of unspoiled wildernesses and endless horizons centred around human subjects who are especially attuned to the natural world, finding themselves in reverent communion with charismatic animals and landscapes much like explorers of yore. While these visions adapt to local histories and geographies, to illustrate this point I focus in this chapter mainly on how the African continent, and specifically the dream of 'wild Africa', figures in the conservation imaginary. This dream is juxtaposed to a nightmare of destruction and despoliation driven by the insatiable hunger of populations near and far for land, fuel, food and wildlife products such as rhino horn that can fetch astronomical prices on the illegal market. The crisis presented by the imminent violence of the poacher legitimizes the fortress and shifts resources to further militarization. Within capitalist logic, the costs of the urgent work of protecting nature against violent people are covered through philanthropic and business processes, where tourism and development reduce the lure of poaching, making the natives slightly less menacing, while at the same time paying for the guns and the fences. Because these logics historically and in the present day require the enclosure of land previously held in common, biodiversity conservation thus works as a kind of colonial frontier, peeling back the historical connections to land and place of local communities and replacing them with a mode of extraction tied in to the global tourism market, as well as institutions of green capitalism such as carbon finance. The ideology that structures contemporary conservation thus naturalizes the mode of production that causes biodiversity loss in the first place, drawing boundary lines around wildernesses imagined as reliant on markets for their survival. As a number of political ecologists have pointed out, the fact that this model is unsustainable and ultimately self-defeating does not seem to have reduced its power to hegemonize thinking about the protection of 'Nature'. Meanwhile, the human cost of colonial conservation in terms of lives and livelihoods lost is considerable. But it is definitely not the only option available to humanity, however hegemonic it may seem today, and pathways out of its violence will thus be discussed in the concluding section of this chapter. First, however, it is important to see how the current model is articulated in discursive practice.

## On the 'frontlines' of biodiversity conservation

In early November of 2024, William, Prince of Wales, was in Cape Town for the Global Summit of 'United for Wildlife', a conservation NGO he founded in 2013 to fight the illegal wildlife trade.[2] In his address to the conference, he launched a new initiative to support game rangers:

> We cannot protect our planet without a well-supported workforce. Currently, fewer than 300,000 rangers are working globally. But estimates suggest we need five times that to properly meet our conservation commitments. That is why I am proud to announce that United for Wildlife, in collaboration with the Game Rangers Association of Africa and Tusk Trust, is launching a new life insurance product specifically designed for rangers across Africa. This initiative aims to ensure that rangers receive much needed life insurance cover, giving them and their families reassurance that they are protected for the dangerous job they do. [...] A strong and supportive frontline is essential to ending the illegal wildlife trade. [...] What happens over the next five years, especially in relation to

> stopping the illegal wildlife trade, is crucial for the future of life on earth. The urgency is palpable as entire parts of the species chain are being removed. As we look to 2030 and beyond, we have the power and opportunity to change the trajectory. While the challenge may still be complex, the path is now clearer. We must strengthen our networks, support those on the frontlines, and adopt innovative community-focused conservation strategies.[3]

William's remarks trace the generic conventions of business administration. He speaks not in the lofty monarchical register of heir apparent to the British throne, but in the style of a manager who has a business to run. He has a 'workforce' to support, 'commitments' to meet, a 'trajectory' to change and 'strategies' to adopt. Eve Chiapello and Norman Fairclough have argued that management gurus are the 'prophets'[4] of late capitalism. Voicing the figure of the manager performatively accomplishes authority, hierarchy and rationality. From this solid basis, the claim that '[w]e cannot protect our planet without a well-supported workforce' seamlessly integrates biodiversity conservation, military might (the *force* in workforce)[5] and the management of wage labour. What is entailed (but not made explicit) is that without a well-organized and well-supported battalion of workers, things would inevitably fall apart as populations around protected areas descend into orgies of slaughter and despoliation. A classic ingroup–outgroup dyad (or 'ideological square')[6] is thus constructed where the rational management of the modern conservation enterprise with its army of rangers is aligned with environmentally and socially positive outcomes, and the lack of such interventions with savagery and decline. It is thus only logical within this imaginary that a person of William's status would be in Cape Town to sell a 'product' – something to be sold on the market, probably docked in monthly instalments from rangers' pay – as if he were a travelling salesman. At the same time, the actuarial calculations that go into balancing the probabilities of the deaths or disability of individual members of the 'workforce' against the size of the payout their families will receive are produced as necessary for the survival of the earth's ecosystem. Finance thus supervenes on the natural world, providing its very conditions of possibility. Without capitalism, we are led to believe, nature itself would collapse.

The relationship between employer and labourer is produced as a benevolent one through the invocation of 'support' for rangers and 'reassurance' for their families. What William implies but does not spell out in detail is that rangers not only sell their labour to their employers (the 'excess' value of which is considerable) but also put themselves in the way of sometimes fatal harm. The International Ranger Federation recorded that between 122 and 156 of their members were killed on the job *per annum* between 2016 and 2021, with homicide being the leading cause of death in both Africa and Asia by a wide margin.[7] These numbers do not account for all personnel involved in conservation efforts around the world, nor do they include grievous harm and/or disability. That rangers are paid to go to war in the name of conservation is evidenced in William's repeated references to their being 'on the frontlines'. Conservation is thus understood as a battlefield, with park rangers in the trenches.

William further asserts that the war for African biodiversity is raging in the here and now, that the 'urgency is palpable' and fighting back is 'crucial for the future of life on earth'. The stakes are thus constructed as infinitely high. What could possibly be more important than securing the future of life on earth? This sense of urgency is undergirded by the trope of the deadline. Looking 'to 2030 and beyond' references target three of the Global Biodiversity Framework, which sets the goal of conserving '30% of terrestrial, inland water, coastal and marine areas by 2030' – the 30×30 initiative.[8] William's speech thus taps intertextually into a raft of plans at a global level, which determine 'our conservation commitments' which are given solidity through the topos of numbers ('five times'; '300,000') and that constitute a sense of measurability, rationality, seriousness, organization and momentum.

While William sets his sights on 'life on earth' the press release issued by United for Wildlife to report on the launch names a more specific aim in Africa: protecting 'the planet's most iconic species and landscapes'.[9] The fissure between the grandiosity of the former and the limited scope of the latter goal is revealing. In the grand scheme, rangers bravely risk their lives for the future of all life. But in the latter, what they protect are sacred objects, consumed as 'icons' by an unseen viewing public, the satisfaction of whose desiring gazes justifies the risks that they have to take. Another fissure opens up in the prince's text, where he deviates from speaking of rangers as in need of support, to their importance as the 'supportive frontline'. The threat here is that the frontline might turn on their commanders if they do not feel supported themselves, perhaps because they realize that the benefits of the sacrifices they are making are not accruing to them and their families. The ever-present threat of the ranger who colludes with or even becomes the poacher is thus foreshadowed in this slip from 'supported' to 'supportive'.

As the environmental social scientist Aby Sène has argued at length, it would be a mistake to chalk up the involvement of the British royal family in African conservation initiatives to 'innocuous imperial chauvinism'.[10] There are important continuities between the carving up of the continent between colonial powers and what the carnivore ecologist Mordecai Ogada calls the 'second scramble for Africa' constituted by contemporary conservation initiatives. To understand why this is the case, in the next section we will go on safari.

## Dreaming of Africa

'Africa' – as fantasy more than as a landmass – played a central role in the development a white European identity distinct from non-white human Others, and from the non-human 'natural' world. European cultural production historically maintained an uneasy ambivalence about the continent, producing it sometimes as an inhospitable wasteland and sometimes as an earthly paradise.[11] In settler colonial contexts, as 'Africa' became a place that Europeans might call home, the 'myth of wild Africa'[12] as unspoiled and Edenic was entrenched, and the continent was imagined as a 'romanticized wilderness in opposition to the decadent metropole'.[13] For white settlers Africa thus became a 'a place of identity and a site of belonging'[14] as Elsie Cloete argues, citing lines the Italian conservationist Kuki Gallmann's memoir *I Dreamed of Africa* (1991): 'The desire to go to Africa seemed to have been an obscure yearning to return. A nostalgic inherited need to migrate back to where our ancestors came from. It was a memory carried in my genes. The urge to fly home, like the swallows'. Palaeoanthropological evidence for the evolution of *Homo sapiens* on the continent – the so-called Out of Africa thesis, which whites can feel 'in [their] genes' – supports a romanticization of European settlers[15] as potential inheritors of the land and its 'wild' beauty.

The notion of a primordial and universal human belonging in Africa rooted in geological time animates European colonial fantasies.[16] These fantasies project an exogenous image of what Africa 'should' look like onto the landscape in ways that frequently ignore the agency and sovereignty of the people living there.[17] As the political ecologist Roderick Neumann has extensively documented, conservation initiatives on the continent are shaped by an 'Anglo-American nature aesthetic'[18] that delimits the biodiversity worthy of conservation while establishing a moral economy of whose rights should be respected and whose disregarded. These practices of looking have fatal consequences for local people whose struggles against colonial land theft are continuous with their struggles against the creation of protected conservation areas on their ancestral lands. Long histories of human land use and their violent colonial interruption are not allowed to feature in the picturesque, however, which in the colonial imagination is 'unspoiled'. Neumann mobilizes the distinction in Western cultures between landscapes that are seen as *productive* and thus enmeshed in markets and those considered *consumptive*, where leisure, contemplation and emotional connection are the model practices.[19] The African savannah became a 'quintessential landscape of consumption'[20] where the European could signal their class status through communion with sublime nature. In settler colonial contexts such as South Africa and Kenya establishing national parks went hand-in-hand with white nation-building, providing settlers with cultural symbols of the wilds they had tamed.[21] This established a discursive tension around protected areas in Africa, which for settlers and tourists seemed to be precious remnants of threatened nature, while for displaced people they signified a lost 'history in place, symbolized by ancestral grave sites and sustained by a collectively formulated memory of past occupation and utilization'.[22]

With the triumph of neoliberalism in the late twenty-first century, conservation became increasingly privatized and integrated into global markets. The private game reserve emerged in this period as a 'commercial and ideological response to the idea of nature's fragility'.[23] Louise Green identifies the 'new safari aesthetic' as a material rearrangement of space to optimize the consumption of African wilderness in luxury game lodges:

> [The new safari aesthetic] is a design category, but one that goes beyond design to construct a moral and experiential fantasy that centers on the game lodge. It draws its value from an array of sources: a moralizing discourse about the environment, a celebration of the premodern and the traditional, a nostalgic fantasy about the modern.[24]

The 'nostalgic fantasy' is for a stylish *Out of Africa* colonialism where the presence of benevolent whites on the land protects both land and people, and where uniformed Black servants deliver an 'experience of leisure and luxury can be justified because this is merely a side effect of the real work of the reserve, which is to preserve nature for the good of humankind'.[25] As I have shown in discourse analysis of environmental campaigns in South Africa, the production of whites as 'stewards of the land'[26] works to legitimate continued white landholding and socio-economic dominance in the post-apartheid order. For tourists and settlers alike, part of the fantasy of the game lodge is of an unbounded wilderness outside of national borders and thus above the messy complexity of individual nations and peoples. The aestheticization of (political) nature works to oppose a 'real' Africa to modern consumer society, and thus to include actual African people only if they fit the mythos of the premodern and prehistoric indigene.[27] Actual African people, societies, economies and histories are brushed over in a moral imagination that stretches from Cape to Cairo, much in the mould of historical colonizers.[28]

An important element of the dream of wild Africa is the inclusion by conservationists of charismatic animals in their 'moral community'.[29] This granting of moral standing to rhinos, elephants and gorillas tends to happen in concert with the dehumanization of the poachers who threaten them. A colonial regime of value is activated in cases where love of animals is associated with Western cultural evolution and enlightenment, and disregard for them with African barbarism and backwardness. The racist implications of this moral hierarchy are resisted in popular critiques of private conservation. During the Covid-19 pandemic, for example, calls by a game lodge owner to lift lockdown so that tourists could return with the cash needed to sustain expensive anti-poaching tactics were lampooned on social media for their blatant disregard for the human cost of reopening the economy. The conservationist in question (whom I shall introduce below) is a white South African émigré cricketer who has restyled himself as a protector of rhinos. He was told in no uncertain terms that if he and his rhinos were such great pals, perhaps he should think of taking them with him when he leaves the country again.[30]

The dream of wild Africa carves out a special role for the white conservationist, whose subjectivity is valorized as being at the centre of protecting the 'unassailably positive value'[31] of nature. Alignment with this subject position gives whites in the postcolonial period a 'reason' for being in Africa, as the anthropologist David McDermott Hughes argues.[32] Numerous ethnographies of the racial dynamics of African conservation have identified how profoundly its habitus is structured by whiteness, and how white conservationists legitimate their belonging in Africa through communion with and love for the natural world.[33] Where the guiding value of the conservationist is claimed as non-human nature, opportunities for good relations with other people to secure an ecologically sound future are shut down. This insight retraces critiques of Arne Naess's 'Deep Ecology'. The historian Ramachandra Guha observed that deep ecology's shift from an 'anthropocentric' to a 'biocentric' ecological ethic retraces an unworkable and impractical dichotomy.[34] Anthropos is always already enmeshed in bios and vice versa. Humans who speak for nature as if they have privileged access to moral insights disentangled from the human and more-than-human world arrogate to themselves the right to correct the naturecultures of the Other.[35] When conservationists turn to communion with the non-human world to anchor their moral projects, they too often turn away from fellow humans. The 'non-relational' character of whiteness is thus reproduced in a conservation sector that remains a 'white man's game'.[36]

The prominent involvement of white celebrities such as Prince William in African conservation confirms the idea that if it weren't for Western intervention continental actors would mismanage ecosystems and squander their natural wealth. The need for 'white saviours' to rectify this situation is reproduced in media texts that, for example, entirely exclude African conservation expertise and activism in favour of a series of white talking heads.[37] The characterological figure of the white conservationist in Africa is very much modelled by William, and his brother Harry, who in 2022 appeared in ranger uniform alongside the more formally dressed Rwandan president Paul Kagame.[38] These contrasting wardrobe choices, where the royal attire is not ermine but khaki, bespeak a readiness to get down to the hard work of conservation while the African president will keep his hands (and his suit) clean, indexing shifting regimes of value in late modernity. Involvement in conservation moves value up the hierarchy towards the celebrities who associate themselves with it, sanctifying their unearned privileges or obscene wealth. Conservation may also be marshalled to 'philanthrocapitalist'[39] ends. A video clip released by Amazon owner Jeff Bezos's 'Earth Fund', for example, portrays the world's second richest man and his then-fiancée Lauren Sánchez surrounded by smiling conservationists in the Democratic Republic of Congo, discussing how the fund's \$110-million investment in the Congo basin would help protect the area. 'Lauren and I are very committed to nature, we're committed to conservation, we're committed to restoration,' he says. But when this investment is compared with the other things he spends money on, such as his \$500-million luxury yacht Koru, which emits over 7,000 tonnes of carbon dioxide every year and racks up an annual running cost of \$25 million, the sincerity of this 'commitment' is cast in doubt.[40] By playing the part of the conservationist, Bezos is able to enjoy all the kudos of saving wild Africa while at the same time leading the kind of lifestyle that disproportionately contributes to environmental decline at a global level. The environmental sociologist Stasja Koot argues that conservation philanthrocapitalists derive *jouissance* from 'saving' nature in no small part through lingering over the gory details of just how threatened it is, and therefore how important they are.[41] As we shall see in the next section, this is nowhere more apparent than in the 'war' on poaching.

## Green violence: 'It's raw. It's war'

Kevin Pietersen, renowned batsman and one-time captain of the English cricket team, was born and raised in South Africa but left for the UK in 2000 citing the racial 'quota system' as the reason for his move.[42] In 2017 he announced that he was retiring from international cricket in order to focus on saving the remaining rhinoceroses of Africa. Styling himself as a rhino conservationist in the midst of a 'poaching crisis'[43] was met with almost universal acclaim in the media. A ten-part podcast produced by the BBC in 2019 entitled 'Kevin Pietersen: Beast of Man' portrayed the retired cricketer as the hero Africa needed to save the rhino from the people hunting them to extinction. The tagline of the podcast runs as follows:

> The rhino. An icon, all but extinct. A maverick who never backs down says he'll save them. It's raw. It's war.[44]

While promising to save the rhino from extinction, Pietersen is the owner of the luxury Umganu Lodge 'an idyllic retreat built on the banks of the magnificent Sabie River, providing uninterrupted vistas of the world renowned Kruger National Park'.[45] The return of a wealthy white settler to the land of his birth to engage in a 'war' funded by a tourism business is put in a positive light through its association with the value of protecting nature. The idea that meeting biodiversity objectives requires violent conflict is affirmed in the anaphora 'It's raw. It's war'. The 'beast' in the title meanwhile refers to the rhino itself[46] while also hinting at the violent actions a beast of a man such as Pietersen must engage in to save it, and the beastly people ranged against him. It is within this triangle – the charismatic animal, the conservationist and the poacher – that 'green violence' is legitimated.

As Bram Büscher and Maano Ramutsindela show, 'green violence' has material, social and discursive facets.[47] It works through processes of division. Communities are divided into '"good citizens" (informants) and "bad citizens" (poachers and those working with poaching syndicates)'.[48] Territory is materially divided in ways that create 'spaces of exception' where suspected poachers may be shot on sight. This necropolitical complex relies on a moral economy that divides those who must be protected from those who may be killed. Charismatic species are included in the former category, and thus granted a moral standing which is denied to the humans who would kill them, who become subject to extrajudicial killing.[49] The discursive production of the killable human not only impacts materially on people suspected of poaching in the African context, but more broadly on Black rural men whose torture and dismemberment are earnestly longed for in forums devoted to the topic.[50] This discursive violence is vented in gory detail, coalescing into a 'politics of hysteria'[51] in no small part animated by more general feelings of white rage at the postcolonial African order.

The overpopulation imaginary, with its racist and colonial undertones, multiplies the figure of the poacher into a seething mass of hungry subhumanity red in tooth and claw. Kevin Pietersen was not the only one to sound the alarm during the Covid-19 pandemic that fewer tourists meant less money, and less money meant fewer patrols, fewer drones, fewer guns and fewer boots on the ground. In news reportage of Covid-era poaching, jam-packed communities are pictured pressed up to the fences of protected areas as an ominous presence.[52] Conservation biologists worried about the increased pressure on protected areas from hungry people during lockdown argued that the lack of photographic tourism and the $200-million per annum trophy hunting business posed 'increased threats to wildlife and ecosystems'.[53] The distinction here between hunting for food as a crime perpetrated by local poachers and hunting for sport as a public good performed by conservation-minded foreigners is particularly stark. When the pandemic was over, normal programming resumed, and the same biologists started to fret about the impact of 'rapid human population growth'[54] on biodiversity. A review of the evidence for this claim conducted by the conservation biologist Alice Hughes and colleagues however found that there is no evidence that high population density is the direct driver of biodiversity loss, calling it a 'long-held but overly simplistic and inaccurate assertion'.[55] As these authors argue, a focus on population diverts energy from the actual causes of biodiversity loss (which as discussed in the previous chapter are caused by rich-world consumption) while justifying underdevelopment and neocolonialism.

From the periphery of empire to its centre and back again, the figure of the 'poacher' is continuous with the enclosure of land traditionally held in common.[56] While the things poachers and hunters do are more or less identical, the 'hunter' is valorized while the 'poacher' is vilified. The distinction between the two is a simple matter of whether they are seen to participate in capitalism in the correct way. Trophy hunting, for example, supports 68 per cent of Tanzania's protected areas.[57] For the value of killing African wildlife to be high enough to pay for 'fortress conservation' the supply must be controlled. The global elites who can afford the high prices come to Africa to kill animals which are already their property: they have bought the rights to the animal and its carcass, they have purchased the thrill of the hunt.[58] Poachers disrespect these rights: they steal the property that rightly belongs to others, they treat the abundance of nature as a common resource when in fact it is an enclosed commodity reserved for those wealthy enough to enjoy it, whether armed with cameras or rifles.

This commodification of nature, which I will unpack in more detail in the next section, both necessitates and subventions the fortification of conservation.[59] A protected area is a protected commodity given value in an economic system, and the developmental objectives of African states thus make protecting spectacular natures that will draw international tourists a priority. It is not all about money, however. As has been documented in a number of different contexts, geopolitical and security objectives often merge with conservation projects in 'green militarization'.[60] The borderlands of South Africa, Zimbabwe and Mozambique, for example, or the Caprivi strip, where Namibia borders Angola and Zambia to the north, Zimbabwe to the east and Botswana to the south, are epicentres of biodiversity conservation as well as of historic insurgency and counter-insurgency.[61] In Tanzania, Kenya, Malawi, South Africa, the Central African Republic and Rwanda 'the militarization of conservation has overlapped with state repression of minority ethnic groups and liberation movements'.[62] There is a 'convergence' of conservation and security personnel[63] as rangers become better trained and armed and start to resemble soldiers, and as soldiers are deployed to patrol for poachers and start to resemble rangers. The casualties of this war are not only those who get shot on sight for trespass, but their families and their broader communities, who are given the choice either to affirm fortress conservation or become part of a human developmental problem in need of intervention. Guns are of course not the only form of violence. When it comes to accessing water, fuel and food fences themselves are a kind of 'infrastructural violence'.[64]

The construction of the poaching crisis has discursive effects that renew colonial practices, which cleave as much to gender as they do to race. As Tommaso Milani and I have shown, the killability of poachers is established through their construction as over-sexualized and hypermasculine savages in need of lynching and castration.[65] A gendered 'conservation labour geography'[66] meanwhile typically exposes Black male rangers to deadly risks that impact heavily on the women left at home, which is in turn mirrored in the extreme levels of deprivation evident among the families of dead poachers.[67] An exception that proves the rule is the proliferation of all-woman anti-poaching units, reconfirming the idea that Black men really can't be trusted.[68] Media representations of women fighting poaching, of men pinned down and being arrested, or graphic images of butchered animals make these conflicts the most visible form of violence at play in conservation areas. This politics of the spectacular makes other social dynamics, such as the 'slow violence' of historical dispossession and poverty, invisible.[69] What is made visible and what invisible is intimately tied up in colonial ways of seeing, and with the hegemonic regime of value in late capitalism.

## Tourism, climate finance and 'neoliberal conservation'

While the rest of the world was entertaining themselves during the Covid-19 pandemic with what Kai Bosworth sardonically calls 'bad environmentalism', sharing memes claiming that 'nature is healing' due to the reduction in economic activity,[70] tourism operators and wildlife entrepreneurs such as Kevin Pietersen, along with some of the world's largest conservation NGOs, arrived at a very different conclusion: that far from healing, nature was being torn apart by the *lack* of economic activity. As discussed above, this logic rests on a particular model of conservation, where fences, patrols, drones, rifles and rangers are paid for with the proceeds of trophy hunting and tourism. A 'schematic' proposed in the *Nature* sub-journal *Ecology and Evolution* by a group of ecologists and NGO actors models how this works.[71] As economies shrink, money from philanthropic donations, tourism revenue and state coffers reduces. This decline in turn impacts negatively on the food security of local communities, the perceived benefits of protecting nature and the 'effective management' of protected areas, resulting in these communities losing faith in the conservation model and encroaching on the natural resources of the park. This encroachment then starts to eat away at African biodiversity, diminishing its 'global existence value' and thereby the need for its conservation, potentially resulting in a vicious circle of accelerating decline. The authors argue that this imminent threat should spur the development of 'new, resilient conservation' models built on diversifying revenue streams, including with income from 'wildlife-friendly' agricultural land use and green financial instruments such as carbon trading schemes, and 'debt-for-nature swaps'.[72] These models would deviate from 'fortress conservation', promising to embrace innovations that connect wildlife protection more closely to global financial markets, and thus hewing more closely to what has been termed the 'new conservation'[73] or, less vaguely, 'neoliberal conservation'.[74] New conservation aims to unlock climate financing for biodiversity protection through assigning values to ecosystem services.[75]

The account in *Nature* of the inevitable decline of ecosystems in the context of an economic downturn is doubtless not far from the truth if we accept that 'it is easier to imagine the end of the world than it is to imagine the end of capitalism'.[76] What the current mode of conservation presupposes and thus naturalizes is the actually rather strange idea that nature *depends* on capitalism for its continued existence, rather than the other way round. In this imaginary, the commodity value of a particular element of an ecosystem needs to be set within an ecological market that ultimately determines the 'existence value' of biodiversity. If we just get these markets right, we will 'transition' from fossil to green capitalism. As discussed in Chapter 3, this fantasy is ignorant of history for two reasons: first, because capitalism by necessity 'cannibalizes' the conditions of its own existence, and secondly, because in human history there never has been an 'energy transition' – only accelerating use of new agglomerations of different forms of energy.[77] These structural dynamics of capitalism, especially in its late neoliberal form where 'competition' is imagined as the essential organizing principle of natural and social relations, results in what the ecologist Robert Fletcher refers to as neoliberal conservation's tendency to 'fail forward'.[78]

In Fletcher's account, it is a mistake to imagine neoliberal conservation actors as craven profiteers. As he points out, actual profits from neoliberal conservation models have remained elusive. Rather, we should understand that these actors ardently desire a form of capitalism that is indeed sustainable, where 'people, planet, and profit'[79] are in perfect balance. Four of the world's largest conservation NGOs – Conservation International (CI), The Nature Conservancy (TNC), the World Wildlife Fund-US (WWF-US) and the Wildlife Conservation Society (WCS) – have thus populated their boards with bankers and other financiers who guide their decision-making.[80] The inevitable failure of these schemes is taken not as evidence that their assumptions were wrong, but that they just need to try harder, invest more and (paradoxically given the nature of neoliberalism) that states need to introduce more legislation to facilitate green capitalist transitions. Failing forward is caused by the 'fictitious' nature of financial capitalism and animated by a 'fantasy structure promising an eventual success that is continually deferred into the future'.[81] Investors in neoliberal conservation thus fall prey to 'ergoic' reasoning, where an explanatory framework (in this case neoliberal capitalist rationality) is held inviolate in the face of contradictory experience.[82] Because capitalism *must* be the correct way to think about the world, we will continue to throw our time and energy into neoliberal conservation models until the world conforms to our theory of the way things *should* work. The fundamental contradiction at the heart of capitalism – that 'profit' is derived from finite social, political and ecological externalities which if fully internalized cancel it out – dictates that building infinite economic growth through the commodification of the non-use of natural resources is a non-starter. But as a soothing just-so story that provides a glimmer of hope for the future of capitalism, there is no amount of energy that it is not worth pouring into it.

While the grand schemes of neoliberal conservation may not be especially profitable, this does not mean that nobody is making money. Spectacular conservation 'successes' in one area provide cover for extractivist sacrifice zones in others.[83] But conservation itself is a profitable mode of production that enables accumulation, especially for landowners and operators of tourism businesses.[84] Tourists flock to the wilds of Africa from places far afield, sustaining intercontinental luxury travel and requiring networks of roads and automobiles to ferry them to and through nature. The 'conservation labour geographies'[85] surrounding game lodges rely on racial, class and gender inequalities that ensure that the exquisitely luxurious wilderness experiences are facilitated by local personnel who are only too grateful for the money being brought into their otherwise poor and precarious lives from faraway places. Even in instances where local groups have managed to secure title to ancestral lands, the people who profit most from conservation concessions on the land are too often white settlers, or the urban middle classes.[86] In this sense, contemporary conservation is 'fossilized' both in its reliance on hydrocarbons and on unjust social structures historically petrified in colonial power relations that facilitate Euro-American consumption of exotic Southern landscapes.[87]

## Conservation as colonial frontier

In December 2022, Fiore Longo, a European activist at the Indigenous rights organization Survival International shared a video clip on her Twitter feed in which the former chief scientist of the World Wildlife Fund, Eric Dinerstein, is presents a bold plan for the preservation of planetary biodiversity. Dinerstein states:

> The pope of conservation biology, E.O. Wilson, has come out with a new book called *Half Earth* which calls for putting half of the terrestrial realm and half of the marine under some form of conservation management. What we've done is to apply this idea of 'Nature Needs Half' to the 846 ecoregions that will be in this new map called Ecoregions 2.0. Now the idea of taking like the tall grass prairie of Illinois, which is 99% soybeans and corn, would probably require the end of the civilized world to ever get back to nature […] But the surprising and positive finding is that there are many ecoregions, perhaps a third already, that either have the ability to protect, that have already protected half or close to it – 40% – or could do so, so that it's not as far-fetched as we would think. […] If we are to leave behind a living planet, we have to undertake the exercises to plan, to

zone for conservation and zone for intensive human development in the right places so that we leave space for the tigers and rhinos and elephants of Bardia that I had the great chance to study when I was 21 – and leave things for future generations as well.[88]

The 'Nature Needs Half' movement enjoys the support of scores of biologists, ecologists and policymakers around the world. In support of the movement, the prominent Canadian conservationist Harvey Locke has written that 'it is time for conservationists to reset the debate based on scientific findings and assert nature's needs fearlessly'.[89] Dinerstein and his collaborators around the world have produced detailed maps of planetary biodiversity that identify areas where half or nearly half of 'nature' is already protected, and those where 'nature' is decisively 'imperilled' and therefore unlikely to recover.[90] Wilson's *Half Earth* meanwhile takes aim both at green capitalist schemes to monetize nature and at academic critiques of the nature–culture dichotomy, arguing forcefully for the preservation of 'untrammeled wilderness'.[91] His definition of wild areas is that they are largely 'self-willed' without human intervention, though they 'have often contained sparse populations of people, especially those indigenous for centuries or millennia, without losing their essential character'.[92]

While establishing new protected areas may in certain cases serve the interests of local communities, political ecologists point out that they are just as likely to cause conflict and dispossession.[93] Top-down, map-based approaches concentrate power in the hands of powerful conservation NGOs and scientific elites, who are unlikely to prize the livelihood and cultural goals of local human populations above what they construct as the ecological integrity of a specific territory.[94] Wilson's mention of 'sparse' populations is salient here: as long as people are melting into the wild background, they can be tolerated, but if they become too visible or too numerous, they become a problem for the 'essential character' of the landscape. In Dinerstein's presentation, we get a clear view of how these ideas could cleave to colonial divisions. The farmers of Illinois have so successfully annihilated the local flora and fauna that they will be left alone, while the comparatively low ecological impact of people who live alongside 'tigers and rhinos and elephants' will have their lives disrupted by conservation interventions. Interfering with commercial agriculture is constructed as a threat to civilization itself, while displacing populations in far-flung places is glossed over. As a case in point, Bardia – the national park in Western Nepal mentioned by Dinerstein – was created through the displacement of 9,600 people over three decades.[95] If a dichotomous human–nature logic dictates the zoning of vast swathes of the world as wilderness, the result might well be the disproportionate dispossession of rural populations in the global periphery, over whose heads ecoregions developed in the rich-world will be imposed.

The scientific energy behind Nature Needs Half has fed into advocacy for initiatives to meet Target 3 of the Global Biodiversity Framework – the 'ambitious commitment to conserve 30% of terrestrial, inland water, coastal and marine areas by 2030'[96] – or 30×30 for short. Activists, not least of all on the African continent, thus faced off against major NGOs ahead of the UN Biodiversity Conference (COP15) in Montreal in December 2022 to oppose 30×30. When in August of that year a major player in African conservation, the Frankfurt Zoological Society (FZS) tweeted an image of forested hills with the hashtag #FAQfriday and the caption 'Why does FZS protect wild areas?' with text explaining their commitment to 'protect 30% of the earth by 2030' the Kenyan environmentalist Violet Matiru shot back: 'Evicting indigenous people to create areas for the exclusive enjoyment of rich white people. You are colonizers & white supremacists! Who gave you the mandate of "protecting" "wild areas" from whom & for who? What you call "wild areas" are homes of people who co-exist with wildlife.'[97]

The 30×30 target has been championed both by 'neoprotectionists' and 'new' conservationists who embrace the neoliberal absorption of nature into markets.[98] Whether a wild area is protected for its 'ecological services' or for its own sake, because of its potential impact on peasant and Indigenous landholding, Survival International warned that the target constituted 'the biggest landgrab in history'.[99] In 2019 they put together a coalition of environmental justice and Indigenous rights groups behind a campaign called 'No Deal for Nature'.[100] Twenty Indigenous organizations in Asia released a report in February 2022 arguing that 'new conservation modalities are needed to end exclusionary approaches, embrace human rights-based strategies, and advance the recognition of the land, forest, water, and territorial rights of Indigenous Peoples'.[101] Amnesty International meanwhile launched a petition to include 'robust protections for the rights of Indigenous Peoples, and for the rights of subsistence land-users' in the 30×30 target.[102] The Kenyan ecologist Mordecai Ogada published a series of no-holds-barred editorials arguing that 30×30 was the '2nd scramble for Africa'[103] and a ploy to 'enable Western capitalism to annex over 80 per cent of Africa's landmass'.[104] By December when the 30×30 goal was adopted in Montreal it had been amended to exclude mention of 'strictly protected' areas and included references to the importance of Indigenous rights.[105] What it did not engage with, however, was the actual driver of biodiversity loss worldwide: rich-world consumption.[106]

Suspicion of ambitious conservation initiatives is backed up by historical experiences of displacement and dispossession. From its very inception in the nineteenth century, the idea of 'unspoiled' nature in national parks involved the violent displacement of native peoples. The US environmentalist John Muir, who co-founded the Sierra Club and was the leading activist behind the establishment of the Yosemite National Park, famously claimed that dirty, lazy and ugly 'Indians' had 'no right place on the landscape'.[107] The 'Yosemite model' was organized along military lines[108] and was exported to other contexts, including South Africa, where it guided the formation of the Kruger National Park from which thousands of local people were displaced in the early twentieth century.[109] Displaced people eked out an existence in the areas surrounding the park. In the successful tourist towns around the Kruger National Park, for example, Black labour is still confined to former apartheid 'homeland' areas. Tourism and conservation workers have to commute into the 'white' and wealthy towns where the tourists are, in a repetition of historical apartheid divisions.[110] The Western logic of enclosure, which results in the 'forced incorporation into a capitalist economic underclass'[111] thus advances racial capitalism.

While establishing protected areas has a proven impact on the protection of endangered species, there is very little evidence that the displacement of people itself plays any protective role.[112] There is another, nastier, logic at play. Asserting the necessity of 'spaces' for biodiversity that require the partial or total evacuation of human populations is a way of thinking about the future of biodiversity that political ecologists have called 'conservation *Lebensraum*'.[113] It was no accident that Muir saw no space for the Indians on the landscape, nor that Bernhard Grzimek, founder of the FZS, was an enthusiastic Nazi Party member who joined the *Sturmabteilung* in his twenties.[114] His vision for African conservation consistently excluded actual African people from the scene. The 'pope' of biodiversity conservation, E.O. Wilson, meanwhile, had an enthusiasm for eugenics that lasted well beyond the Second World War.[115] The division of the world into spaces of productivity and order, where wealth is generated, and spaces of leisure, where wealth is enjoyed, leaves little room for poor people constructed as less-than-human on the wild landscape.

Mainstream conservation too often follows the script of colonialism. It degrades the humanity of people in far-flung places as a pretext for expanding the frontier of its own influence, transforming previously sovereign peoples into precarious wage labourers. Aby Sène thus recently called attention to the Guyanese pan-Africanist and Marxist thinker Walter Rodney's coinage of the term 'Wildlife Republic' to refer to the coloniality of conservation.[116] Rodney wrote that imperialism was

> threatening to transform Kenya, Uganda and Tanzania into Wildlife Republics. Every effort was made to attract tourists to look at the animals, and the animals assumed priorities higher than human beings. [...] Certainly, tourism in all its aspects is proving to be one of the new areas of expansion of the imperialist economy. It is a new way of confirming the dependence and subjugation of Third World economies

Two ongoing cases in East Africa illustrate Sène's and Rodney's point. Maasai people have used the pastures of what is now the Ngorongoro Conservation Area (NCA) and its surrounds in modern-day Tanzania for hundreds of years. When the NCA was first established in 1959 there were roughly 8,000 Maasai in the area, but their numbers have since grown to about 100,000.[117] A quest for foreign investment and economic growth has seen the Tanzanian government play a leading role in removing them from their ancestral lands. As the Maasai scholar and activist Leiyo Singo observes, Maasai are being forced to 'leave our homelands so that others, be they white tourists, Emirati royals or wealthy urbanites, can use our grazing lands as their playgrounds'.[118] The Maasai are under enormous pressure both from private hunting and from traditional conservation as advocated for by the FZS, which holds enormous sway in Ngorongoro. In August 2017 the four villages in the Loliondo area, where a private company based in the United Arab Emirates had invested in a hunting lodge, were forcibly removed, leaving 20,000 people homeless.[119] More land has since been taken, cattle have been confiscated and all legal recourses have failed. The Maasai are meanwhile routinely represented in Tanzania as backward savages standing in the way of national progress. In August 2024, the Tanzanian government delisted all of the wards in Ngorongoro, effectively disenfranchising the Maasai from 2025 local elections and boding ill for their future in the region.[120] Their plight has become so urgent that even the otherwise elite-friendly *The Atlantic* featured the story in their May 2024 edition.[121]

The ferocity of the Tanzanian state's response to Maasai land use is to a large extent explained through their adoption of an economic rationality in thinking about national territory. People who stand in the way of state planning are constructed as 'economic saboteurs', while conservation projects often open the way for extractive ones, such as the Russian uranium mine inside the Selous

Game Reserve.[122] These partnerships between authoritarian states are more smoothly facilitated if relations to land can simply be overridden. But even apparently democratic and inclusive conservation projects in the region have had violent consequences. Roderick Neumann noted that it was already 'de rigueur' by 1998 to speak of 'community involvement' in conservation.[123] The nature of this 'involvement' is however severely limited by prevailing discourses and power relations.[124] Community 'benefits' in these models are frequently stated as family planning and education about the importance of conservation. Constructing Indigenous Africans as lacking moral and civilized relations to the natural world repeats well-worn racist canards. The existing 'landed moral economy' is furthermore distorted in ways that disrupt inter-group relations.[125] In Kenya's flagship community-based conservation project, the Northern Rangelands Trust (NRT), cattle rangers work together with conservation figures and wealthy landowners to manage a vast constellation of protected areas – 7 per cent of Kenya's entire area – in parts of which limited agricultural activity is allowed. While these spaces seem to unlock some progressive forces, they have been critiqued as 'green grab by bricolage'[126] due to the difficulty of negotiating mutually advantageous relationships between indigenous pastoralists and the conservation goals of white settlers, 'contributing to land shortages that have precipitated violence'.[127] The decision-making structure of the NRT brings outside authority to bear in local disputes, with often highly disruptive effects.[128] Political elites capitalize on changing land tenure arrangements in ways that exacerbate existing ethnic tensions.[129] The easy availability of guns in the area, and a trigger-happy Kenya Wildlife Services (KWS), contributes to an environment in which dozens of people have been killed, and government agencies are 'accused of abducting, disappearing, and torturing people in the name of conservation'.[130]

As the peace studies scholar Kennedy Mkutu Agade argues, the community conservancy model used in the NRT does bring some benefits to communities, but it also tends 'to formalize land claims, militarize communities and sometimes exacerbate conflict'.[131] In an area where conservation is militarized by both public and private actors, accountability and transparency become key problems.[132] The local dynamics of the NRT are furthermore turbocharged by the broader investment and conflict context of the region. The Lamu Port, South Sudan, Ethiopia Transport Corridor (LAPSSET) is one of the continent's most ambitious development programmes, and runs directly through the NRT area. LAPSSET's promise of jobs, money and development for local people raises the stakes for local communities. In a region already known for skirmishes between pastoralists, the intervention of the NRT as a well-armed

NGO seeking to keep 'peace' between ethnic groups is fraught with difficulties. Conrad Schetter and colleagues thus classify the NRT as a 'frontier NGO'.[133] The 'regime of exception' that emerges at the frontier demands violence and, even if the forces martialled by the NRT technically report to the police, they very often act independently. What is emerging in the north of Kenya, therefore, is the expansion of new violent frontiers of dispossession whose point of equilibrium is as yet unknown.

There is extensive empirical evidence that where Indigenous people have sovereignty over land, biodiversity is protected.[134] And yet, in a typical colonial reversal, mainstream conservation too often constructs protected areas 'as threatened by land hungry populations rather than as being historically complicit in the creation of conditions of land and resource scarcity'.[135] Too much of the spirit of conservation is similar to the impetus that drove historical imperialism, that 'the modern and civilized world must continue to save Africa from the Africans'.[136] Aby Sène points out that sustaining this skewed and unjust picture of what it will take to protect biodiversity requires the operation of a non-stop, ubiquitous 'propaganda machine'.[137] The content of this propaganda is the colonial conservation imaginary.

## Summarizing the colonial conservation imaginary

Now that the various strands of critique of mainstream conservation have been laid out, a sketch of what they share – which I am calling the 'colonial conservation' imaginary – may be offered. In this section, I will outline the elements of this imaginary, while in the next, I will present two pieces of textual analysis. As with all of the other empirical work in this book, the summary (see Table 3) and the analysis that follows are of necessity incomplete and partial.

It may seem foolhardy to reduce the complexities and contradictions of contemporary conservation debates to a single imaginary. Whereas neoprotectionism and 'fortress conservation' enforce a strict separation between 'wilderness' and human settlements, the 'new' conservation with its embrace of neoliberal 'fictitious' capitalism is happy to blend the economic with the ecological, identifying 'ecological services' as fungible, tradable commodities. However, as I showed in the discussions above, what these modes of conservation share is a tendency to extend the influence of rich-world conservationists into territories in ways that disrupt the web of human and more-than-human relationships already in place. Many conservationists are critical of rich-world

**Table 3** Summary of the colonial conservation imaginary.

| THE COLONIAL CONSERVATION IMAGINARY | |
|---|---|
| Ecological protection requires enclosure, commodification and capitalism. The unassailably positive value of the 'wild nature' thus protected motivates a primary relation to protecting it and not people who may harm it – and to an ongoing extension of the protected horizon. | |
| **SUBJECT ROLES**<br>• The ranger (supported/supportive)<br>• The poacher (gendered and sexed)<br>• Good citizens (informers) vs bad citizens (economic saboteurs)<br>• Backward savage populations who are land-hungry and do not love nature<br>• The noble (white) conservationist (dressed in khaki)<br>• The eco-mapping ecologist<br>• The green financier | **INSTITUTIONS**<br>• Development (economic; national and international; financial)<br>• Scientific (ecoregions, mapping and ecological services)<br>• Military (drones, rifles, guns, fences, shoot-on-sight, etc.)<br>• Moralistic (mavericks; luxury lodges; ritual communion with nature) |
| **CHRONOTOPES**<br>• Timeless unspoiled Eden – the Anglo-American nature aesthetic<br>• Lands currently unprotected as conservation *Lebensraum*<br>• Wilderness as battlefield, zone of exception<br>• Sparse indigenous communities turn into a blight on the landscape | **AFFECT**<br>• Love, awe and reverent communion with nature<br>• Fearless assertion of nature's needs<br>• Earned enjoyment of luxury<br>• Unmitigated rage against poachers<br>• Ardent desire for 'people, planet, profit' |

consumerist lifestyles and recognize their devastating impact on the planet. And yet some neoprotectionists imagine that they can stop this impact at the fence of the protected area, taming capitalism through using the proceeds of photographic and trophy hunting tourism to pay for rangers, rifles and drones. Colonial conservation is ultimately articulated around the sacrosanct value of a 'nature' that must be brought into modernity through enclosure. Its protection is thus imagined as relying on capitalist economic activity. As we have seen in other imaginaries, capitalism is naturalized as the only viable model for human–nature relations. Disagreements between conservationists tend to centre on the types of value extraction that can happen – whether low-impact photographic and leisure tourism, trophy hunting, or higher-impact mixed-use models that include agricultural, extractive or ecological services – rather than the commodification of nature to begin with. Across this spectrum, ecological systems thinking converges with economic rationality to offer the promise of a world in balance, where (green) capitalism can go on in metropolitan centres while wild Nature is preserved in the periphery.

The colonial dynamic of conservation can be thought of as a shifting frontier that sometimes realigns human–nature relations with the conservation mode of production, but more typically severs them entirely, evicting people from their ancestral lands. This frontier is realized in discourse when displaced people are constructed as land-hungry threats to wildlife rather than as people made hungry by enclosure and dispossession. The progress of this frontier is constructed as the only viable way to save nature, and its failure as guaranteeing inevitable decline. Lands not governed by the conservation mode of production are lands at imminent risk of despoliation by irresponsible local populations that have not learned to attach value to nature in the proper way. To prevent this from happening, conservationists are called on to think like business administrators or, if need be, military commanders. They must set targets, pore over maps, establish boundaries and, if necessary, declare war. Because (as per Prince William) the very 'future of life on earth' is at stake, a moral economy is established that valorizes the actions of the conservationist and their backers no matter what, while pushing 'saboteurs' into spaces of exception where they can be cleared away or subjected to extrajudicial killing. Because who can be treated in this way so closely cleaves to historical racial and colonial differences, Ruth Wilson Gilmore's definition of racism as the 'production and exploitation of group-differentiated vulnerability to premature death'[138] is again salient. Colonial conservation is thus of a piece with the global system of racial capitalism.

## Colonial conservation subject roles

The controlling image of the poacher as a greedy, backward savage is in dynamic tension with the figure of the ranger, who is constructed as a reliable member of the workforce as long as he is well-support*ed*. If he is not well-supported, he becomes less support*ive*, and in an atavistic fashion may revert to his racial-sexual type as a cruel and relentless poacher. The exception that proves the rule in this case is the Black woman ranger, who is constructed as not having the same capacity to murder and maim iconic animals, and of being a natural protector of Mother Nature. While white male trophy hunters are a necessary evil, a throwback to older modes of conservation that must be tolerated in order to bring much-needed foreign currency into protected areas, the poacher is absolutely and always intolerable. The poacher may be shot on sight, whether armed or not.[139] The ranger on the other hand, is a subject position that increasingly converges with that of the soldier, and who is given license to exercise lethal violence. Both ranger and poacher are drawn from populations defamed as hungry for the flesh of treasured (and sometimes nearly human) animals. As has been extensively documented, the supposed lack of concern for the environment evinced by Black and working-class people is merely an ideological effect of the nature of conservation itself, and its racial and class dynamics.[140] A preoccupation with why the Other does not love nature appropriately is ubiquitous within colonial conservation. A focus is thus often placed on educating the peasantry to form the correct relations with the natural world, to become good citizens who work with the rangers and not the bad ones who support the poachers. A further distinction is made here between communities that stay (as per E.O. Wilson) 'sparse' on the landscape, harmonizing with it in aesthetically acceptable and preferably premodern ways. Should these populations grow too large or too modern, they invite removal.

In the colonial conservation imaginary, 'iconic' animals are often granted moral standing. This status reflects on the animal-lovers who form close bonds with them, whose moral excellence is thus established. Loving 'nature' is the essence of the khaki-clad conservationist who is rough and ready to protect it. The tones of their outfits help them blend into the bush, establishing that they belong there. Whether historical settlers or newly arrived tourists, the subject position of the nature-loving conservationist is available as a mode of being that implies that one is at home in the wild, while conferring an enlightened moral tenor on otherwise problematic activities such as luxury travel, or extrajudicial violence. The sanctifying mantle of conservation valorizes the rawness of 'war' and the willingness of the 'maverick' to stop at nothing to save animals from slaughter. Other subjects also build on this license. Financiers design and trade market-based instruments that maintain ecological integrity, billionaires and celebrities lend small fractions of their wealth and time to environmental causes in return for fawning press, and conservation biologists and other scientists communicate objective truths about what must be done in order to save the planet. Together, in a fashion not dissimilar to historical imperial projects, nature-lovers who are typically white, Western or wealthy are lauded as the pioneers and visionaries who have taken up the Conservationist's Burden of saving the world's wild places from the untrustworthy and damaging people currently living there.

## Colonial conservation institutions

The dominant genres of colonial conservation are developmental, scientific, moralistic and militaristic. The developmental genre corroborates the establishment of protected areas as building blocks of the progress of nations, where an economic rationality predominates. The foreign currency brought to wild areas by international tourists is especially important in this regard, as are the investments of foreigners in hunting lodges, and the unlocking of the 'existence value' of nature and its related market-based instruments. Conservation is seen as opening up development corridors in previously isolated backwaters, attracting investment and modernization that promise to benefit surrounding communities and reduce poverty, while more typically enriching elites, but in any event growing the national GDP. The developmental register institutionalizes nature's commodity value, linking it to a supranational network of regional, continental and global sustainable development objectives, to the extent that a member of the British royal family can talk of 'our conservation commitments' in Africa without fear of contradiction. For the commodity value of nature to be realized, it must be enclosed to some extent – whether through the formalization of stricter rules to govern land use in community conservancies, or the building of high fences and access control measures in 'fortress' conservation models – pushing the frontiers of a racialized property regime born in the West to the corners of the world. Developmental discourse is complemented by the institutional heft of science, which informs where boundaries are drawn and which land use forms are acceptable. Scientists aid in mapping 'ecoregions' and their relative health, identifying which ones should be targeted for conservation. They may interpret data from areas such as conservation ecology and climate science for commodification, helping to assign precise values to carbon sinks and other 'ecological services' and thus serving as bridges between scientific and financial institutions. The moral register of colonial conservation meanwhile institutionalizes a particular form of hierarchy where conservation 'visionaries' (including billionaire philanthrocapitalists and other celebrities) are granted the license to advance frontiers, to occupy the subjectivity of 'mavericks' who will never back down. In the moral universe of African conservation specifically, impossibly luxurious lodges, pre-dawn game drives with champagne breakfasts and late-night conversations in the boma around a well-fed campfire are part of a ritual that enables modern people to contemplate their modernity in communion with nature. This luxury, and its reliance on cheap wage labour, is no more embarrassing to conservationists than the exquisite interior of a cathedral is to clergy. Together, the institutionalization of conservation as developmental, scientific and morally unassailable underpins its militarization. On the green frontier, private landowners, conservation NGOs and governmental police and military personnel blur the lines between public and private, legal and illegal, 'just' war and war crime. Global conservation audiences bay for extreme violence to be meted out to people suspected of poaching – torture, dismemberment, summary execution – and the moral economy of the frontier thus becomes a zone of exception. 'Shoot-on-sight' and 'shoot-to-kill' are given the tacit blessing of the moral, developmental and scientific bastions of colonial conservation.

## Colonial conservation chronotopes

'Iconic species and landscapes' are central to the colonial conservation imaginary. These 'consumptive' landscapes are imaginatively contrasted to 'productive' landscapes – the decadent metropole or the corn and soybean fields of Illinois without which 'civilization' would collapse. The archetypal colonial chronotope is the unspoiled Eden outside of time. As I have shown in South Africa building on the work of David McDermott Hughes in Zimbabwe, white settlers construct their belonging on the landscape in part through connection with its prehistory, imagining dinosaurs grazing around recently constructed dams, or pointing to the traces of inland oceans that they are uniquely positioned to see.[141] These richly imagined primordial landscapes promise an encounter with sublime, untouched, awe-inspiring Nature. The dream of extending these landscapes, of animals being able to move freely through unpolluted and unpeopled space, invokes a *terra nullius* – no man's land, the 'coloniser's dream'[142] – onto which the extension of 'conservation *Lebensraum*'[143] may be projected, along with a place for (white) conservationists on the landscape. In this chronotope an eternal and essential nature is consumed by its ecologically enlightened protagonist, in a wilderness that is above national borders and outside of human history. In this nature there is no society, no politics, no history. There is thus also no complicity in colonialism or historic violence, nor need for any shame about one's current or future enjoyment of luxury. Where local people are included in this scene they are kept in place through the notion of 'community' – conservation may be 'community-focused', 'community-based' or occur in 'community conservancies'. 'Community' is acceptable if it is sparse, premodern and close to nature. The ecological protagonist relates to this community through a 'nostalgic fantasy of the modern'[144] – where, for instance, uniformed servants recruited from the 'community' happily serve gin and tonic in the lodge on one's return from safari. In neoliberal conservation imaginaries there is always a future landscape in which the non-use of nature is the basis for economic growth and continued accumulation of wealth. This vast and soothing chronotope however recedes from view when it is imagined as under threat. When poachers are on the horizon, wild nature becomes a battlefield and militarized conservation a frontier that must be extended at all costs to save the few remaining iconic animals whose butchered and bloody carcasses on the landscape are evidence of the savagery of the poacher. Wilderness as battlefield is a place outside of civilization where the normal rules no longer apply and thus space of exception where 'It's raw. It's war'.

## Colonial conservation affect

The emotional core of conservation lies in love of the wild and the development of warmly affectionate relationships with landscapes, plants and animals; a reverent state of communion with nature which involves spiritual awe at its grandeur; and an affinity for rational comprehension of the delicate balance of ecosystems and the importance of all of their elements. For neoliberal conservationists, these emotions are combined with the ardent desire that capitalism is compatible with this love, and that profits will not be harmed by care for the planet. These emotional dispositions participate in coloniality when they cohere around a hierarchy of humanity that is non-relational. Where one way of being affected by nature is understood as the only correct way, and other ways not merely as different but as antagonistic, the subject experiencing the correct emotions is placed at the pinnacle of humanity. Their moral and intellectual superiority leaves them in no need of relating to the world-views and practices of ecological Others, which are irreducibly alien and inherently problematic. Non-relationality is an arrogation of power enabled by material and symbolic capital. This power manifests in colonial conservation through the work of protection. This über-human conservationist is above all a protector, and it is only right that the hard and caring work of protection is rewarded with luxury, which luxury is also an enabling condition for the experiences of the sublime that sets them apart from other people to begin with. This affirming affective circle is given considerable energy through another central emotion: rage. Conservationists have to be 'fearless' in their work to protect wild nature, and their work can be 'raw' precisely because they are antagonized by others. Sharing images of animals butchered by poachers fills the reservoirs of rage needed to either perform or legitimate lethal violence. In dwelling on the terrible and painful harm done to beloved creatures, anti-poaching activists experience the *jouissance* of identification as the antagonists of the poachers, and the agents of their destruction.

## Textual analysis

In this section, I present two cases where elements of the colonial conservation imaginary are made visible.

### Case 1: A 'Looming Crisis' in Laikipia

In July 2017, *The New York Times* published a lengthy feature article on land pressure on the African continent, using the Laikipia region of Kenya as a case study. The headline reads 'Loss of Fertile Lands Fuels "Looming Crisis" across Africa' while the subhead explains that 'Climate change, soil degradation and rising wealth are shrinking the amount of usable land in Africa. But the number of people who need it is rising fast'.[145] It is thus the looming threat of overpopulation that provides this narrative with some of its drama. Laikipia's importance as a conservation area is however asserted as a crucial aspect of the 'looming crisis' – as can be seen in the excerpt below.

> Kenya has a land problem. Africa itself has a land problem. The continent seems so vast and the land so open. The awesome sense of space is an inextricable part of the beauty here – the unadulterated vistas, the endless land. But in a way, that is an illusion. Population swells, climate change, soil degradation, erosion, poaching, global food prices and even the benefits of affluence are exerting incredible pressure on African land. They are fueling conflicts across the continent […] including here in Laikipia, a wildlife haven and one of Kenya's most beautiful areas. […]

While the 3,500-word story focuses on Laikipia as a case study, this small region is offered as a 'microcosm' of the continent, or in the words of a Michigan State University economist, 'the tip of the iceberg'. The reader is thus invited to extrapolate from the very specific dynamics of Laikipia to the continent as a whole. This scale is outlined in punchily statistical terms: 'by the end of this century, there could be as many as four billion people on the continent, about 10 times the population 40 years ago'. The equivalence drawn in the first sentence of the excerpt – that Kenya as well as 'Africa itself' has a land problem – is thus consistently reproduced throughout the piece, revealing a slippage of scale between the local and the continental that persists throughout the feature.'[146]

In grammatical terms, 'African land' is the object (or *patient*) of 'incredible pressure'. 'Pressure' is a conceptual metaphor, which borrows from the domain of mechanics, invoking the 'carrying capacity' of land (as discussed in Chapter 4). In terms of the seven factors listed, when population 'swells' the land must carry more people, while 'affluence' and 'global food prices' require more land to be enclosed in commercial farms, which send food to the cities or overseas. The other factors, 'climate change', 'erosion' and 'soil degradation' reduce the land's fertility and thus its ability to 'carry' the subsistence farmers who rely on it. 'Poaching' thus seems to be the odd term out here, as it is a factor neither in increasing the population nor in making the soil less fertile. Arguably, it is conservation itself that is putting 'pressure' on the land, by making it unavailable for agriculture, and 'poaching' is merely an effect of this pressure. What could be the ideological underpinning for its inclusion in this list?

Except for 'global food prices' what the other items on this list have in common is that they are all the 'fault' of postcolonial (Black) Kenyans. Conservationists are not to blame, as the lands they protect do not increase 'pressure' on land but instead suffer from it. As is mentioned later in the feature, tensions over land in Northern Kenya are fuelled by the fact that 'most of the biggest ranches and wildlife conservancies are owned by a handful of families of European descent'.[147] White conservationists as well as cattle ranchers (white and Black) are portrayed in vivid terms as facing an onslaught of massive proportions: 'Swarms of herders' and 'Waves of young pastoralists' are described as 'shooting people with high-powered assault rifles' and 'AK47s'. The shooting of Kuki Gallmann is related, as is the burning of the eco-lodge of another white landowner. The response of Kenyan authorities is meanwhile constructed in disparaging terms. One white farmer, a 'former British military officer' is shot and killed by trespassers he confronted because the 'security forces were doing little to help him'. A truck supplying land invaders with tea and sugar is 'later traced to a paramilitary police officer'. When the government sends more police officers into the region, landowners complain that 'they've never seen such timid men in uniform'. Landowners in turn respond with their own strategies. When Gallmann was shot, she had 'a contingent of armed wildlife rangers behind her'. A young white farmer patrols his ranch 'in a flak jacket and operates a new drone they bought for surveillance'. When police do arrive at the scene of an attack and start firing 'military-grade mortar rounds at the invaders' the white owner is moved to egg them on, yelling 'Shoot another, shoot another!'. The primary argument of the text is thus that privatized violence is legitimate: a topos of imminent threat to the bodies of (white) conservationists and the safety of precious animals is constructed as unmanaged by the state, and thus the job of 'frontier' actors.

Laikipia – and thanks to the generalizations of this piece, the rest of the continent along with it – is imagined as in the grip of chaos, where the state's monopoly on violence has lapsed and private citizens must take it upon themselves to guard the wild frontier. While according to 'scientists'' solutions exist to reduce pressure on land – agricultural beneficiation of soils and 'smaller families' – the problem with Africa is constructed by a white wildlife activist cited in the piece as 'too many people, too many cattle, and too little planning'. This failure of rationality is thus also at the heart of the failure of Africans to protect the 'Dream of Wild Africa' for future generations of white tourists. What is being retrieved here for the readers of *The New York Times* is thus the dream of the 'awesome sense of space [...] the unadulterated vistas, the endless land'. The protection of a specific nature aesthetic thus lies at the heart of this text, which seeks nostalgically to reconnect with a past in which blameless conservationists and sparsely populated vistas worked together to create an idyll, wishing away the bad citizens of postcolonial Kenya.

## Case 2: The 'Badass Women' taking on poachers

In October 2017, *National Geographic* reported on the 'Black Mambas', an all-female anti-poaching unit launched in the Balule Private Game Reserve in South Africa's Limpopo province. The Black Mambas patrol Balule, which is home to lions, elephants and rhinoceros, entirely unarmed. This is articulated as a conscious decision to avoid the trends of green violence and green militarization outlined above. 'Creating orphans and widows is not the answer to this problem,' argues Craig Spencer, who founded the initiative and is the head warden at Balule.[148] When asked what motivates young women to become Black Mambas, Spencer states that most of them 'started because they needed a job, but now, it has become a source of dignity to be a Black Mamba [...] They get given a break for the first time in their lives'. The fact that they are unarmed seems not to have resulted in any serious injury since their launch in 2013.

Though there are doubtless many things to be celebrated about this initiative, which operates as an NGO and has won numerous international awards, it is suspended in a web of material, racial and gendered power relations that are rarely taken into account in reportage. As the political ecologists Stasja Koot and Frederiek Veenenbos argue, unarmed women in anti-poaching squads seem to function primarily as a 'staged spectacle' that can work to mask militarization in other parts of the conservation enterprise, while serving as important sources of donor funding.[149] They also argue that the project's intervention in racial and gender stereotypes is at best 'unconvincing'. Analysis of how their leader Craig Spencer, a white South African man, talks about the Black Mambas reveals how this plays out discursively:

> I do see myself as a bit of a father figure to them, which is why I'm hard. I call it tough love. I love every single one of them desperately and want the very best for them [...] We're trying to save the rhinos by creating a totally different set of values within the community. These women are very proud to wear the uniform. They are role models. I want to see them grow, build houses, send their children to school [...] I often wonder if I need the Mambas or they need me? I've lost my heart and soul to them. They are my reason for sticking around. It used to be the elephants and the rhinos, but now, it's these women.

The paternalism evident in this text is especially grating given the racial and gendered dynamics at play: Spencer is a white South African man and the Mambas are all Black women. An African need of white tutelage and guidance has been an element of the cultural landscape of the region since the seventeenth century, and surfaces frequently in a range of texts. What is however quite striking in this particular text is how blatantly Spencer articulates that surrounding communities lack 'the values' to protect the rhinoceros. This perspective is shared by the *National Geographic* journalist, who adds notes of girl-power-style success to her reportage. Spencer is here implicitly appealing to the gendered role of women as the carriers of cultural values into families that are constructed as intrinsically backward, oriented against nature and in need of a 'totally different set of values'. The notion that poaching is caused by the ecocidal values of poor communities and not by the process of colonial dispossession and its sedimentation in the institutions of racial capitalism in the first place is a typical reversal, which blames conservation's victims for having the wrong attitude to it. The idea that women would not want to send their children to school if it weren't for Spencer's intervention, and that such intervention is going to result in a total cultural shift away from doing ecological harm to being 'good citizens' on the side of conservation, places communities in the vicinity of conservation in a developmental logic, where their gradual entry into modernity is being facilitated by white largesse and far-sightedness.

## Conclusion: Towards justice in conservation

Protecting biodiversity is an important global project, and critique of models of conservation that dispossess, exclude and harm should not be taken as an argument in celebration of the wildlife trade or of ecosystem degradation. What conservation should however never be is a distraction from the actual systemic factors underlying ecological decline, where the existence of spectacular protected areas in far-flung places assures metropolitan ecological subjects that something is being done about 'saving' the planet. Even people on the left fall into the trap of thinking that ecological health can come from the exclusion of human beings and our systems of social and economic production from vast swathes of the earth's surface. A recent book, *Half Earth Socialism*, for example, makes exactly the same mistakes as colonial conservation, fixating on entirely non-relational modes of engaging with problem-solving, and imagining that half of the world can be set aside for 'rewilding' following the models of scientists and economists who have the god's-eye view of the world.[150] Instead of colonial conservation, what is required is a model of conservation that centres justice and 'hinges not on liberal frameworks, but on its complete disentanglement with global capitalism and the racial order that sustains it', as Aby Sène writes.[151] Bram Büscher and Robert Fletcher have proposed a model they call 'Convivial Conservation' structured around radical equity, structural transformation and environmental justice.[152] Around the world, political ecologists and conservation activists are grappling with ways to make these values come alive in real-world situations where a number of competing interests must be balanced.[153] It is in refining and improving the way we think about our relation to 'protected' or 'wild' nature that (ecointersectional) critical discourse scholars have much to offer planetary ecosystems.

# 6

# Race, nature, nation: Econationalism

Nature is frequently invoked in the name of the nation, whether as resource, landscape, heritage, or as reflecting the essential character of the people. The discursive ties between ethnic groups and specific parcels of territory – between people and place – are forged through historical processes of settlement and conflict that play out at different scales, whether valleys, mountain ranges, deserts, geographical regions or empires. Questions around 'who owned the land, who had the right to settle and work on it, who kept it going, who won it back, and who now plans its future'[1] are answered in narratives of autochthony, divinity or historical settlement, where ecosystems play important roles. Contestations of belonging and legitimate occupation are worked out through the construction of relations of protection or care for the environment, which binds itself to people just as people bind themselves to it. In this chapter, we will consider how nature is used to sharpen the borders between nations both through the logic that excluding ethnic Others is ecologically responsible and through the wilding of borderlands as bulwarks against illegal crossings. We will examine the politics of belonging through the lens of the ethnoscape, which makes nature a historical player in the formation of the people while naturalizing the people's history on the landscape. In settler colonial contexts, we will consider how indigenes are treated as backward and harmful to the land while they pose an active threat to landgrabbers (in the case of Israel/Palestine) but become depoliticized and empty avatars of stewardship once they have been subdued (in the case of the white American invocation of the 'Ecological Indian'). Finally, we will address the spectrum of 'green' nationalisms from left to right in order to sketch the outlines of the 'econationalist' imaginary.

## Natural borders

A tumble of white sand dunes and windswept coastal grasses is pictured against a sky so pale it's nearly white. The sun radiates brightly from a faraway horizon, its rays refracted into perfectly symmetrical beams piercing the scrub below and dissolving into the heavens above. The viewer squints their eyes across a landscape that stretches out to the North Sea and beyond. This 'vastness' is confirmed in the accompanying copy:

> The eastern coast of Calais: Horizon XXL!
>
> From the site of the two seas to Hemmes de Marck via Fort Vert, the vastness is east of Calais. This endless coastline offers its spaces according to your desires. Between sand, green spaces and dune cordon, the landscapes are varied. Fauna and flora open up to contemplation while respecting a few logical rules. The wild and invigorating nature of the east of Calais also has its traces of history.[2]

This website, maintained by Grand Calais Tourism and Culture, invokes feelings of awe at nature's boundless majesty sure to be felt by the visitor should they choose to travel there. The 'few logical rules' to be followed while out in wild nature, as is explained lower down on the page, relate to the regulated harvest of samphire and the protection of birds and seals resting on sandbanks. Within these reasonable constraints lie limitless possibilities along the 'XXL horizon' and 'endless coastline'. This promise of freedom, following your fancy, opportunities for 'contemplation' and communion with a nature towards which one is also responsible and respectful lies at the heart of much communication that invokes 'wild and invigorating nature'. As Anders Hansen and David Machin show, the idea that nature is 'beautiful and tender, that we should seek union with it, that this is a kind of spiritual journey, and certainly a feel-good one'[3] often masks the character of not-so-green products. In the case of Calais, one such obfuscation becomes apparent in the website's map of the region, where icons of cars and numbers with bright green lines radiating out from Calais show that the city is a mere two hours' drive from London or Brussels, three hours from Paris, and three-and-a-half from Amsterdam. Communion with 'endless' nature is thus intimately connected with (auto)mobility: the freedom to cross borders, speed down highways, and march across dunes for the sheer pleasure of it. Travellers bring the money to the city that funds the tourism agency, which will always have more 'XXL' to sell.

Calais's current and historical geopolitical importance is hinted at in the 'traces of history' found in the dunes. Two sites are mentioned on the page: Walde Lighthouse and the Oldenburg Battery. The lighthouse is a rarity, the 'only metal lighthouse at sea on piles'. Though it has not shone since 2001 it is retained for its historic value, and for the fact that it 'marks the boundary between the English Channel and the North Sea'. Oldenburg Battery meanwhile is a 'remnant of the Atlantic Wall'. Initially built by the Nazis in 1940 to facilitate a planned invasion of Britain, it took on a defensive role from 1942 as the tide turned against Hitler. Both lighthouse and battery signify Calais's position at important crossroads, between seas and between nations, and most crucially as mainland Europe's gateway to the British Isles.

Nineteenth-century maritime innovations and the Second World War are not the only 'traces of history' in the dunes east of Calais the website might have mentioned. 'Fort Vert' is a protected area now managed by a coalition of local authorities called 'Eden 62', which is tasked with environmental protection, improvement and education.[4] Walking through these dunes, one might find 'evidence that an injustice had taken place here. Discarded tear gas canisters, a child's shoe, a broken spoon, a pile of asbestos, a spatula, torn fabric caught on a tree, a crushed water jug, shotgun cartridges, a dirty toothbrush'.[5] For Fort Vert is also the site of the notorious Calais 'jungle' that made global headlines during the European migration 'crisis' of 2015–16.[6] In September of 2015, desperate migrants from Sudan, Syria, Afghanistan, Iraq and elsewhere, on their way to the UK and occupying smaller camps which had for years been called 'jungles', were herded by authorities into a 'new' jungle which became their only 'tolerated zone'.[7] Within a year, the population of '*la jungle*' had reached nine thousand. Then, in October 2016, the site was completely demolished by bulldozers and waste workers, and its denizens ejected by riot police to be dispersed throughout France.[8] Fort Vert – which means 'green fort' – was established on the site soon afterwards. Eden 62 does not allow camping for more than 24 hours, an approach to the management of migration known as 'zero-anchor point'.[9] This area now serves as a bulwark against future encroachment by those trying to reach England. As the political geographer Thom Davies and colleagues have argued, at Fort Vert nature is 'not an innocent bystander, but a tool used to exclude, harm, and erase racialised migrants'.[10] The French state's failure to intervene in ameliorating the unsanitary conditions and squalor in '*la jungle*' before it was dismantled turned it into an ecosystem where disease could thrive, conditions that are not tolerated in the post-*jungle* Fort Vert, where ecological health is closely managed.[11] The thriving of migratory birds and sand dune vegetation is now held sacrosanct, and nature works to exculpate both France and Britain – which helped pay for Fort Vert – from harms historic and ongoing.

A number of related terms have been proposed by political scientists, geographers and political ecologists for the use of nature to strengthen a border, whether materially, symbolically or morally. 'Ecobordering' names the discursive practice of constructing national borders as 'the environment's greatest ally'.[12] In Joe Turner and Dan Bailey's analysis of European far-right party materials, ecobordering relies on figuring migration as 'environmental plunder' and/or migrants as 'environmental vandals'.[13] These texts, which draw heavily on the overpopulation imaginary, construct people on the move as racialized Others from the 'Global South' who threaten to overwhelm the environmental integrity of the host nation with their profligate breeding and environmentally unsustainable practices. At Calais, ecobordering is made even more literal, as at Fort Vert migrants are constructed as threats to the ecology of the border itself. These variations on ecobordering are in turn legitimated through what John Hultgren (in the US context) identifies as discourses of 'ecosovereignty'.[14] Ecosovereignty constructs national prosperity and thus consumption as threatened by the unmanageable consumption of the Other, obscuring the facvt that it is rich-world greed that is in fact to blame for poverty, conflict, desperation and displacement in the rest of the world.

Ecobordering and ecosovereignty are advanced by parties across the political spectrum, though they tend to be most clearly articulated by the far right. Calais was no exception. In the months leading up to the corralling of migrants into *la jungle*, which segregated them from the mostly white town, they had been constructed in sensationalist terms by politicians and the media as a worsening blight on the landscape. As reported by *La Voix du Nord* in 2014, Marine le Pen, leader of the far-right *Front National* (now *Rassemblement National*) had complained:

> There is no longer any law that is valid in Calais, no more rule of law, there is only the jungle, violence and the law of the strongest. In short, the opposite of the Republic. […] Immigration is a true poison, the most visible and most shocking consequence of which is Calais. Illegal immigrants are outside of the law, criminals who have no business on our national territory.[15]

This use of 'jungle' as a racialized metaphor in European migration discourse thus anticipates the comparatively liberal Josep Borrell's deployment of exactly the same idea (analysed in Chapter 4) eight years later. The movement of the word 'jungle' from Calais to Brussels is an example of how discursive shifts mainstream far-right ideas: first, Le Pen enacts the lawless jungle in a novel context, then the media perpetuates the idea, until eventually it is so normalized that the EU's chief diplomat sees nothing problematic or racist about it.[16] Two things are however particularly striking in Le Pen's statement. First, she invokes the 'Republic' and 'our national territory' – as opposed to Borrell's 'Europe'. Later in the same report, Le Pen calls for exiting the Schengen agreement, which she claims has resulted in an 'unmanageable' influx of migrants. The erosion of French national integrity, stability and sovereignty is thus blamed on the post-national EU. Secondly, her construction of immigration as a 'true poison' was not normalized by Borrell. If the Republic is being poisoned, France is metaphorized as a living organism susceptible to toxins introduced from the outside. The social body of the nation has thus been violated, its borders breached, by a social substance – an irreducibly foreign Other – that will kill it if left unchecked. The damage done by this poison let in by the EU is 'most visible and most shocking' in Calais. The evidence of national health can thus be read off the landscape; one just needs to *look at Calais* to see what has gone wrong.

Years later while campaigning ahead of the 2019 EU elections, Le Pen articulated her stance on environmentalism:

> Environmentalism [is] the natural child of patriotism, because it's the natural child of rootedness [...] if you're a nomad, you're not an environmentalist. Those who are nomadic [...] do not care about the environment; they have no homeland.[17]

In this echo of the controlling image of the 'rootless Jew'[18] only the ethnic indigene truly belongs in the nation, and people with no fixed abode are a constant threat to the integrity and strength of a socio-ecological body imagined as homogeneous and harmonious. Outsiders – though not those who are in France for tourism, only those who are 'nomadic' and thus associated with poverty and the racialized Global South – are unnatural presences on the land who create disharmony. Ecologies are here imagined as continuous with the patriotic spirit of the people, which is intrinsically and biologically connected to a specific territory. As we shall discuss below, the 'organicist' vision of national identity that is resurgent in European ethnonationalism is built on similar structures of feeling to the *Blut und Boden* ideology of National Socialism. It plays an important role in undermining supranational bodies such as the EU and redrawing the racial lines between Europeans who 'protect' and dark-skinned people who 'poison'.

In the vastness east of Calais, on the windswept white sand dunes, the only footsteps that are welcome are those of the hiker who is given to 'contemplation' and respect for nature. Some 'natural' mobilities are thus prized – the steady step of the nature-lover, the migration of the birds – while others are made illegal and pushed into zones of exception beyond the boundaries of human sympathy. The intimate entanglements of landscapes with national identities dictate that what is left on the landscape reflects the intrinsic relationship between a territory and people. Or as another far-right French party put it, '*Ni minarets, ni charia! France aux Français!*' – 'Neither minarets, nor sharia! France for the French!'.[19] The traces of history left on the landscape – a decommissioned lighthouse, a Nazi battery – are those that may are indexically tied to a French national identity, which the minarets of a mosque will never be. The traces of *la jungle* must thus be erased even as the marks of defunct technology and catastrophic conflict are preserved.

In countries like France, where diverse indigenous ethnicities and languages were gradually and violently formed into a national identity over a period of centuries, claims to patriotic defence of homeland ecosystems may be fairly straightforward. But how do these work in settler colonial contexts? In the next three sections we will consider examples from South Africa, the United States and Israel/Palestine to unpack how this logic works to sever relations between Native peoples and their territory, and in many cases constitutes an attempt to 'cleanse' them from the map entirely while replacing them with settler populations.

## Constructing settler autochthony on the 'ethnoscape'

In August 2015, a member of the public Facebook group 'chase SHELL OIL out of the Karoo!' shared a photograph of an arid, starkly beautiful landscape, adding the text 'This is why!!!'. The image was being proffered to other group members as a reason to 'chase' the multinational company out of the central semi-desert of South Africa, where plans had been gazetted a few years earlier to prospect for natural gas. The group, which was organized by the Treasure Karoo Action Group (TKAG) to oppose 'fracking' – hydraulic fracturing – in the Karoo basin, has members who live in the region, but many more who live in the cities and visit the area either as a holiday destination, or on their way elsewhere. A sample of the comments made about the image speaks to their various reasons for 'treasuring' the Karoo:

1. The Karoo has its own stark beauty and the ever-changing light makes the scenery unforgettable and different every time I drive through

2. It never leaves you .....
3. Definately something intangible – but a feeling so strong.
4. A wide open space of unspoilt beauty ! With the ' quiet' of memories and history !![20]

The Karoo makes up more than a third of South Africa's surface area and is criss-crossed by roads connecting the industrial and agricultural centres in the country's northeast with coastal cities such as Cape Town and Gqeberha in the southwest. The first permanent European settlement at Cape Town established by the Dutch East India Company (VOC) in 1652 gradually expanded its authority and presence eastwards into the interior of the country, establishing the first administrative and legal centre in the Karoo at Graaff-Reinet in 1786.[21] Every wave of Dutch expansion by white *trekboere* (pioneer cattle farmers) came at the expense of the Indigenous Khoe and San people of the region. In 1777, an 'extirpation order' by the VOC authorized the indiscriminate killings of all hunter-gatherer communities, inaugurating what Mohamed Adhikari calls South Africa's 'genocidal moment'.[22] Much of this carnage took place in the arid interior. After the British annexation of the Cape in 1806, a series of intensifying land wars were fought with the amaXhosa around the area of Graaff-Reinet, and these pastoralists were eventually evicted from their grazing land and their herds decimated.[23] English-speaking white settlers arriving at present-day Gqeberha in the eastern Cape Colony from 1820 meanwhile moved inland, introducing merino sheep to the southern Karoo. By 1840, business was booming. In order to make agriculture in the dry region viable, Xhosa and Khoe people were coerced into peonage, and a racial regime of property and agricultural labour emerged.[24] In the less than two hundred years since the VOC first gained a foothold in South Africa, using murder, assimilation and disease, white settlers were thus well on their way to erasing hunter gatherers and turning formerly sovereign pastoralists into sources of cheap labour. The foundations of apartheid and its afterlife in the contemporary racial state were thus well laid.[25]

It is unsurprising that none of this is mentioned by the anti-frackers cited above, who were all white South Africans. Commenter (4)'s assertion that the history of the region is 'quiet' is especially telling of a particular way of thinking about the Karoo that constructs it as 'soul country'[26] for white South Africans. The landscape is 'unforgettable' and inspires a feeling of awe that 'never leaves you' … 'a feeling so strong'. A deep spiritual connection is thus asserted between these online anti-fracking activists and the landscape they aim to protect. In the other comments in this thread, a discussion is initiated by a Black commenter

(5) who is responded to in comments from white group members, one of which (6) is reproduced below:

5. but there's little or no work in the karoo, why are u guyz denying shell the chance of creating jobs in that area? i think it will be for the interest of the ppl in that area. [...] thats selfishness u know. think of the poor ppl of the karoo.
6. The poor poor people of the karoo will have no water if fracking are allowed ... working on a fracking site require competant skilled people ... so no skills. and no water ... where would that leave the poor people ...

In pointing out the desperation of local impoverished people, who would welcome development that helped them feed themselves and clothe their children, commenter (5) draws attention to the unevenness of development in the Karoo, which has continued to benefit white landowners, operators of tourism and private conservation businesses, while leaving (mostly) Black people leading often precarious lives. The response of commenter (6) in turn exemplifies the impasse of post-apartheid South Africa. The racial state cannot be undone under neoliberal capitalist hegemony: one can say one wants to eradicate poverty and racism, but without deep structural change driven by a society's most powerful players and social engineering on a vast scale, poor Black people will remain a permanently unskilled, unemployable underclass and the best that they can hope for is access to clean drinking water.

Asserting a spiritual connection to place in nearly the same breath as upbraiding marginalized Indigenous people to be thankful that they are allowed to remain on the landscape at all is a signature strategy of colonial discourses of belonging. As the cultural geographer Jeremy Foster argues, land is taken not only through the exercise of military and juridical power, but through processes of the 'imaginative appropriation' of landscape.[27] Because lands are newly settled and their former occupants excluded, the cultural production of the landscape may be riven with ambivalence and anxiety. J.M. Coetzee's classic distinction between the pastoral and anti-pastoral traditions in white South African writing speaks to this unease. In the pastoral tradition, where labour is the measure of virtue, the fact that the 'real' work is done by Black people must be obscured. The fact of Black labour 'threatens to deprive the white man of the labours that he, as Africa's new heir, must not only perform but, more important, be seen to perform'.[28] The anti-pastoral tradition, meanwhile, represents the landscape as forbidding, alien and mysterious, a terrain from which monsters may be brought forth signifying 'the return of what is repressed in the poetry of the silent landscape'.[29]

The fiction of *terra nullius* – land owned by 'no one' that may be appropriated by the settler – produces anxiety, and anxiety necessitates repression. The settler feels unsettled, and is therefore preoccupied with managing the way that land is represented. From the earliest days of colonization to the post-apartheid period, the possibility that the land itself would reject its new owners in South Africa is anxiously repeated.[30] Quelling these anxieties through aesthetic and spatial arrangements underpinned evolving racial subject formation. In the gold rush town of Johannesburg in the early twentieth century, for example, European trees lining the streets of the artificial urban forest lent new arrivals a sense of comfort and at-homeness on the reef.[31] A tamed landscape was sufficiently familiar to banish anxieties of unbelonging. Towards the end of apartheid, different anxieties changed the relation to landscape. 'Ecological patriotism'[32] thus started to vie with Europeanization for dominance. Instead of using exotic plants, many white South Africans chose to plant indigenous gardens, though emerald-green lawns have remained markers of 'civilization'.[33] Private housing estates market themselves to well-heeled (often white) clientele through asserting their status as 'eco-villages', which are typically gated communities that offer a retreat from the perceived spatial chaos of a multiracial, politically fractious and threateningly hybridized society – ecobordering at a different scale. Commitment to knowing and working to protect local biodiversity provides a convenient pretext for the privatization of segregationist practices.[34] As Renugan Raidoo argues, experiencing natural 'Africa' in the confines of a gated community creates a liveable affective world in which the anxiety-inducing memory of Black labour and displacement, and ongoing environmental and economic injustice, can be temporarily excluded.[35] Creating and caring for indigenous landscapes lends the descendants of white settlers some claim to autochthony through their claims to knowledge, and the proximity afforded through landholding.

Whether on the suburban or regional scale, white belonging in South Africa coalesces into what Christi van der Westhuizen has termed 'enclave nationalism'.[36] Well-manicured indigenous gardens or the vastness of the Karoo with its picturesque rural towns are landscapes turned into screens for the projection of vestigial white nationalism. They serve as 'ethnoscapes' that naturalize settler histories through the territorialization of memory, at the same time as making nature an actor in the history of the people whose character it was instrumental in forming.[37] As Anthony D. Smith argues, the foundational myths of an ethnic group 'are vital "evidence" of territorial "title-deeds" in that they associate the people with a specific terrain'.[38] In an example analyzed by the literary scholar Johan Geertsema, the white supremacist poet Dan Roodt uses the 'Karoo as a metonym for South Africa and the toughness of white Afrikaners who, he thinks, are able to use land that would be considered worthless in other parts of the world'.[39] Roodt's claim to nativeness rests on the uniqueness of the relationship of the Afrikaner people to a specific territory, the roughness of which formed the character of the *volk* just as the *volk* imprinted the signs of their belonging onto the terrain.

The landscape of the Karoo, which is described as a 'wide open space of unspoilt beauty' by commenter (4) above, is often referred to by its protectors as 'pristine', 'God-given' or 'perfect'.[40] And yet it is covered with the marks of settler agriculture: barbed wire fences with signs that intruders will be shot, sprawling farmhouses, concrete farm dams, sheep dips, tarred highways and – above all else – windpumps. Windpumps are ubiquitous in the Karoo, not only on the landscape, but in ornamental wire miniatures sold on the side of the road or in gift shops, in the signage of tourist accommodation, or in the logos of local organizations, including the TKAG. The windpump is so iconic of the Karoo that it is easy to forget that it is a sign of dispossession and plunder, a sentry on the horizon marking the advance of colonial agriculture. Coming from white South Africans, therefore, celebrating the 'unspoilt' Karoo as having a 'quiet' history is about as straightforward a claim to autochthony as one can get: there was no genocide, there were no wars; we have always been here and the agricultural marks we have left on the land are part of pristine nature. Indigenous Khoe and San people are occasionally invoked as benevolent spirits on the land, stuck in an eternal past, rather than as contemporaries demanding justice. As we shall see in the case of another settler colony, the United States, to which we now turn, ecological nationalism and white ethnoscapes often go hand in hand with the discursive depoliticization of Indigeneity, making it available for settler appropriation.

## The 'Ecological Indian'

To the beat of warlike drums, we see a Native American man through the underbrush on the shore as he rows his canoe down a broad river. In the next shot we see him close-up, and recognize Iron Eyes Cody, star of Paramount's 1948 blockbuster *Paleface*, headlined by Bob Hope and Jane Russell. As Iron Eyes glides steadily down the river, horns signal that something is wrong: now we see that the river is filled with litter. As he rows on, he is increasingly surrounded by factories and smoking chimneys. The music reaches a brassy crescendo as he pulls his canoe up onto the shore and walks to a road, where the driver of a passing car throws a plastic packet of fast-food detritus at his feet. A sombre voice intones: 'Some people have a deep, abiding respect for the natural beauty that was once this country. And some people don't.' As the camera slowly zooms in, Iron Eyes turns to face the audience, and a single tear rolls down his cheek.[41] 'People start pollution, and people can stop it,' the voice concludes.

This advertisement, for the 'Keep America Beautiful' campaign, was so popular after its 1971 launch that television stations were constantly asking for new copies as they had worn through the old reels.[42] A generation of Americans grew up with the 'Crying Indian' as the avatar of environmental protection, an embodied indictment of pollution and national defilement. Picking up on the counter-cultural meanings that had been attributed to Native American culture by the flower power generation, the Crying Indian retraced the outlines of an older myth, of the 'Noble Savage' who is perfectly attuned to the natural world and untainted by modernity. While this would at first blush seem to be a good thing – a positive representation that resists racist depictions of the bloodthirstiness and brutishness of a savage Other – in the hands of white people, it reinscribes the trope of what has been termed the 'Ecological Indian'. This trope works as a controlling image that depoliticizes indigeneity while flattening the cultural diversity and complexity of Indigenous nations. The idea that Native Americans were the 'original ecologists and conservationists'[43] of North America – which the anthropologist Shepard Krech argues gained currency in the United States in the early 1970s, around the same time the Crying Indian was making his TV debut – ignores evidence of historical overhunting and environmental harm.[44] This does not mean that Indigenous nations do not have ecological epistemologies that balance their impact on the earth with renewal, maintaining good relations between the human and the more-than-human worlds. Indeed, some Native American scholars reject the scientific basis for Krech's claim that their ancestors were what the Sioux historian Vine Deloria calls 'Pleistocene hit-men' who hunted large fauna to extinction.[45] As the ecocritic Greg Garrard argues, however, what must be analysed is not necessarily the trope's accuracy but rather how it 'is propagated and contested under specific historical and geographical conditions'.[46] The Ecological Indian does symbolic harm through relegating Native peoples to a space outside of history, an eternal and unchanging past where nothing new is ever learned and all peoples think the same. This is most commonly the case when invoked by white people doing their own identity work.

The fact that 'Iron Eyes Cody' was not the name of an actor, but that of a character played in redface by the Sicilian-American actor Oscar de Corti, puts the Crying Indian in his appropriate socio-political context.[47] At a time of Native

American resurgence, with the rise of 'Red Power' and the American Indian Movement (AIM), as well as the occupation of Alcatraz, there was no shortage of Indigenous activists for Keep America Beautiful to call on. For a mainstream TV advertising campaign, however, actual Native American activism was the last thing on their mind. Iron Eyes Cody is thus little more than a two-dimensional picture of how white Americans think an Indian should look, talk and behave; as such, he is a projection of their own desires. He is the 'racial epidermal schema'[48] of a 'good Indian' who lacks interiority or agency. The salient settler desire is for belonging on the land that is rooted in ecological indigeneity.[49] Whereas genocide and dispossession may be erased through the naturalization of the ethnoscape as outlined above, here the people the settlers tried to erase return to sanctify their moral right to their possession of the territory, to 'Keep America' – whether beautiful or not. That this campaign was funded by the packaging industry in a ploy to shift responsibility for ecological protection to consumers rather than producers adds a further twist to the tale.[50] What the Ecological Indian was thus summoned to sanctify was the white right to possess America and the system of capitalism *tout court*.

The trope of the Ecological Indian creates problems for Indigenous nations around the world. They may face 'racist vitriol and even violence'[51] from environmentalists when they engage in traditional activities such as whale hunting or pastoralism. In the mainstream white popular imagination Indigenous people are *not supposed* to behave in any way construed 'unsustainable'. Head of agriculture and food security at the African Union, Simplice Nouala, warns that rich-world environmental advocacy against traditional pastoralism threatens to rob already vulnerable herders of their major source of food and energy, arguing that, against the backdrop of historical Northern responsibility for climate change, African livestock is a form of 'climate justice'.[52] In the well-documented case of Sápmi (the homelands of the Sámi people, which stretch from Norway through Sweden and Finland to Russia) the development of supposedly 'green' energy has for more than a century impacted on reindeer herding culture. What Sámi scholars refer to as 'green colonialism' involves the effective enclosure of common land, the killing of their reindeer, and the increased risk to life and health that herders must face in navigating hydroelectric and wind turbine terrains.[53] The Sámi are constructed by the majority populations to the south as standing in the way of the green energy 'transition', and criticized for using petrol-powered ice scooters in their husbandry work. Colonial definitions of 'environmentalism' and 'conservation' that rely on fantasies of how Indigenous people ought to act are thus wielded in universalizing ways that malign the livelihoods of actual Native nations.

The political scientist Claire Jean Kim argues that Indigenous ecological ontologies are 'no more "constructed" or provisional than Western ones, but no less so either'[54] and are thus not frozen in time, either as essentially 'bad' or essentially 'good'. As we have seen across imaginaries, environmental advocacy too often universalizes its remedies as necessary for all humans, and alliances between mainstream environmentalists and Indigenous nations have often been unstable.[55] When Indigenous ideas such as granting personhood to rivers or forests are taking up within majority-white nations as ecological remedies, their implementation can work to 'straitjacket' Indigenous concepts, leaching them of their radical potential.[56] As the Indigenous studies scholars Eve Tuck and K. Wayne Yang famously argue, real (as opposed to metaphorical) decolonization 'eliminates settler property rights and settler sovereignty. It requires the abolition of land as property and upholds the sovereignty of Native land and people.'[57] These radical demands on the settler state are thus tamed in a piecemeal approach to Indigenous lifeways that appropriates them as elements of settler desires for guilt-free futures of green capitalist consumption.

For Indigenous identities to be appropriated by settlers as romanticized avatars of responsible land stewardship that endorse and even sanctify the colonial state, they must first be discursively available for depoliticization. If they have not been sufficiently subdued or eradicated, their potential to destabilize settlement and resist expropriation is settlers' dominant concern. The historian Miles Powell has shown in the context of nineteenth-century US conservation discourse that it was only once the displacement and genocide of Native Americans was nearly complete that the 'noble Indian' could be recruited as an environmental avatar. The anxieties of white men about their 'virility' and the strength of their 'stock' attended the disappearance of a wild frontier on which they could prove their rugged manhood.[58] Where Native peoples represent an active threat to colonizers, however, they cannot be used to buttress settlers' rights to be there in the first place. This, as we shall see, plays out in Israeli conservation discourse, which constructs Palestinians as living ecological threats.

## Eco-orientalism and the ethnic cleansing of Palestine

*Foragers*, an award-winning 2022 film by Jumana Manna, follows the stories of Palestinian citizens of Israel[59] who forage on open land for the wild herb *za'atar* and the spiny delicacy *akkoub*, which is similar in flavour to an artichoke. The camera dwells on moments of gentle connection with the earth, which take up much of the 64-minute runtime. A forager speaks lovingly with his two dogs as we see their soft golden pelts rub up against the leaves and bright flowers of an overgrown footpath. Later he carefully lifts a prickly *akkoub* out of a dense patch of herbs. A woman in a black bathing suit steps carefully down into a stone pool, the water lapping languidly against her skin as she calmly chats with her husband who sits on a nearby rock. We see the same couple (Manna's parents) carefully harvesting herbs from a verdant hillside. Later, two women sit in a kitchen slowly and deliberately paring the *akkoub* with small knives, laughing quietly as they talk about people living and dead.

The scenes of foraging are lush and intimate, even fragrant. They are however interspersed with scenes that have a very different tone. In a series of reconstructions of court appearances, individual Palestinians prosecuted for the crime of foraging – which has been a restricted activity under environmental laws since the 1970s[60] – are filmed against stark backgrounds of stone and cement, their exchanges in Hebrew with the unseen legal personnel playing out in dialogue form (Figure 4). For example:

> Man: You're a thorn in my gut and you say you're trying to help?
> Woman: There are laws in this country. You're here in a court of law!
> Man: But this law is shit! Banning us from foraging food …
> Woman: I see there is no sense in talking to you
> Man: What protected species? This land isn't yours! Neither is the plant!

**Figure 4** Still from 'Foragers' (Jumana Manna, 2022).

Some of the scenes of the film, where foragers hide behind bushes from park officials who patrol the terrain, or are then caught and forced to engage in ridiculous evasions and justifications, are wryly comic. Even the courtroom scenes have a touch of the absurd to them. The petty bureaucracy of Israeli officialdom is subtly lampooned by Manna's directorial choices, while at the same time the logic of environmental 'protection' is revealed as a cynical ploy. Manna includes archival footage from the 1970s of Israelis discussing how to make money from farming *za'atar* instead of allowing it to be picked in the wild. Palestinians are pictured navigating overgrown hillsides beneath the concrete Israeli infrastructure of highways and separation walls. Long aerial shots from cinematic drones show that the land being foraged is adjacent to ruined villages. Throughout, closeness to the land and its verdant bounty is associated with those who pick herbs, while harsh light, drone's-eye view, and Kafkaesque chicanery attach to conservation agents and their legal enforcers. Palestinians caress the land, while Israel lands heavy blows.

Environmental logics are routinely used by the State of Israel 'as a tool to erase the memory of previous Palestinian culture, marginalize and curb the development of Palestinian communities, and even uproot them, while justifying these policies by portraying them as progressive environmental policies'.[61] An important aim of environmental policymaking is the remaking of the Palestinian landscape to fit the self-image of the State of Israel. The ecofeminist scholar Edna Gorney recounts how key Zionists projected the antisemitic trope of the 'rootless Jew' onto Palestinians, who were said to be a nomadic, desert people with no deep roots in the land, in contradistinction to Jews, who proved the depth of their roots in *haaretz* (the land) through the planting of trees, thereby 'greening' Israel.[62] The roots of the hundreds of millions of pine trees planted with support from the Jewish National Fund (JNF) would fix the soil and make it fertile, whereas Palestinians were like the sands of the desert: unproductive and unfixed, ready to be blown away. *Haaretz* was imagined as empty – a land without people for a people without land – and Jewish settlers constructed themselves as returning indigenes who would prove their autochthony through restoring the 'desert' into a Garden of Eden.[63] Pine trees are thus 'fighting partners and brothers in arms'[64] as well as living 'flags'[65] that establish settler control. Trees have been named as casualties of conflict, with pines burnt in arson attacks constructed as 'martyrs' for Zionism, and olive trees uprooted by settlers in 'a tree for a tree' retaliation.[66] To the extent that pine trees manage to cover over the marks of conquest – of ruined villages, uprooted olives, and formerly cultivated land – they form a landscape that, as Edward Said observed, promotes '"collective", if selective, forgetting'.[67]

Some early settlers made sense of the evidence of Palestinian agriculture in olive groves by attributing their planting to Alexander the Great.[68] Constructing people as unworthy inheritors of a land once occupied by far greater civilizations is a signature move of orientalism, which, as Said defines it, always locates a 'good' Orient in the ancient past, while the 'bad' Orient lingers on.[69] Much Israeli conservation thus focuses on restoring what are imagined as biblical landscapes, connecting the 'original' species of animals and plants on the land with Jewishness and conservation, while problem species and environmental decline are associated with Palestinians. This harking back to a civilizational and ecological Golden Age is what Roey Gafter and Tommaso Milani term 'eco-orientalism'.[70] Israeli eco-orientalism constructs Palestinians as 'backward, less than human, and therefore threatening [...] to nature itself'[71] making them 'enemies of the ecological state'.[72] This 'conservation-colonialism nexus'[73] has very material impacts on the lives of Palestinians both in Israel and in the occupied territories, as Irus Braverman has extensively documented. Conservation areas are tools for land expropriation and zones of militarization that are used to pressure residents of the occupied West Bank. Jewish settlers articulate the 'green case' for expropriating Palestinian land, often successfully.[74] Conservation meets a number of different objectives in addition to land grabs, as the Palestinian environmental justice scholar Ghada Sasa argues, including preventing the return of Palestinian refugees, Europeanizing the landscape in ways that wipe away evidence of Palestinian identity, and covering up its perpetration of apartheid with an ecologically conscious image.[75]

A not uncommon claim of fundamentalist Zionists is that Palestinians lack a language, spirituality and culture distinct to the region, and that they therefore fail to meet the definition of an Indigenous people.[76] This makes Israeli Jews the only 'real' indigenes, legitimizing interventions such as the 2018 Nation State Law, which declares Israel the homeland of Jewish people alone. The critical sociolinguist Muzna Awayed-Bishara has shown how Israel uses key state institutions to 'de-Palestinize' its Arabic-speaking citizens.[77] Indeed, many Israelis refuse to use the word 'Palestinian' at all, and will only use the term 'Arab'.[78] Jumana Manna's film is thus a stunning visual refutation of this attempt at deracination. For Palestinians, as the legal scholar Rabea Eghbariah argues, foraged *za'atar* and *akkoub* are what 'make a household Palestinian'.[79] The criminalization of this practice by the Israeli state in the name of conservation is a strategy to ethnically 'cleanse' the land of Palestinianness, and to assert its essential Jewishness. In the face of these attempted erasures, as Awayed-Bishara explains, Palestinians mobilize the ideology of *sumud* – steadfastness – which is ultimately the tone of Manna's film. Ghada Sasa also identifies *sumud* as a core principle of Palestinian environmentalism, alongside *a'wna* (collaboration) and *a'wda* (return). As is made clear in Manna's film, *a'wda* plays a profoundly important role for foragers as they range across the ruins of ancestral villages, and their *sumud* in returning again and again, resisting the State of Israel's right to decide (in Manna's words) 'what is made extinct and what gets to live on'.[80] The idea that the nation-state must extinguish within itself lifeways constructed as hostile is the kernel of a kind of environmental thought that constructs the Other as a kind of pollution, a topic to which we turn in the next section.

## 'Black trash': From econationalism to far-right ecologism

In August 2018, the right-wing political commentator Tucker Carlson offered his personal definition of 'environmentalism'. Greens should stop banning straws and other things that ordinary folk enjoy, he opined, and instead focus their energy on the most important thing: securing the border. 'I actually hate litter, which is one of the reasons I'm so against illegal immigration: It produces a huge amount of litter – a *huge* amount of litter.'[81] As we saw at Calais, the association between racialized migrants and dirt is central to ethnonationalist imaginaries. The philosopher Charles W. Mills has even argued that in white-majority societies, while 'white trash' is understood as a slur, 'black trash' is merely redundant.[82] The implicit racial message of Carlson's 'environmental' commitment is thus arguably contained in the emphasis he places on how '*huge*' the amount of litter is. It's so *huge* it seems to include not only the objects strewn on the ground, but the people who had left them there; human rubbish fouling up the landscape, undeserving of sympathy or support.

Racial regimes are often held in place through the threat that the national body will become polluted or infected. While the 'environmentalism' of Carson remains undeveloped in this example, the 'profound political ambivalence'[83] of ecological thinking makes it perfectly compatible with right-wing ethnonationalism. Political scientists have however differed over whether some form of 'green nationalism' might actually be a progressive force in averting environmental catastrophe.[84] Arguing that nationalisms have developed the powerful and long-lasting institutional bases that make environmental action feasible, Daniele Conversi and Lorenzo Posocco identify a range of factors in countries such as Norway, Sweden, Denmark, Switzerland and Germany that facilitate ecologically responsible policymaking.[85] In a world as divided as it is, they argue, the only way *out* of environmental catastrophe is *through* the nation. Others have noted that minority nationalist parties often advance green causes through articulating the 'ecological patriotism' that aligns the health of the people with the health of nature.[86] Competition between nations to be 'greener' apparently reconfirms, and even strengthens, national identity, creating a potentially virtuous cycle between national feeling and ecological protection.[87] In a world where nationalism is resurgent, and the nation doesn't seem to be fading away, perhaps it would be foolish *not* to welcome putting national sovereignty to ecological use. I will refer to this broad category of political discourses as 'econationalism'.[88]

John Hultgren observes a split in the critical environmental politics literature between scholars who see the 'green state' as an enabling condition for 'ecological democracy' and those who view national sovereignty as by definition an exclusionary 'death knell for the emancipatory potential of environmental thought'.[89] And indeed there is plenty of evidence that econationalism buttresses rather than replaces older forms of resource nationalism, and is far from a harbinger of radical change.[90] When nationalist parties 'domesticate' climate issues that are by definition global, they tend to obfuscate the real issues and frustrate urgent, globally coordinated action.[91] Nationalism respects political borders before ecological ones, and asserts sovereignty in ways that limit the enforceability of international agreements, potentially making econationalism an inconsistent and unstable proposition.[92] Solving ecological problems may require suspending commitment to the 'national interest', and weighing the prosperity of faraway people as equal to that of citizens. How can nationalist governments plausibly sell this to their people? Authoritarians may make green commitments to grow their international profile and legitimize their hold on power in the eyes of the world,[93] while in democracies environmental policies have to first build popular support. The question posed by the environmental historian Stephen Brain is highly salient: '[I]s environmentalism a popular ideology only when it requires personal sacrifices of others, and not of its putative adherents?'[94]

An answer to this question becomes apparent in how ecological themes in nationalist discourse tend to hinge on shifting blame for environmental problems to non-national Others. In just one of the examples of 'good' econationalism cited by Conversi and Posocco, Switzerland, the right-wing populist Swiss People's

Party – which became the largest party in the Federal Assembly in 2024 – states that 'if you want to effectively protect the environment in Switzerland, you must fight mass immigration'.[95] There has been evidence for some time that white supremacist social movements align themselves with environmental causes in a ploy to gain legitimacy and a place in the mainstream.[96] This is in some cases a kind of 'rebranding'[97] connected to a very limited political repertoire, as Andreas Malm and the Zetkin Collective argue, though there are also many examples of environmental ideologies that exceed mere window-dressing. An oft-cited case of a fairly well-developed fascist environmentalism is that of Nazi Germany, where the focus on the intimate and essential relationship between the Aryan 'race' and their 'homelands' – between *Blut und Boden* – prioritized the conservation and beautification of the landscape. Though the Nazis eventually reneged on many of their environmental commitments, a relatively strong 'green wing' within the party successfully promulgated environmental policy.[98] This tendency within Nazism survived the war, resurfacing in 1980s neo-Nazism and persisting into the digital age, where closeness to nature is articulated with Nazi authenticity.[99] The phenomenon is thus neither merely incidental nor particularly new.

Much has been written about 'ecofascism' since white supremacist terrorists offered environmental justifications for their acts of mass murder.[100] Scholar of fascism Graham Macklin argues that modern ecofascism combines *Blut und Boden* ethnonationalism with an anti-human ecology inspired by mysticism, anti-technologism, and a misanthropic strain of deep ecology that longs for the decimation of human population.[101] Ecological thought offers the extreme right a compelling account of the harmonization of humanity and territory premised on the expulsion of a polluting Other. As antifascists Sam Moore and Alex Roberts argue, 'fascism is intensely interested in the interface between humans and the natural world, and the ordering of social relations according to nature's laws'.[102] The fascist critique of modernity as degenerate and polluted fails to amount to a systemic critique of capitalism as it latches onto race instead of capital, and thus sees nature's 'bleak lessons of scarcity, competition and dominance' in social relations, affirming 'the "natural" character of racial struggle and the superiority of its own race within it'.[103] Terrorism scholar Kristy Campion meanwhile defines ecofascism as a 'reactionary and revolutionary ideology that champions the regeneration of an imagined community through a return to a romanticized, ethnopluralist vision of the natural order'.[104] In this context, 'ethnopluralism' refers to the separateness of peoples based on fundamental characteristics which degenerate if racial 'stock' is allowed to mix.

The environmental sociologist and critical discourse analyst Balša Lubarda argues that ecofascism in practice can be hard to pin down, as it overlaps with other right-wing ecological visions. He argues instead for a broad approach to 'far-right ecologism' (FRE) that takes into account its derivation from fascism, conservatism and national populism.[105] Lubarda builds on political scientist Jonathan Olsen's analysis of eco-naturalism (where social relations are modelled on 'nature'), eco-organicism (where nature and society are thought to form an organic whole) and eco-authoritarianism (where democracy cannot halt ecological decline)[106] to offer an account of FRE as structured by distinct values (paraphrased here):

- *Naturalism.* The nation is at one with nature and national life is governed by natural laws. Respecting nature is respecting the nation. Non-nationals are by nature out of place, unable to respect the nation or its laws.
- *Spirituality and mysticism.* A mystical connection between nature, homeland and nation is governed by some kind of spirituality, whether religious or not.
- *Authority.* Democracy has a degenerating effect on nature/nation, which only strong, centralized decision-making can prevent.
- *Organicism and autarky.* Animals, plants, people and the land in which they are rooted form a homoeostatic self-sustaining sociobiological system that is opposed to city life and threatened by invading alien 'species'.
- *Nostalgia.* Past organicity and autarky have been lost with urbanization and modernity, but they are also our future once the nation is reborn.
- *Manicheanism.* Absolute antagonism is constructed between the nationalist and the globalist, the traditional and the cosmopolitan, the insider and the outsider.[107]

Lubarda warns that these values are not intended as the basis for 'box-ticking'.[108] Specific values may be emphasized to varying degrees in different geopolitical and historical contexts. In Hungary, for example, values of rootedness and localism structure a unique relationship to the Carpathian basin that justifies an exclusive Magyar right to the management of the area, which is constructed as the foundation of their prosperity.[109] The localism of Hungarian FRE helps to explain how closeness to nature contrasted to an unfeeling 'globalist' order gives an ethnonationalist form to what is essentially a critique of neoliberalism and globalization. FRE is also evident in contexts outside of Europe.[110] In neo-Nazi forums in the United States, for example, genocidal fantasies are entertained by ecofascists who envision extreme population control measures at home while the rest of the world is left to collapse.[111] This 'accelerationist' position sees environmental catastrophe as a welcome event that will enable the phoenix-like rise of a new world order from the ashes of the old, and where racial purity and ecological harmony will secure a permanent utopia.[112] A less apocalyptic environmental vision is meanwhile advanced by the prime minister of India, Narendra Modi, whose *Bharatiya Janata Party* (BJP) constructs Hindus as environmental protectors and minorities as despoilers. In Hindutva ideology, Lord Krishna is an environmentalist god who is offended by the practices of Muslims especially, who do not respect Hindu dietary laws, and cruelly slaughter and devour cows.[113] The BJP claims a 'spiritual environmentalism'[114] aligned with national development. Their form of green nationalism uses wind energy development as a tool of conquest to dispossess Muslims, confirming that the space of India is an ethno-religious one in which upper-caste Hindus have the right to dictate development.[115] The BJP in power acts cynically, moreover, rejecting environmental policies that may reduce friction between Muslim and Hindu communities as such conflict serves their long-term interests.[116]

The gendered aspects of FRE have so far been underexplored. Fascism arguably hinges as much on gender, sexuality and reproduction as it does on race and nation.[117] While the importance of 'petro-masculinity' to 'fossil fascism' has been theorized[118] and the emergence of 'eco-modern masculinities'[119] can be thought of as a reaction to the predominance of fossil-friendly subjectivities in macho cultures, distinct far-right sexual anxieties are also clearly visible in a range of FRE discourses. The environmental scientist Giovanna di Chiro's pathbreaking analysis of anxiety about environmental toxins in the important volume *Queer Ecologies: Sex, Nature, Politics, Desire* is an excellent starting point for such analysis. Di Chiro shows how industrial pollution is constructed as feminizing men who are 'swimming in a sea of estrogen'.[120] These discourses are entangled in a kind of 'eco(hetero)normativity' that pathologizes LGBTQ people as 'unjustly impaired by chemical trespass'.[121] Environmental texts that fixate on 'gender-bending, homosexual, and emasculated frogs, fish, birds, and alligators' – the last of which are said to have 'teeny weenies'[122] – thematize emasculation and declining sperm counts as issues vastly scarier than the other effects of environmental toxins. A further dimension to this discourse is described by gender scholar Mel Y. Chen, who shows how US cultural production about the 'Yellow Peril' in the twentieth century has endured in the twenty-first in 'anxieties about intoxications, mixings, and Chinese agents'.[123] The idea that feminizing agents in food and drinking water are part of a programme of bioterrorism aimed at weakening men is a popular one on the 'dissident' or 'alternative' right. Whether it is the Chinese, feminists or Jews who are scapegoated, the refrain remains constant: men's virility is being stolen from them; they are growing weaker, less fertile and more feminine; and the only option is to fight back.[124] Far-right conspiracy theorists such as Alex Jones routinely argue that 'they' are intentionally 'putting' chemicals in the water, making the 'frogs gay' and by extension leading to an increase in homosexuality and trans identities among humans.[125] The influential far-right ideologue Bronze Age Pervert meanwhile describes a world controlled by the feminized globalist 'bugman' who 'seeks to bury beauty under a morass of ubiquitous ugliness and garbage ... The waters are polluted with birth control pills and mind-bending drugs emitted by obese high-fructose-corn-syrup-guzzling beasts'.[126] The far right's latent critique of capitalist consumption is unable to rise above ethnic and sexual fantasies, choosing instead to respond to an increasingly toxic world through articulating racism, transphobia and anti-feminism as saving the future of humanity.[127] In this world of absolute differences between man and woman, white and Black, boundary crossings are obsessed over as sources of impurity. Whether viruses[128] or toxins, the FRE landscape is dotted with racialized threats to white identity that adopt an 'immunological and anticipatory register'.[129] In the far-right imaginary the Other is a poison, a disease that must be expelled at any cost from the national body.[130] The register of environmental health is thus a particularly useful one for the expression of sexual and gendered anxieties about race and purity.

One of the forces making FRE so insidious in the present moment is that the environmental crisis has already begun, and the far right thrives in places where ecological crises are ongoing.[131] In the United States especially, the idea of an impending ecological collapse meshes well with religious apocalypticism, which forms the basis for coalition-building and ideological entrepreneurship on the far-right.[132] Droughts, hurricanes and devastating fires leave people asking questions that are easily answered through scapegoating or conspiracism. State authoritarianism is furthermore often welcomed by people in the wake of a natural disaster.[133] Environmental communication strategies that invoke the seriousness of climatic threats have been shown to increase negative attitudes towards ethnic and religious out-groups.[134] The possibility of their own extinction may make people less likely to cooperate with each other, and more likely to see their own survival as depending on the absence of others, and so fascist recruiters build their appeal in proportion to the size of the crisis and the perceived lack of solutions, where any extreme action, including terrorism, may be justified in order to survive.[135] As ecological breakdowns intensify, many scholars and activists fear, it becomes harder and harder to resist 'the ecofascist creep'.[136]

## Summarizing the econationalist imaginary

There are many strands of econationalist thought that might be included in a mapping out of this imaginary. In this section, I propose elements that connect nationalism to environmentalism, ecology to race, and politics to biology. I then present three case studies showing how the econationalist imaginary plays out in real-world texts. Again, I must remind the reader that this summary (outlined in Table 4) is merely a starting point for future analysis, and far from the final word on the full dimensions of the econationalist imaginary.

We might think of the econationalist imaginary as a flexible formation that adapts to local dynamics, animating social visions from banal nationalism to full-on fascism. At its heart is the imaginative equation of the nature of a territory with the nature of its people. The foundational myths of how the nation came into being inform which landscapes are protected, which events memorialized

**Table 4** Summary of the econationalist imaginary.

| THE ECONATIONALIST IMAGINARY | |
|---|---|
| Articulated around ecosovereignty, naturalism and organicism, where the national ethnic body is naturalized on the land, and must therefore be protected from pollution by outsiders | |
| **SUBJECT ROLES**<br>• The ecological patriot<br>• The green traveller<br>• The 'Ecological Indian'<br>• The migrant<br>• The savage<br>• The shadowy puppet master | **INSTITUTIONS**<br>• The 'green state'<br>• Heritage and nature conservation<br>• Law: mobility, property, immigration<br>• Militarized borders |
| **CHRONOTOPES**<br>• The Heroic Age<br>• Decline (and Fall)<br>• Shithole countries<br>• Pastoral idyll<br>• Retrotopia | **AFFECT**<br>• Love, connection, belonging<br>• Tranquillity and spiritual connection<br>• Disgust<br>• Anxiety<br>• Hatred<br>• Sorrow and nostalgia |

and which natural features lionized in national cultural production. In settler colonial contexts, especially, nature must submit itself quietly as *terra nullius* to its conquerors, whose violence is strategically forgotten. Given this rich and important heritage, the nation is rightfully mistrustful of ceding any sovereignty to supranational or extra-national entities. Where global issues such as climate change threaten the health of the land and thus of the people, they are domesticated and used as pretexts for the centralization of authority. The health of the race/nation is metonymized as the health of the (gendered) body susceptible to penetration, poison and disease. In a naturalized world system imagined in ethnopluralist terms as separate territories with separate peoples, the bleak lesson of nature is that each country must look out for its own interests if it is to survive. This same naturalism constructs 'shithole countries'[137] as producing (racialized) human shit, which the national interest demands be kept at bay. Lines between the national and the non-national may be Manichean or drawn less distinctly. In general, there is a suspicion of the body of the Other on the landscape, which is read as a sign of environmental decline, litter, pollution and the entry of a toxin. What is promised in the econationalist imaginary is the future re-harmonization of the people and their land, which will be achieved once the polluting Other has (in certain cases) been assimilated into the national body and thus culturally neutralized, or more typically either once they have been expelled or extirpated.

## Econationalist subject roles

The central subject role of the econationalist imaginary is the ecological patriot, who understands that their duties to nature and the nation are aligned. The ecological patriot is rooted in the landscape; they connect to their 'soul country' with a 'feeling so strong' it 'never leaves you'. In colonial contexts, the settler's indigeneity is asserted through their ecological patriotism but also through a process of selectively forgetting and edulcorating history, allowing for the alignment of their own subjectivity with two-dimensional Indigenous avatars of environmental purity, the controlling images of 'Ecological Indians'. A fellow subject role to the ecological patriot is the green traveller, whose visits to the homeland to hike and explore the countryside give important confirmation of the nation's beauty. These similarities between the patriot and the traveller, whose character is imagined as shaped by their own beautiful homeland, are important for the development of transnational racial and class. These figures are diametrically opposed to the controlling figures of the migrant, the savage and the puppet master. The migrant is not a traveller, but a pest. They are the product of a 'shithole' and have no fixed abode; their deleterious effect on lands to which they have no connection is inevitable. The savage meanwhile is the Janus-face of the Ecological Indian whose enduring presence on the land is a mistake in need of rectification. The savage Palestinian, for example, does not truly belong on the land, as they are constructed as lacking the signs of a unique culture that binds them to it, which only the settler (who is also constructed as the truly autochthonous indigene) can claim to have. The final figure, the shadowy puppet master, is the root cause of global environmental problems. Whether it is a threateningly toxic China, blue-haired feminists whose birth control poisons the water or Jewish 'globalists' trying to inaugurate a New World Order, other people's hunger for power and profit is what threatens our nation, and it is them who will have to be stopped if our homeland is to be spared an environmental catastrophe.

## Econationalist institutions

The core institutions of econationalism are the institutions of the nation-state – like Le Pen's 'Republic' that is in tension with 'Europe'. Because the exercise of national sovereignty is constructed as the only viable path to a green future, the hegemony of the nation as the appropriate unit of human socio-political organization is affirmed. The specific institutions that develop within these nations will prioritize environmental concerns that advance the narrow interests of the 'green state'. Their focus is therefore on tourism, picturesque cultural landscapes, national parks, stronger or better borders, and regimes of forgetting and remembering that tell a unified national story. The landscape may be rewilded, covered in pines, fenced off or subsidized in ways that mirror the imagined national character. Special institutional attention is paid to property, ownership, and the right to own, do business, camp or subsist on particular patches of land. Imposing a national vision on nature and the people in it may be a more or less violent process. Historical and ongoing genocides, where people are eliminated from the old *Heimat* or new 'homelands' as if they were dirt, have not been without their environmental justifications. More typically, however, 'cleansing' the land through forced removal and the criminalization of traditional human–nature relations enacts a cultural genocide that seeks to permanently sever the connection to territory and right of return. Laws that govern foraging, hunting and fishing and other forms of subsistence supplement laws that govern camping, trespass and private property in an attempt to make certain ways of living go extinct. Ecologists, conservation authorities, police, the courts and the military collaborate to give nature this essential national character. The selective provision of healthcare is another strategy that exploits the 'vulnerability to premature death'[138] of racialized migrant and Indigenous populations. The signal institutional apparatus of econationalism is however the border regime. Separation walls dotted with armed turrets that have one-way glass and loaded guns; immigration controls and fortified concrete walls covered in spikes and razor wire; wilderness areas with warning signs that camping is not allowed; police agents decked out with automatic weapons and Kevlar; visa regulations that impose enormous costs on every (undesirable) non-national wishing to cross a border: this is the dominant institutional face of econationalism.

## Econationalist chronotopes

Nationalist myths typically identify a 'Heroic Age'[139] where the first people, founding fathers or feisty pioneers established the true character of the nation in a process inseparable from the characteristics of the territory, which becomes their 'ethnoscape'. In relation to the Heroic Age, nature is both a site where (selectively preserved) 'traces of history' such as landmarks or ruins serve as title deeds to the land and a force that formed their national or ethnic character in the first place. In Israel, the nation may be imagined in a biblical chronotope while in the United States, it is often the 'wild frontier' that maps onto both conservation practices and the national self-image. Whereas resource nationalisms produced nature in a 'Promethean' chronotope of endless exploitation, econationalism tends to target the 'idyllic'.[140] A return to some rural idyll, a mixture of the wild and the pastoral, is thus a frequent theme. A more natural life outside of cities, which are problematized to various degrees, is constructed as tranquil, simple or otherwise 'quiet'. This tranquillity is especially important to assert in settler colonial contexts where land was taken violently. Historical threats that might invalidate the claim to autochthony are either repressed or hollowed out of their political content. The idyll is however very much threatened in the present by ethnic Others who are both the evidence and underlying causes of a 'Fall' from the Heroic Age. These threats grow within the nation in a chronotope of decline. Dirt is visible everywhere; sometimes because people are flowing out of their shithole countries and into our idyll, or because they represent the degenerated 'bad Orient' so starkly different from the 'good'. These people bring the 'jungle' with them – and it is a jungle of constant nomadic motion which can take root anywhere it is left to its own devices. It erodes the rule of law; it becomes a ghetto that poisons the national body. They blew in here like sand, and now they must be dispersed again, cleared away, like so much human detritus. Only once their threat has been neutralized will the future truly belong to us once again. A new Golden Age yet to come will be realized through a new generation of heroes, whose ecological patriotism will rid the land of what poisons it, and restore the lost idyll in the future, in what Zygmunt Bauman calls a 'retrotopia'.[141]

### Econationalist affect

Love and hatred in equal measure animate the econationalist imaginary. Love of the ethnoscape with its nature and traces of history, a 'spiritual' connection with nature's majesty, and knowledge of belonging and at-homeness that is proved in knowledge of the fauna, flora or topography of the homeland. Connection with the ethnoscape becomes an essential part of one's identity; it is a feeling that 'never leaves you'. When one is in communion with the homeland, one enjoys tranquillity, feeling at peace and truly at home. Sorrow, disgust, anxiety, rage and hatred attend contemplation of how the homeland is being ruined by invading Others, as they have already ruined their faraway territories. The dirt itself is disgusting; it brings disease and decay, rotting the national body from the inside. The ever-present threat of infection, toxicity or retaliatory violence provokes anxiety about its location and trajectory, which must at all costs be managed. Rage at what They have already done, and hatred of Them for having done it, energizes the people for violent action to expel or hold back the threat. Shared nostalgia for the lost idyll is the 'cohesive' force that sticks the social body together. This sorrow, like the tear that rolls down Iron Eyes Cody's cheek, is a spur to action: to repair what has been broken, to stop the pollution at its source and to work together to purify the homeland.

## Textual analysis

In this section, I present three cases where elements of the green capitalist imaginary are made visible.

### Case 1: The Green Party's 2018 election campaign in Sweden

The Green Party in Sweden (*Miljöpartiet de gröna*) has been contesting elections since 1982, with limited success. Its best performance was in 2010, when it won 7.3 per cent of the national vote and was thus the third-largest party, with twenty-five seats in the Swedish *Riksdag*.[142] At the most recent election in 2022, it scraped into parliament with just over 5 per cent of votes. Though the Greens were in the so-called red-green ruling coalition with the Social Democrats (*Socialdemokraterna*) and the Left Party (*Vänsterpartiet*) between 2014 and 2021, they have traditionally attempted to stay aloof from categorization as either a left- or right-wing party, on the logic that the environment should be everybody's concern, and not only that of the proletariat or the bourgeoisie.[143] In the run-up to 2018 national elections, the Greens launched a campaign that spelled out a 'green ideology' redefining 'solidarity', a key nodal point of Sweden's social democratic system. It was spelled out as follows: 'solidarity with animals, nature and the ecosystem; solidarity with coming generations; solidarity with all the people of the world'.[144] This 'solidarity without borders'[145] was a bold, if risky strategy to adopt at a time when the 2015 'migration crisis' was still fresh in the minds of the electorate. In the webpages of their online manifesto, which included sixty-three images, eleven short videos and over 60,000 words of text, the Greens presented a reassuring picture of how they understood this new, borderless world. In Figure 5, a farmhouse surrounded by trees and covered in solar panels is shown. This particular style of building, with its white trim, and walls painted in a shade of red known as *Falu rödfärg*, is immediately recognizable as iconic of both of the Swedish countryside and of Swedishness itself. The Swedish self-image of always being close to a rural, 'authentic' way of life, is intimately connected with the 'little red cottage' – which may be returned to from the city for long summer holidays.[146] *Falu rödfärg* is a paint made

**Figure 5** Still of a farmhouse taken from the Swedish Green Party's 2018 election campaign video.

through a method developed at the copper mines of the town of Falun in the sixteenth century. 'Red earth' that contains 'silicic acid, iron ochre, copper and zinc, among other things'[147] is fired in a kiln until the iron oxidizes, resulting in a distinctive red hue. Its production is thus intimately interwoven with histories of extraction and the gradual industrialization of Sweden, a mineral-rich country that famously kept the Nazis supplied with iron throughout the Second World War.[148] Iron extraction is also one of the major forces driving historical and ongoing dispossession in Sápmi.[149] This faint echo of resource nationalism in the image combines with non-fossil, renewable energy indexed in the solar panels, and a voiceover that promises a 'solar revolution' with 'larger investments, shorter wait times, and simpler regulation than ever before' – in reference to a 30 per cent rebate offered by the government for the installation of photovoltaic energy, which the Greens proposed should also be available to people living in apartments. The green 'transition' is thus effected visually as aligned with Swedish identity, where a continuous national character in a specific cultural landscape is maintained even as the technology is shown to be changing.

In Figure 6, another icon of Swedishness is mobilized: 'the dip' (*doppet*). Jumping into the crystal-clear water of a Swedish lake[150] for an invigorating swim signifies the alignment of the health of the people with the health of their environment.[151] In choosing this specific activity, the Greens thus represent a moment of great affective significance for Swedes as ecological patriots. The

**Figure 6** Still of people jumping into a lake taken from the Swedish Green Party's 2018 election campaign video.

voice of erstwhile party co-chair Gustav Fridolin however adds a note of unease to the idyllic scene:

> This dip was taken on the 25th of May 2018. But I can't stop thinking about how, more and more, we're swimming with small particles of plastic. Clean lakes and living oceans can no longer be taken for granted. If we don't fight for nature, soon nothing will be.

The reference to the date on which the *dopp* was taken connects with the twenty-second clip's tagline, which is superimposed in large text at the end: 'Now: The climate cannot wait'. While the connection between the climate and microplastics is not elaborated, the choice of the *dopp* powerfully invokes how the Swedish rural idyll is threatened by invisible forces which have managed to penetrate right into the heart of the nation, disrupting healthy communion between the people and the land, indexed by the two young white people who are filmed whooping as they jump into the lake. Fridolin's claim that he 'can't stop thinking' about microplastics performs his own anxiety, an affect that contrasts to the delight of the swimmers jumping into what they 'take for granted' will be clean water. The anxiety of Fridolin thus becomes the anxiety of the ecological patriot, who perhaps now will also be unable to stop thinking about the hidden pollutants and toxins in even the most idyllic scene. Whereas waste tends to be hidden or suppressed in everyday discourse, here it returns to haunt the Swedish electorate as they engage in treasured activities, intruding into the very heart of their ethnoscape.[152]

In a third scene (Figure 7), the scale of oceanic plastic pollution is outlined in the campaign's main (one-minute-forty-three-second) campaign video. In this scene, which shows what incongruously appears to be stock footage of a yacht sailing next to a bank of seaweed, Fridolin again narrates the scene: 'Now there is a continent of plastic the size of Mexico floating in the Pacific'. Fridolin's claim is traceable to news reportage at the time. In 2017, a flurry of press told of the discovery of a garbage patch in the South Pacific during a six-month scientific mission led by the oceanographer Charles Moore and his Algalita project.[153] Reporters cited Moore's project updates on ResearchGate. Moore did not mention Mexico, however, instead stating that the 'southern patch could be as big as a million square kilometers, 1.5 times the size of Texas'.[154] While some stories stuck with the comparison to Texas, the southward drift to Mexico was inexorable. *Göteborgs-Posten* (The Gothenburg Post), for example, the largest newspaper in the west of Sweden, which picked up the story from CNN, equated the patch to Mexico.[155] Because large numbers can be hard to comprehend,

**Figure 7** Still of a boat near a brown mass taken from the Swedish Green Party's 2018 election campaign video.

spatial similes are often useful. Similes are interdiscursive ties that scaffold new information on knowledge which may be presupposed. The statement 'X is cold as ice' presumes knowledge of how cold ice is, for example. In a report on *IFLScience*, which has a large British readership, the patch is said to not only be the size of Mexico but also 'four times the size of the United Kingdom'.[156] The assumption here is that the audience is more familiar with how large the UK is than they would be with Mexico. The problem, however, is that while it is accurate in relation to Moore's original claim to say that the UK's surface area is about one quarter of the million-square-kilometre polluted area, and that Texas is two thirds of a million square kilometres and would therefore fit into the garbage patch one-and-a-half times, Mexico covers two million square kilometres, and is thus *double* the estimated size of the garbage patch. For a website that claims to 'Fucking Love Science'[157] to fail at basic arithmetic (the UK would fit into Mexico eight times, not four) is revealing enough on its own. That the Swedish Greens stick to the Mexican simile without so much as a fact check, on the other hand, may reveal presumptions about what a 'continent' of trash is similar to. As the education scholar Malin Ideland shows, 'dirt' and 'pollution' in Swedish environmental textbooks are words that 'cling to bodies of colour' while white people are represented as 'world champions of recycling who sort their garbage with enthusiasm and joy'.[158] Arguably, the reason that the inaccurate simile of Mexico sticks so well is because it confirms ideas about where plastic waste is coming from: definitely not us, the responsible Swedes, who are embracing the green transition and connecting with nature, and Texans may be big polluters, but the real faraway problem people who litter, who don't recycle, and who merely throw their trash into the nearest body of water, must be brown. Or in other words: the Mexico simile presupposes the redundancy of 'black trash'. Solidarity with 'all the people of the world' is thus perhaps more aspiration than guiding value in this particular 'green ideology'.

## Case 2: Heresy against Mother Nature

The second case is also from Sweden, though from a different location on the political spectrum: the far right. 'The Golden One' is the alter-ego of the Swedish esoteric masculinist, fitness coach and neo-Nazi Marcus Follin. While many far-right influencers stoke the rage and disappointment of young white men, The Golden One 'remakes "white pride" as motivational language'.[159] His own spin on 'red pill philosophy'[160] – the 'Glorious Pill' – involves bodybuilding, dieting, masturbation abstention and playing fantasy role-playing games (RPGs). While he has a number of digital platforms on which he posts updates and photographs of his scantily clad 'fizeek' (physique) as well as advertisements for his nutrition supplements, fitness clothing and coffee brands, his YouTube channel hosts the bulk of his content. The cultural historian Simon Strick has termed the niche occupied by The Golden One on YouTube 'Nordic kitsch', which he defines as 'somewhat humorous fascist camp'.[161] The Golden One is a character played by Follin, who performs a vaguely ancient martial masculinity that is sometimes aligned with Nordic mythology, sometimes with ancient Rome, and always pulled off in heavily accented English. What holds these subject roles together is their connection both to a supposed European Golden Age and to the world of RPGs, which mixes science fiction, history and fantasy together. When The Golden One speaks, it is in a stilted voice that aims to mimic this genre, while also sounding otherworldly because of the fact that English is a foreign language for him, as can be seen in the extract below (e.g. 'intake' instead of 'take in'; 'a countryside' instead of 'the countryside'). This video was published in February 2019 and entitled 'I Cannot Endorse This Barbarism With My Silence – Heresy Against Mother Earth'. I combine the spoken transcript with screenshots of the images used by Follin to illustrate his argument.

Follin's archaic register, and the illustrated panel of a cloaked Golden One sitting in a library with vaulted ceilings reading a book with esoteric insignia (Figure 8), index a time-space of the fantasy past, a chronotope that (much like

Figure 8 Still from a video published on the YouTube channel The Golden One showing a cloaked figure reading a book.

Greetings my dear subscribers! Today I thought to show you something of utmost blasphemy, and this will hit you right in the feels if you are a true friend of Mother Earth. I will play a clip right now and you can avert your gaze if you are not ready to intake this, this heresy…

Figure 9 Still from a video published on the YouTube channel The Golden One showing a truck dumping trash into the Amazon River.

[…] so what we have before us is a truckload full of trash which is being emptied right into the Amazon River. And the reason for me showing this is first and foremost because […] I want you all also to agree that this is extremely wrong, this is not worthy, this is not how anyone should treat Mother Earth. And also a secondary reason for me showing this: This is from Peru, and I am quite sick and tired of hearing how bad Europeans are for destroying Mother Earth […] Sweden is extreme as in all cases. So what they do here is that they tax the people out in a countryside making it ever more expensive to own a car, and to drive a car, because […] the politician class in Stockholm they think that – you know – Sweden can save the environment with these measures punishing the people in the countryside.

Figure 10 Still from a video published on the YouTube channel The Golden One showing a crowd of identical Wojacks.

The people who are polluting tremendously, the people who are treating earth in this way, yeah it's not Europeans: this is in Peru! […] So you can show this next time someone tells how bad white people are. 'White man bad'!

the RPGs Follin plays) conflates the details of diverse historical periods and geographies. This is a retrotopian chronotope of vaguely premodern adventure, heroism and male violence that is constructed by its enthusiasts as predating the empowerment of women, the advent of multiculturalism and 'political correctness'. RPGs are often starkly 'ethnopluralist' in that they imagine clear lines between (human/non-human) 'races' or (historical) 'peoples' with different traits that lend them different essential characteristics in competition with each other.[162] Follin furthermore mixes codes: being hit 'right in the feels' (i.e. being deeply affected) is taken from everyday online youth culture, while 'utmost blasphemy' borrows clumsily from a religious register. These acts of bricolage are subcultural cues that establish Follin's authority to inhabit the character of The Golden One, and in turn, for The Golden One to act as an ideological and spiritual guide to his followers, who learn from him how to distinguish between what is and what is not 'worthy'.

Speaking from a fantasy-historical chronotope lends The Golden One the subcultural capital and thus the authority to adjudicate other salient chronotopes in a regime of value: the endlessly polluting South, urban vs rural Sweden and 'Europe'. In the construction of Peru in this text, we see (again) how 'black trash' travels. Follin goes on to explain that the trash will end up in the ocean, and in the bellies of 'sea creatures'. Peru has thus been caught in the very act of poisoning the planet (Figure 9). At one point (not pictured) as he talks of 'the people who are treating earth in this way' he includes a graphic from a *Scientific American* article illustrating the top ten riverine sources of plastic pollution on earth, all of which are in Asia and Africa.[163] South America is thus connected to Asia and Africa as a Global South endlessly haemorrhaging trash into the oceans, and his followers enjoined to feel disgust at this 'heresy'. Disgust attaches to both the 'heretical' people and the earthly poison/infection/pollution with which they are aligned. The racial dynamics of this heresy are clarified through the progressively larger demonyms he employs in the extract as he moves from 'Swedes' to 'Europeans' and eventually to 'white people'. A chain of equivalence is thus established, where Swedishness is Europeanness and Europeanness is whiteness. Mandisi Majavu's 'white international'[164] – the global network of power initiated during slavery and colonialism that sustains white supremacy – is thus invoked. The memetic combination of a sea of identical Wojacks (Figure 10) and the text 'white man bad' index the perceived conformity of liberal culture, where everybody supposedly gets in line to beat up on white people.[165] Voicing this culture using a non-standard grammatical construction suggests intellectual deficiency and thus participates in a eugenic discourse of racial inferiority and biological 'degeneration'. Using Lubarda's FRE framework, it is clear that a Manichean division between (intelligent/white) 'Europeans' and a (degenerate/racialized) ethnic Others from the Global South is constructed, and furthermore that it is sustained by spiritual/mystical semiotic choices and the threat of violence. In the opening frame, The Golden One is dressed in a cloak like an old-fashioned holy man or knight, whose job it would have been to put 'heretics' to the sword or on the pyre. Though this violence is for the most part kept veiled, he does at one point say that 'if I had the reins of Sweden at least I would have no problem at all stationing soldiers to make sure that no one treats the environment in this sort of way'. Authority is thus also a clear guiding value of his ecological vision. Follin's 'ideological square' between polluting Others and ecological whites is thus far more explicit than that of the Swedish Green Party.

This text also 'domesticates'[166] the environmental crisis by intervening in a local controversy. Far-right parties in Sweden have built substantial support in rural communities dissatisfied with the rising costs of petrol due to a government tax intended to promote the use of sustainable energy.[167] Opposition to this tax, which pits 'the countryside' against the Stockholm 'politician class', cleaves to the familiar populist opposition between the people and the elite, and the (chronotopic) conservative opposition between the modern, degenerate town and authentic, idyllic countryside.[168] By juxtaposing this issue with Global South pollution, which is constructed as the *real* crisis, Follin's far-right ecologism thus combines opposition to local interventions to reduce greenhouse gas emissions with passionate devotion to 'Mother Nature'. The rural idyll – populated by 'true' Swedish people who cannot afford the taxes imposed on them by a cosmopolitan and race-mixing elite – is constructed as threatened by globalism, while the real threat of Global South pollution perpetrated by racial Others is allowed to continue unchecked. Ethnic bodies are thus made the targets of disgust and rage, as is the conformity of dominant elite culture, which together conspire to destroy 'Mother Earth'.

Because The Golden One tends to be associated with positive affect, he apologizes towards the end of this video for giving his followers a 'black pill' and promises to correct this in future with more positive content. Follin's performance of The Golden One is to get young men 'white-pilled' about nationalism: in other words, to see white supremacy and the future of the pure white nation in a positive light. Being 'black-pilled' on the other hand, means being depressed and despondent, convinced of the inevitability of immigration, feminism, 'globalist' hegemony and, in this case, pollution. Seeing the severity of the environmental crisis in Peru was a moment of 'black-pilling' that devotion to white homelands will counterbalance. Ecologism is thus produced as in alignment with the popular fascist slogan 'Tomorrow belongs to us'.[169]

## Case 3: Returning to the Land in the United States

The final case in this chapter stays on the far-right side of the political spectrum, and examines a white separatist homesteading initiative in the United States known as 'Return to the Land' (RTTL) set up by Eric Orwoll, a moderately popular YouTuber, podcaster and entrepreneur.[170] RTTL is a 'Private Member Association' that buys up stretches of inexpensive rural land, which is owned by a Limited Liability Company (LLC).[171] In their promotional videos, Orwoll explains that while the 1968 Fair Housing Act prohibits racial discrimination in selling or hiring property, no such restrictions apply to the admission of a new member to an LLC. RTTL is thus an entirely legal whites-only property development opportunity. It is currently only operational on 158 acres of the Arkansas Ozarks, but Orwoll has big plans for expansion. Buying into the LLC costs $6,000 an acre, and the people who move there build their own homes, secure their own utilities, and collaborate on educating each other and their children. Their aim is to bring together 'groups that share common ancestry' in 'independent and fully-functioning parallel societies'.

Though they do not explicitly articulate an ecological philosophy, their website is filled with pictures of moss-covered forests, and their vision, mission and goals are backgrounded by an image of a white woman's hands holding a small growing fern. The articulation of 'community' as something that will 'grow' to be independent is thus metaphorized in the small plant she holds in her hands, which is also typical of the iconography of the environmental movement, discussed in Chapter 1, where it signifies care or protection. 'Nature' also features prominently in the promotional videos. In the first extract below, Orwoll (whose tag on YouTube is @aarvoll_, see Figure 11) has been walking through a forest, explaining the legality of the scheme and the Christian white separatist utopia it represents. He then turns to some very practical matters:

> It's also just a nice way to get cheap land and live in a more sustainable way. I mean [...] we could see the end of the world as we know it in the near future, for all we know. We could have a nuclear war, we could have some kind of catastrophic natural disaster that collapses our economic system. Or there's just a financial collapse. A lot of bad stuff could happen, and I think your priority should be living in a way where you are resilient, where even if the broader

**Figure 11** Still from promotional video of Return to the Land showing Eric Orwoll walking through a forest. https://returntotheland.org/.

> society goes down you and your local community can make it. [...] That's what we're doing out here. We're trying to live sustainably, live in a way where we can hold each other accountable and aspire towards virtue.

Orwoll is trying to cast his net wide in appealing to a variety of different white audiences as potential investors in his segregationist scheme. The possibility of societal collapse is a preoccupation both of 'preppers'[172] and other survivalists who view some kind of drastic disruption of life as we know it imminent and of Evangelical Christian sects that prepare for the end of the world by modelling their lives on the gendered performances of entertainers and celebrities in American popular culture.[173] Orwoll's repetition of the word 'sustainable' and his reference to 'natural disaster' are also evidence of the fear of ecological collapse. As he mentions the word 'sustainable' a shot of a woman picking raspberries in front of a bank of solar panels appears (Figure 12). Green energy is thus part of RTTL's definition of 'sustainable' and 'resilient'. There are also numerous drone shots here of the forested Ozarks, and a canyon with a winding river in it, suggesting that part of the experience of returning to the land will be enjoying wild natural landscapes.

These implicit links are made explicit in another promotional video where Orwoll interviews LLC members who are living on the Ozarks property about their reasons for having invested in RTTL.

**Figure 12** Still from the promotional video for Return to the Land showing solar panels in the distance.

> *INTERVIEW 1*
> This is a part of who we are, and I believe it's written in our very nature. Denying the beautiful aspects of our nature slowly kills us, because it harms our souls. Spiritual damage leaves us prone to making lifestyle choices that further remove us from alignment with God and kinship. I joined because I want us as a people to heal, rebuild, live prosperously, fully and freely, and I want us to know who we are.

Here, the word 'nature' is used to express social and spiritual harmony connected to the idea of an organic community whose naturalistic connections to territory are spiritually determined. Nature is inside of people and ordained by God: it is 'written' by Him, and it is distance from Nature that causes harm to us. Nature is thus made equivalent to God's design, and God's design is imagined as an ethnically pure community whose connection to each other ('kinship') and to the internal 'beauty' of each member of the community secures spiritual health. If this community between 'souls' and their Nature is disrupted by modernity (i.e. race-mixing) then the soul is slowly 'killed' as it becomes disaligned from God and nature, and polluted by 'bad lifestyle choices'.

Another interviewee was even more explicit with the links he saw between closeness to Nature and the importance of a whites-only community:

> *INTERVIEW 2*
> I've always wanted to do homesteading and build my own house, grow all my own food, and live off the land, be as self-sufficient as possible, and live a little

> closer to nature like our ancestors did. I kind of wish I was born during the pioneer period of America. Living where I was before I just felt like my soul was slowly chipping away, and I didn't feel like my children were gonna have a good future in a society where schools and media and pretty much every institution in society is pushing false and misleading anti-white narratives, pushing mental, physical and emotional propaganda designed to harm our people. I wanted to live around people who want to promote healthy lifestyles and have my children's best interests at heart, and my people's best interests at heart.

'Closeness' to nature in this extract is equated to pioneer lifestyles, and this interviewee is thus invoking a retrotopian fantasy of an originary natureculture. He longs to be a pioneer, and the fact that he lives in the degenerate modern age results in his feeling as if his soul is 'slowly chipping away'. Like the 'spiritual damage' referred to by the other interviewee, damage to the soul is equated with a disalignment with nature, and with spiritual unhealth. This interviewee however adopts a slightly less spiritual style in describing the effects of this: the harms are mental, physical, and emotional and they arrive through 'anti-white narratives' in 'propaganda'. The solution to an *unnatural* world in which the descendants of pioneers are forced to endure the propaganda of a degenerate culture is thus to return to a life close to nature and to the values of organicism and naturalism that are imagined as having sustained previous generation of white settlers.

## Conclusion: 'Earth needs no borders'

It may at first seem as if some of the Indigenous land ontologies and the cultural commitment to rootedness and place that I have referenced above conform to a similar structure to the kinds of exclusionary ethnonationalisms that I have critiqued in this chapter. Leanne Betasamosake Simpson, a Michi Saagiig Nishnaabeg scholar, clarifies that visions of Indigenous resurgence are of 'nation building, not nation-state building'.[174] Indigenous nation building is a rejection of the violence of colonialism and its claims to be able to dictate what should be done with territory, determining the horizon of the socio-political order. Relations to land, place and people in ways that are 'nonhierarchical, nonexploitative, nonextractivist, and nonauthoritarian'.[175] These imaginaries, though they also thematize connections between people and place, are thus diametrically opposed to the values of FRE and to the ecocidal violence of colonialism. Indigenous environmental justice is rooted in a properly historicized account of people and place, and in non-proprietorial ethical regimes, and thus should not be confused with violent and exclusionary ecological ethnonationalism.

Organized resistance to econationalism has taken many forms. The No Border collective in the United States, for example, draws attention to the coloniality of the border, and the historical and ongoing violence of militarized border regimes. The cultural studies scholar Joseph Pugliese has noted a 'convergence' in this kind of anti-colonial and anti-racist activism with environmental discourses that understands the global nature of environmental decline as requiring solutions that cross borders.[176] Building these kinds of coalitions may help to repair the relations that have been damaged by both extractivist colonial racial capitalism and imperialist environmentalism.

# Conclusion: Critique, despair and critical hope

This book has argued that CDS has much to offer political ecology and that political ecology has much to offer CDS. The result of a more robust engagement between these scholarly traditions is an approach to discourse analysis that strives to articulate the full dimensions of socio-ecological problems in all of their complexity, and that therefore aims to be *ecointersectional.* Such analysis is built on the historicization of present crises, on tracing the roots of discursive formations *genealogically* to understand the power relations that forged them. To the extent that such an approach deserves to be called critical, it rejects the notion of critique as pure negativity and instead sees it as necessary for 'strengthening our political hand'.[1] It is important to reiterate that the strength of ecointersectional critique is not merely an academic matter, but a path to better praxis and coalition-building. Ecointersectional critique is about standing in solidarity with those who are harmed by the racial-colonial-capitalist world system and taking action to realign societal relations in ways that promote human and more-than-human thriving. The slogan often attributed to the slain Brazilian environmental activist Chico Mendes – that 'ecology without class struggle is just gardening' – is apposite.

Among the spate of executive orders signed by Donald Trump on reassuming the US presidency in January 2025 was one that froze funding for the Inflation Reduction Act (IRA), the bipartisan infrastructure law that channelled funding to green energy. While Trump's second term is without a doubt going to have a massive and irrevocably negative ecological impact, rushing uncritically to the defence of the IRA would be a mistake, as Daniel Cohen and Thea Riofrancos warned in a recent *New York Times* op-ed.[2] The law was primarily a boost in the arm for the green capitalist consumption patterns of rich people. The wealthiest 25 per cent of households captured 66 per cent of the IRA's tax credits ($5.5 billion) while the poorest 25 per cent received only 0.4 per cent ($32 million).[3] Poor households in the United States struggle to pay rent and energy bills, and many do not have adequate healthcare. Imagining that the Establishment as Americans know it is magically going to deliver relief is one of the key reasons why Kamala Harris and her enthusiasm for large corporations were unable to secure the trust of voters.[4] The IRA fell so far short of 'green economic populism' that it in fact entrenched the appeal of the ecological disaster that Trump represents. Failure to think critically, genealogically and ecointersectionally when designing environmental policy can have catastrophic effects.

As I write, Elon Musk, the world's richest man who was (like me) born to a life of privilege in apartheid South Africa, is currently dismantling large parts of the US federal government in a chaotic performance before an audience that has been led to believe that progressive values and public spending are their enemies. Musk has entrenched himself for the moment in a position of unthinkable power to harm the very people who think he's helping them. In December 2022, he tweeted an artwork by the Japanese artist Shigeru Komatsuzaki.[5] In a reimagining of the biblical story of Noah's Ark, Komatsuzaki depicts recognizably African animals – elephants, lions, giraffes, rhinoceroses – proceeding two by two into the 'Ark of Space'. Their movement into the ark is cordoned off by a row of men in blue uniforms pointing rifles at a crowd of dark-skinned people, a seething mass of humanity, hands clawing the air, desperate to be let onto the ark. Perhaps Komatsuzaki meant to depict a dystopia, but as shared by Musk, whose plans to colonize Mars as 'Earth 2.0'[6] are broadly celebrated by his cult-like following, this image signifies a threat that is scarily real. The idea that a mass must be kept back by the military while billionaires take wildlife to Mars combines Musk's brand of green capitalism with the overpopulation imaginary, colonial conservation and – if not a narrowly econationalist vision – some kind of intergalactic ecofascism. Musk's tweet encapsulates how capitalists create the conditions of impoverishment and environmental degradation that people seek to flee and then 'blame those fleeing, attempting to victimize them yet again'.[7] The fact that Musk was for many years the face of the transition from the internal combustion engine to the electric vehicle is all the more alarming.[8]

Thinking ecointersectionally about the environmental crisis in an age of oligarchy and unfettered greed can seem overwhelming, given that our problems come not singly but in battalions. Contemplation of how capitalism, racism, sexism, ableism and other systems of oppression are enmeshed in ecological destruction may even be cause for despair. This is precisely the opposite of what good ecointersectional critique should do. The point is never merely to understand the crisis, but to change its course. And we will not be able to do that if we refuse to see problems in their entirety and instead try to advance single-issue politics.

The mistake of intellectuals, as Gramsci warned, is believing they can *know* without *feeling*. Such intellectuals become a 'caste, or a priesthood'[9] disconnected from the everyday passions of the people, and they can thus have no influence over the formation of a bloc powerful enough to change the course of history. It is only once intellectual pursuits align with popular sentiments that the potential of their knowledge may be realized. The intellectual 'caste' is also frequently middle-class and, as Rebecca Solnit points out, this makes the stakes of despair in the face of ecological crisis very uneven. 'For those of us whose lives are already easy', she writes, 'giving up means making life even easier, at least in terms of effort. For those directly impacted, it means surrendering to devastation'.[10] The environmental despair of the intellectual, especially the comfortable middle-class intellectual, is thus 'like bringing poison to the potluck'.[11] Learning from history, from struggle, and seeing how the arc of the moral universe bends towards justice when people work together to make things happen, should result in hope, not despair. Solnit cites the environmental lawyer Julian Aguon, who writes that the Indigenous intellectual tradition has 'a unique capacity to resist despair through connection to collective memory'.[12] This tradition 'might just be our best hope to build a new world rooted in reciprocity and mutual respect'. Aguon thus echoes Kyle P. Whyte, who, as I mentioned in the introduction, argues that consent, trust, accountability and reciprocity are 'qualities of relationships that are critical for justice-oriented coordination across societal institutions on any urgent matter'[13] and are thus indispensable to meeting the environmental challenge.

In economies of 'hope' powerful actors will always sell themselves as securing the future in ways that promise to make our problems go away. They will align trust in them with hope for a better future. It is thus of crucial importance that scholars engaged in environmental debates reject the solutions of 'settler environmentalisms'[14] that undermine good relations with human and more-than-human worlds through space-clearing gestures and top-down, map-based, half-earth, *terra nullius* approaches to conservation, carbon capture and the next big green scheme. And so it is not our responsibility merely to hope, but to always critique.

If environmental politics asks us to accept racial and economic inequality, accept the world system as it is, forget about coloniality, forget about justice, and to focus instead on building a warm and personal relationship with 'nature' then we have every reason to reject it. To borrow Flavia Dzodan's phrase: our environmentalism will be ecointersectional or it will be bullshit. Initiatives that heighten a sense of crisis without offering any action outside of wiser consumer choices distract us from having the conversations we need to be having, building good relations throughout the web of life and beyond, and forming coalitions across lines of difference to stop the cannibalistic world system in its tracks. Connecting with nature while remaining disconnected from each other falls into the same old binaristic trap where we can choose one or the other. We must remember that the choice between Humanity and Nature is a false one, and that the only solution to the challenges we face lies in strengthening and deepening our relations with each other and with the world that sustains us.

# Notes

## Preface

1 Malm, *The Destruction of Palestine Is the Destruction of the Earth.*

## Introduction

1 Rayner, 'How to Eat an Elephant'.
2 Crenshaw, 'Demarginalizing the Intersection of Race and Sex', 167.
3 Crenshaw, 'Mapping the Margins', 1252.
4 Tuana, *Racial Climates, Ecological Indifference.*
5 Alvarado, 'Index of Winds'.
6 Hayasaki, 'A Year Ago, He Ran into the Flames'.
7 Wang, 'Lahaina, Site of Incalculable Native Hawaiian Importance'; Lyte, 'Lahaina Wildfire Death Toll'.
8 Lyte, 'Lahaina Wildfire Death Toll'.
9 Yeginsu and Chung, 'Travel in 2023'.
10 Kolden et al., 'Wildfires in 2023', 240.
11 Rush, Borenstein, and McDermott, 'Maui's Fire Became Deadly Fast'.
12 Flavelle and Andreoni, 'How Climate Change Turned Lush Hawaii into a Tinderbox'.
13 Rush, Borenstein, and McDermott, 'Maui's Fire Became Deadly Fast'.
14 Flavelle and Andreoni, 'How Climate Change Turned Lush Hawaii into a Tinderbox'.
15 Flavelle and Andreoni.
16 Kevany, 'Non-Native Grass Species Blamed'.
17 Rush, Borenstein, and McDermott, 'Maui's Fire Became Deadly Fast'.
18 *The Economist*, 'Why the Fires in Hawaii Have Been So Bad'.
19 Rush, Borenstein, and McDermott, 'Maui's Fire Became Deadly Fast'.
20 Rush, Borenstein, and McDermott.
21 Kevany, 'Non-Native Grass Species Blamed'. In the original article, silver bullet is glossed as 'solution' and GHG as 'greenhouse gas'.
22 Flavelle and Andreoni, 'How Climate Change Turned Lush Hawaii into a Tinderbox'.

23 Rush, Borenstein, and McDermott, 'Maui's Fire Became Deadly Fast'.
24 Flavelle and Andreoni, 'How Climate Change Turned Lush Hawaii into a Tinderbox'.
25 *The Economist*, 'Why the Fires in Hawaii Have Been So Bad'.
26 Yam, 'Why Hawaiian Sovereignty Has Undeniable Context'.
27 Wang, 'Lahaina, Site of Incalculable Native Hawaiian Importance'.
28 Takaki, '"An Entering Wedge"'.
29 Trask, *From a Native Daughter*.
30 Klein and Sproat, 'Why Was There No Water'.
31 Wang, '"Occupied by the US"'.
32 Hiraishi, 'Lahaina Fires Reveal Ongoing Power Struggle'.
33 McKay, 'Native Hawaiians Fear Maui Wildfire Destruction Will Lead to Their Cultural Erasure'.
34 Wang, 'Lahaina, Site of Incalculable Native Hawaiian Importance'.
35 Klein and Sproat, 'Why Was There No Water'.
36 Ing, 'The Climate Crisis and Colonialism'.
37 Wang, '"Occupied by the US"'.
38 Klein and Sproat, 'Why Was There No Water'.
39 Wang, 'Lahaina, Site of Incalculable Native Hawaiian Importance'.
40 Wang, '"Occupied by the US"'.
41 Katy Tur Reports [@KatyOnMSNBC], 'Lahaina Is Not for Sale'; McKay, 'Native Hawaiians Fear Maui Wildfire Destruction Will Lead to Their Cultural Erasure'.
42 Wang, '"Occupied by the US"'.
43 Hawai'i Alliance for Progressive Action (HAPA), 'Stop Land Grabs'.
44 Wang, '"Occupied by the US"'.
45 Klein and Sproat, 'Why Was There No Water'.
46 Hiraishi, 'Lahaina Fires Reveal Ongoing Power Struggle'.
47 Hiraishi.
48 Klein and Sproat, 'Why Was There No Water'.
49 Hiraishi, 'Lahaina Fires Reveal Ongoing Power Struggle'; Klein and Sproat, 'Why Was There No Water'.
50 Eddins, Board of Land and Natural Resources v. Crabtree, West's Hawai'i Reports and Pacific Reporter.
51 Hiraishi, 'Lahaina Fires Reveal Ongoing Power Struggle'.
52 Hiraishi.
53 Klein, 'The Rise of Disaster Capitalism'.
54 Klein and Sproat, 'Why Was There No Water'. The post-tsunami land-grabs are discussed in detail in Cohen, 'Tourism and Land Grab in the Aftermath of the Indian Ocean Tsunami'.
55 Klein, 'The Rise of Disaster Capitalism'.

56 Klein and Sproat, 'Why Was There No Water'.
57 'Hawaii Governor Calls on People to Visit West Maui When It Reopens in October'.
58 Wang, 'Lahaina, Site of Incalculable Native Hawaiian Importance'.
59 Kippen, 'Mai Hilahila: Uncomfortable Conversations'.
60 I am using 'frame' here in the tradition of media studies, picking out the modality in which a story is told that provides its audience contextualization cues, e.g. 'this is a story about the science' or 'this is a story about the history of Hawai'i'. See D'Angelo, 'Framing'. Teun van Dijk has recently argued that these broad definitions lack theoretical and methodological precision. See van Dijk, 'Analyzing Frame Analysis'. It is however not my purpose to enter into these debates here.
61 Rush, Borenstein, and McDermott, 'Maui's Fire Became Deadly Fast'.
62 Hiraishi, 'Lahaina Fires Reveal Ongoing Power Struggle'.
63 Ing, 'The Climate Crisis and Colonialism'.
64 Andrea Mayr argues that as large bureaucracies, news organizations intrinsically privilege the institutional authority of official sources embedded in the establishment over those of other versions of stories that might challenge the societal status quo. See Mayr, *Language and Power*, 79–89.
65 Darrah-Okike, 'Theorizing Race in Hawai'i'.
66 Arvin, *Possessing Polynesians.*
67 Fraser, *Cannibal Capitalism*, xv.
68 McKay, 'Native Hawaiians Fear Maui Wildfire Destruction Will Lead to Their Cultural Erasure'.
69 Heron, 'Capitalist Catastrophism and Eco-Apartheid'.
70 Kumar, 'Energy Geographies in/of the Anthropocene'.
71 Cretney and Nissen, 'Emergent Spaces of Emergency Claims'.
72 Bonilla, 'The Coloniality of Disaster'.
73 Nixon, *Slow Violence and the Environmentalism of the Poor.*
74 Grove, *Green Imperialism*, 1.
75 Grove, 15.
76 See also Anker, *Imperial Ecology: Environmental Order in the British Empire, 1895–1945.*
77 The idea that natural laws determine agricultural output, which in turn determines the wealth of a nation. See Grove, *Green Imperialism*, 9.
78 Argyrou, *The Logic of Environmentalism.*
79 Argyrou, 123–4.
80 The condition this was understood to cause was 'neurasthenia' – a debilitating and degenerative condition that saw men lose their potency and vigour. See Powell, *Vanishing America*, 53.
81 Powell, 5.
82 In 1962, see Carson, *Silent Spring.*

83 Taylor, 'American Environmentalism', 47.
84 Eder, 'Risk, Environment and Modernity'.
85 Khan, 'Environmentalism in South Africa', 172.
86 Shaw, *Liberalism and the Challenge of Climate Change*.
87 Shaw, 22.
88 Legum, Zekeria, and Crosby, '1500 Environmental Lobbyists Are Double-Dealing with the Fossil Fuel Industry'.
89 Lee, 'Big Oil Helped Shape Stanford's Latest Climate-Research Focus'.
90 Altvater et al., *Anthropocene or Capitalocene?*
91 Parenti, *Against Empire*; see also Davis and Todd, 'On the Importance of a Date, or, Decolonizing the Anthropocene'.
92 Haraway, 'Anthropocene, Capitalocene, Plantationocene, Chthulucene', 162 n.5. These reframings are however resisted by the scientific mainstream. The working group developing the concept of the Anthropocene, despite overwhelming evidence to the contrary, insists on dating its start to the 1950s, many years after colonialism and capitalism started to alter the planetary system. As was evident on Maui, the shifting of temporal frames is an important strategy to exculpate the nations, companies and people who today continue to reap the benefits from historical land theft and dispossession. See Ellis, 'Why I Resigned from the Anthropocene Working Group'.
93 Mirzoeff, 'It's Not the Anthropocene, It's the White Supremacy Scene: Or, the Geological Color Line'.
94 Wynter, *Sylvia Wynter*, 23. Original emphasis.
95 Lorde, *Sister Outsider*, 138.
96 Davis and Todd, 'On the Importance of a Date, or, Decolonizing the Anthropocene', 771.
97 Hall et al., *Policing the Crisis*, 394.
98 Pulido, 'Geographies of Race and Ethnicity II'.
99 Bryant and Mohai, *Race and the Incidence of Environmental Hazards*; Bullard et al., 'Toxic Wastes and Race at Twenty'; Mohai, Pellow, and Roberts, 'Environmental Justice'; Taylor, 'The Rise of the Environmental Justice Paradigm Injustice Framing and the Social Construction of Environmental Discourses'.
100 Nixon, *Slow Violence and the Environmentalism of the Poor*, 1.
101 Mpanya, 'The Dumping of Toxic Waste in African Countries: A Case of Poverty and Racism'.
102 Gilmore, '*Golden Gulag: Prisons, Surplus, Crisis, and Opposition in Globalizing California, 28*'.
103 Tilley et al., 'Race and Climate Change'.
104 Buckingham and Kulcur, 'Gendered Geographies of Environmental Injustice'; Lewis, Reddy, and Mafofo, 'Transnational Perspectives on Food, Ecology and the Anthropocene'.
105 Oswald et al., 'Global Redistribution of Income and Household Energy Footprints'.

106 Millward-Hopkins, 'Inequality Can Double the Energy Required to Secure Universal Decent Living'.
107 Sultana, 'The Unbearable Heaviness of Climate Coloniality'; Sultana, *Confronting Climate Coloniality*; see also Bhambra and Newell, 'More Than a Metaphor'.
108 Liboiron, *Pollution Is Colonialism*, 5.
109 e.g. Bresnihan and Millner, 'Decolonising Environmental Politics'; Dunlap, 'Decolonizing Energy Justice'; Pulido and De Lara, 'Reimagining "Justice" in Environmental Justice'; Tornel, 'Decolonizing Energy Justice from the Ground Up'.
110 Whyte, 'Too Late for Indigenous Climate Justice', 2.
111 Collins, 'Intersectionality's Definitional Dilemmas', 2.
112 Collins, 17.
113 Collins, 16.
114 Tuana, *Racial Climates, Ecological Indifference*, 19.
115 This is what Amitav Ghosh calls 'terraforming'. Ghosh, *The Nutmeg's Curse*.
116 Tuana, *Racial Climates, Ecological Indifference*, 5.
117 Tuana, 11.
118 Robbins, *Political Ecology*, 13.
119 Neumann, *Making Political Ecology*, 11.
120 Fletcher, *Failing Forward*, 14.
121 See, for example, Bosworth, 'The Bad Environmentalism of "Nature Is Healing" Memes'; Büscher, 'Rhino Poaching Is out of Control!'; Fletcher, Breitling, and Puleo, 'Barbarian Hordes'; and Lunstrum, 'Feed Them to the Lions'.
122 Fraser, *Cannibal Capitalism*, 2.
123 Naess, 'The Shallow and the Deep, Long-range Ecology Movement' cited in Stibbe, *Ecolinguistics*, 12.
124 Fairclough, *Language and Power*, 3.
125 Fairclough, *Discourse and Social Change*, 12.
126 Rampton, 'Gumperz and Governmentality in the 21st Century'.
127 Foucault, *Power/Knowledge*, 36; cited in Rampton, 'Gumperz and Governmentality in the 21st Century'.
128 Halberstam, *Female Masculinity*, 13.
129 Klein and Sproat, 'Why Was There No Water'.
130 Wang, '"Occupied by the US"'.
131 Yam, 'Why Hawaiian Sovereignty Has Undeniable Context'.

## Chapter 1

1 Bearak, 'It's Not Just Willow'.
2 Tweet by @ECOWARRIORSS (https://x.com/ECOWARRIORSS/status/1645069701612011520) posted 9 April 2023, last accessed 10 September 2024.

3 Tweet by @SaintPalermo (https://x.com/SaintPalermo/status/1645084141753434114) posted 9 April 2023, last accessed 10 September 2024.
4 Houghton, Jenkins, and Ephraums, *IPCC First Assessment Report*, vi.
5 Pickard, 'Young Environmental Activists Are Doing It Themselves'. On *Letzte Generation*, see Euronews Green with Reuters, 'Climate Activists Stop Berlin Traffic by Gluing Themselves to Roads'.
6 This motivational quote has been attributed both to Helen Keller and to Edward Everett Hale.
7 This point is made powerfully by Malin Ideland in her book on environmental education in Sweden. Ideland, *Den KRAV-märkta människan*, 27.
8 Burns, 'To a Mouse, On Turning Her Up in Her Nest with the Plough'.
9 Phillipe Descola has argued powerfully that the firmness of the nature–culture divide is a relatively recent Western invention which serves as an obstacle to understanding our place in the world. See Descola, *Beyond Nature and Culture*.
10 Amitav Ghosh, for example, writes of the indifference of 'monstrous Gaia' to the survival of humanity. Ghosh, *The Nutmeg's Curse*, 85.
11 Dryzek, *The Politics of the Earth*, 14–17.
12 Dryzek, 235.
13 Heise, 'Introduction: Planet, Species, Justice – and the Stories We Tell about Them', 2.
14 Bennett and Chaloupka, *In the Nature of Things*, xii.
15 Bennett and Chaloupka, xv–xvi.
16 Hannigan, *Environmental Sociology*, 79.
17 Hannigan, 59.
18 See also the useful typology developed in Swaffield, 'After a Decade of Critique'.
19 Stephens, *Critical Environmental Communication*, 5.
20 Milstein, 'Greening Communication', 161.
21 Political ecologists are especially interested in the flow of representations within media ecologies structured by 'platform capitalism'. See Büscher, *The Truth about Nature*.
22 Milstein, 'The Performer Metaphor'.
23 Milstein, 242.
24 Dwyer and Jones III, 'White Socio-Spatial Epistemology', 2000.
25 Discussed in Fairclough, *Language and Power*, 215.
26 Cox, *Environmental Communication and the Public Sphere*, 12.
27 Cox, 'Nature's "Crisis Disciplines"', 8.
28 See, for example, Olbermann, Mayer, and Schramm, 'Greenfluencing through the Power of Emotions?'
29 See Stephens, *Critical Environmental Communication* for an overview.

30 Joosse et al., 'Critical, Engaged and Change-Oriented Scholarship in Environmental Communication. Six Methodological Dilemmas to Think With', 760.
31 Milstein, 'Greening Communication', 162.
32 Cox and Depoe, 'Environment, Communication and Environmental Communication: Emergence and Development of a Field'.
33 See Milstein, 'Greening Communication'; and Stephens, *Critical Environmental Communication.*
34 Garrard, *Ecocriticism*, 7.
35 Nixon, *Slow Violence and the Environmentalism of the Poor.*
36 Slovic, Rangarajan, and Sarveswaran, *Routledge Handbook of Ecocriticism and Environmental Communication.*
37 Lousley, 'Ecocriticism', no page.
38 Buell, *The Environmental Imagination*, 104.
39 Stanley and Phillips, 'South African Ecocriticism'; see also Nixon, *Slow Violence and the Environmentalism of the Poor* for a vindication of material postcolonial analysis.
40 See for example Gosine, 'Dying Planet, Deadly People'; Gosine, 'Non-White Reproduction and Same-Sex Eroticism: Queer Acts against Nature'; Jenkins, 'The Dystopian World of Blade Runner: An Ecofeminist Perspective'; Sturgeon, *Environmentalism in Popular Culture*; Sturgeon, *Ecofeminist Natures.*
41 Fairclough and Wodak, 'Critical Discourse Analysis', 259.
42 Augé, *Metaphor and Argumentation in Climate Crisis Discourse*, 8.
43 Augé, 161–2.
44 See, for example Alexander, *Framing Discourse on the Environment.*
45 Kurtz, 'Reflections on the Iconography of Environmental Justice Activism'.
46 See Maran, *Ecosemiotics.*
47 Schmitt, 'Anthropomorphism, Anthropocentrism, and Human-Orientation in Environmental Discourse'.
48 Stibbe, 'An Ecolinguistic Approach to Critical Discourse Studies', 120.
49 The term was originally coined by the Norwegian linguist Einar Haugen in 1972 as an approach to the study of languages in multilingual contexts. Throughout the 1970s and 1980s, it was used largely to analyse how languages change in response to complex environments, which include speakers, other languages, and language styles and registers, as well as other symbolic and material forces. Ecolinguistics thus also theorizes as 'ecological' the interrelationships between language, society and speakers in multilingual contexts. It was only in the 1990s that the 'real' (as opposed to metaphorical) ecology of the natural world as constructed in language became a focal point for analysis. For overviews of the history, see Fill and Mühlhäusler, *Ecolinguistics Reader*; Fill and Penz, *The Routledge Handbook of Ecolinguistics.*
50 Fairclough, *Discourse and Social Change*, 3.

51 Halliday, 'New Ways of Meaning: The Challenge to Applied Linguistics', 81.
52 Halliday, 89.
53 Mühlhäusler and Peace, 'Environmental Discourses', 469.
54 Alexander and Stibbe, 'From the Analysis of Ecological Discourse to the Ecological Analysis of Discourse'.
55 Cameron, *Verbal Hygiene*.
56 Okri, Birds of Heaven, 21 cited in Stibbe, *Ecolinguistics*, 3.
57 Stibbe, 2.
58 Martin, 'Grace: The Logogenesis of Freedom', 51–2.
59 Martin, 'Positive Discourse Analysis: Solidarity and Change', 183.
60 Stibbe, 'Positive Discourse Analysis: Re-Thinking Human Ecological Relationships', 165.
61 Stibbe, *Ecolinguistics*, 27.
62 Available at https://www.ecoling.net/.
63 Nash, 'Ecolinguistics and Placenames'.
64 Poole, 'A Corpus-Aided Ecological Discourse Analysis of the Rosemont Copper Mine Debate of Arizona, USA'; Poole, *Corpus-Assisted Ecolinguistics*.
65 Kinefuchi, 'Critical Discourse Analysis and the Ecological Turn in Intercultural Communication', 213.
66 And yet, given what is at stake at a planetary level, the ecolinguistics toolkit has not been taken up by many discourse analysts. In terms of the field-defining journals of CDS, *Critical Discourse Studies* has published only four articles in the past ten years that use an ecolinguistic approach, while *Discourse and Society* has published one and *Social Semiotics* two. The term has never appeared in the *Journal of Language and Politics*.
67 Chilton and Wodak, 'Preface'.
68 Fairclough, *Analysing Discourse*; Krzyżanowski, 'Discursive Shifts and the Normalisation of Racism'.
69 See, for example, Bauer et al., 'Gender Equality in the Name of the State'; Brotherton, '"We Have the Best Gays, Folks"'; Gal, 'Gender and the Discursive Authority of Far Right Politics'. Stibbe in fact discusses an excellent example of this tendency, showing how the discourse of Indigenous rights is taken up by the anti-immigrant, far-right British National Party; Stibbe, *Ecolinguistics*, 44.
70 Stibbe identifies William Wordsworth as a rich source of beneficial discourses. It is interesting to note that after a youthful flirtation with revolutionary social change, Wordsworth's politics turned profoundly reactionary. He came to believe that both the English working class and African slaves were 'quite wrongly destroying the bond of mutual dependence and obligation which had existed between groups'. Paul-Emile, 'William Wordsworth: An English Romantic Poet's Response to Colonialism', 631. Wordsworth's revolutionary fervour was however contemporaneous with his best nature writing, so perhaps Stibbe has a point after all.

71 Brown, 'Revaluing Critique', 470–1.
72 Interview with Ruth Wodak in Kendall, 'What Is Critical Discourse Analysis?' no page.
73 See, for example, Flowerdew, 'Critical Discourse Analysis and Strategies of Resistance'; and Milani, 'Sexual Citizenship'.
74 Fairclough, 'A Dialectical-Relational Approach to Critical Discourse Analysis in Social Research', 163f.
75 I am sure that I am not the only academic who has received feedback on an unsuccessful proposal stating that the committee preferred work that was 'solutions-focused'.
76 Levitz, 'Blaming "Capitalism" Is Not an Alternative to Solving Problems'. Though Levitz avers that one problem with ultra-left critiques of policy initiatives such as the Green New Deal in the United States is that 'it is annoying' he does admit that the bigger one is that 'the compulsion to denigrate the stakes of ordinary politics and the merits of actual reforms tends to blur the analytic vision of otherwise insightful intellectuals. There is value in understanding how our economy's undergirding logic might circumscribe or subvert particular avenues of reform. Inattention to such dynamics can undermine policy design. Dyspeptic Marxist intellectuals can therefore serve as useful critics even if they wish to keep mundane "bourgeois" politics at arm's length.'
77 Lamb et al., 'Discourses of Climate Delay'.
78 Stephens, *Critical Environmental Communication*, 1.
79 In his discussion of the role of intellectuals in establishing hegemony, Gramsci observes that the '"spontaneous" consent given by the great masses of the population to the general direction imposed on social life by the dominant fundamental group [...] is "historically" caused by the prestige (and consequent confidence) which the dominant group enjoys because of its position and function in the world of production.' Gramsci, *Selections from the Prison Notebooks of Antonio Gramsci*, 12.
80 Fairclough, *Language and Power*, 216.
81 O'Regan and Gray, 'The Bureaucratic Distortion of Academic Work: A Transdisciplinary Analysis of the UK Research Excellence Framework in the Age of Neoliberalism'.
82 Butler, 'What Is Critique?', n.p.
83 Hook, 'Genealogy, Discourse, "Effective History"', 7.
84 Bhaskar, *Enlightened Common Sense*, 109. Both of these citations of Bhaskar are also found in O'Regan and Gray, 'The Bureaucratic Distortion of Academic Work'.
85 Bhaskar, 109.
86 Butler, 'What Is Critique?'
87 It is important not to instrumentalize critique as only being worthwhile if it does indeed lead to solutions. As Wendy Brown writes, while critique may help us to 'comprehend the constitutive elements and dynamics of our condition' it

'does not elaborate alternatives to the order it illuminates and only occasionally identifies possible strategies for resisting the developments it charts. However, the predicaments and powers it illuminates might contribute to the development of such alternatives and strategies, which are themselves vital to any future democracy'. Brown, *Undoing the Demos*, 28.

88 Hannigan, *Environmental Sociology*, 94.

89 Peluso, 'Coercing Conservation?'; Peluso and Lund, 'Introduction'; Peluso, 'Racial Fictions at Work in a Regional Plantation Complex'; Peluso and Watts, 'Violent Environments'.

90 As detailed in Büscher, *The Truth about Nature*.

91 Salleh, 'Class, Race, and Gender Discourse in the Ecofeminism/Deep Ecology Debate'.

92 Sturgeon, *Environmentalism in Popular Culture*.

93 Mortimer-Sandilands and Erickson, *Queer Ecologies*, 3.

94 See Burnett and Milani, 'Fatal Masculinities'; and Burnett, '"Without White People, the Animals Will Go!"'

95 Mortimer-Sandilands and Erickson, *Queer Ecologies*, 34.

96 Gosine, 'Non-White Reproduction and Same-Sex Eroticism: Queer Acts against Nature'.

97 McWhorter, 'Enemy of the Species'.

## Chapter 2

1 Tuana, *Racial Climates, Ecological Indifference*, 47.

2 Many of the ideas I present in this section are primarily associated with Critical Discourse Analysis (CDA). According to Michał Krzyżanowski and Bernhard Forchtner, CDS refers to the expansion of CDA into a broader field of research since the 2000s, to include novel concepts and theories. See Krzyżanowski and Forchtner, 'Theories and Concepts in Critical Discourse Studies', 254.

3 Fairclough, *Discourse and Social Change*, 3.

4 Fairclough, 4; Fairclough, *Language and Power*, 21–2.

5 Fairclough, *Discourse and Social Change*, 65.

6 Krzyżanowski and Forchtner, 'Theories and Concepts in Critical Discourse Studies', 255; see also Jørgensen and Phillips, *Discourse Analysis as Theory and Method*, 4.

7 Fairclough, *Language and Power*, 4.

8 Fairclough and Wodak, 'Critical Discourse Analysis'.

9 Macgilchrist, 'Fissures in the Discourse-Scape', 269.

10 Allen, *The End of Progress*, 9.

11 Bhambra, 'Decolonizing Critical Theory?', 78.

12 Allen, *The End of Progress*, 205.

13 Ahmed, 'Political Discourse Analysis', 141.
14 Wynter, 'Unsettling the Coloniality of Being/Power/Truth/Freedom'. The concept of 'coloniality of power' is Aníbal Quijano's; see Quijano, 'Coloniality of Power and Eurocentrism in Latin America'.
15 Macgilchrist, 'Media Discourse and De/Coloniality: A Post-Foundational Approach', 391.
16 Thurlow, 'Queering Critical Discourse Studies', 487.
17 Thurlow, 491.
18 Thurlow, 503.
19 I am choosing to refer to the theoretical tradition associated with Ernesto Laclau, Chantal Mouffe and the 'Essex School' of discourse analysis as post-foundational discourse theory as opposed to discourse theory (as it is often named) in an attempt to particularize and parochialize it. Referring to 'discourse theory' has the unfortunate effect of implying singularity and universality.
20 Laclau and Mouffe, *Hegemony and Socialist Strategy*, xxi.
21 Marttila, *Post-Foundational Discourse Analysis*, 140.
22 Macgilchrist, 'Fissures in the Discourse-Scape', 264.
23 Jørgensen and Phillips, *Discourse Analysis as Theory and Method*, 18.
24 As one of the readers of this manuscript observed, this is a position that PDT shares with feminist standpoint theory. See Harding, *Whose Science?*
25 Macgilchrist, 'Fissures in the Discourse-Scape', 269.
26 Rorty, *Contingency, Irony, and Solidarity*, xv.
27 Importantly, Macgilchrist goes on to argue that post-foundationalism can 'move CDS on from post-positivist debates about objectivity and bias, in order to embrace surprise, irony and transgressive validities'. Macgilchrist, 'Fissures in the Discourse-Scape', 273.
28 Marx and Engels, *The German Ideology*, 42. Original emphasis.
29 Laclau and Mouffe, *Hegemony and Socialist Strategy*, 94.
30 Laclau and Mouffe, 57.
31 Marttila, *Post-Foundational Discourse Analysis*, 37.
32 Laclau and Mouffe distinguish their notion of articulation from Althusser's: articulation is not 'like clockwork'. In their view, when moments in a discourse are articulated 'the presence of some in the others hinders the suturing of the identity of any of them'. Laclau and Mouffe, *Hegemony and Socialist Strategy*, 91.
33 Strictly speaking, once a discursive element is articulated within a discourse, practitioners of PDT would call it a 'moment' of that discourse, as distinct from the elements that are *not* articulated within it. I have decided to dispense with this distinction as it does not play a decisive role in textual analysis.
34 Laclau and Mouffe, *Hegemony and Socialist Strategy*, 91.
35 Laclau and Mouffe, 95.
36 Laclau and Mouffe, 92.

37 Jørgensen and Phillips, *Discourse Analysis as Theory and Method*, 28.
38 Leurs, 'The "Chain of Equivalence". Cultural Studies and Laclau & Mouffe's Discourse Theory'.
39 Bartlett and Montesano Montessori, 'Towards Webs of Equivalence and the Political Nomad in Agonistic Debate'.
40 Laclau, *New Reflections on the Revolution of Our Time*, 39.
41 Laclau, 60.
42 Althusser, 'Ideology and Ideological State Apparatuses (Notes towards an Investigation)'.
43 Macgilchrist, 'Fissures in the Discourse-Scape', 265.
44 De Cleen et al., 'Moving Discourse Theory Forward', 30.
45 Marttila, *Post-Foundational Discourse Analysis*, 38.
46 Stavrakakis, *The Lacanian Left*, 81.
47 Strictly speaking I should refer here to 'sinthome' and not 'symptom'. Stavrakakis is explaining the shift from the Lacanian 'symptom' which is understood as within language to the 'sinthome' which is both embodied and social. See Stavrakakis, 79–81.
48 Stavrakakis, 81.
49 Marttila, *Post-Foundational Discourse Analysis*, 41.
50 De Cleen et al., 'Moving Discourse Theory Forward', 30–3.
51 A closed identity, where X=X, is impossible because all meanings are contingent on the structuring of the discursive field. In PDT the sign is 'the name of a split, of an impossible suture between signified and signifier'. Laclau and Mouffe, *Hegemony and Socialist Strategy*, 113.
52 Marttila, *Post-Foundational Discourse Analysis*, 90.
53 Marttila, 95–6.
54 McWhorter, 'Foucault's Genealogy of Homosexuality', 49; cited in Tuana, *Racial Climates, Ecological Indifference*, 8.
55 Mouffe, *The Return of the Political*, 76.
56 Blommaert, *Discourse*, 126.
57 Marttila, *Post-Foundational Discourse Analysis*, 118.
58 Fairclough, *Discourse and Social Change*, 93.
59 Freeden, 'Discourse, Concepts, Ideologies', 54.
60 De Cleen et al., 'Moving Discourse Theory Forward', 29.
61 De Cleen et al., 26.
62 This is Jack Halberstam's concept, and was discussed in the Introduction. See Halberstam, *Female Masculinity*.
63 Fairclough, *Analysing Discourse*, 124.
64 Norval, *Deconstructing Apartheid Discourse*, 105. Following Laclau, Norval understands the imaginary as a 'surface of inscription' which 'facilitates a displacement from its initial association with a specific set to wider social demands.

[…] If the dimension of specificity predominates, there will be little possibility for an expansion of the logic of the myth and its particular principle of reading to wider areas. […] An imaginary thus emerges insofar as there is a movement away from the specificity of the demands of a particular group to the very form of the metaphorical itself.'

65 Laclau, 'Identity and Hegemony: The Role of Universality in the Constitution of Political Logics', 50.
66 Laclau, *New Reflections on the Revolution of Our Time*, 64. Laclau goes on to observe that there is constant interpenetration between the myth of the Gramscian 'corporate class' and the imaginary of the 'hegemonic class', 65.
67 Machin and Mayr, *How to Do Critical Discourse Analysis*, chapter 6.
68 See Thurlow, 'Queering Critical Discourse Studies'.
69 Laclau, *New Reflections on the Revolution of Our Time*, 65.
70 Machen, Davoudi, and Brooks, 'Climate Imaginaries and Their Mediums', 2. Lilie Chouliaraki and Norman Fairclough famously argue that these overlaps are symptomatic of late modernity, where 'boundaries between social fields and therefore between language practices have been pervasively weakened and redrawn, so that the potential seems to be immense, and indeed hybridity has been widely seen as a characteristic of the "postmodern"'. Chouliaraki and Fairclough, *Discourse in Late Modernity*, 13.
71 Stavrakakis, 'Green Ideology', 271.
72 Stavrakakis, 273.
73 Stavrakakis, 271. More recently, Giovanni Bettini and Lazaros Karaliotas used PDT to analyse 'peak oil' discourse, which they argued is articulated around the biophysical quantities and qualities of oil, rather than the socio-economic relations involved in its extraction and distribution. They argue that this serves to sediment rather than dislocate prevailing conservative and neoclassical conceptions of the relations between resources and the economy. This imaginary depoliticizes and naturalizes a phenomenon the political character of which is vital for socialist strategy. See Bettini and Karaliotas, 'Exploring the Limits of Peak Oil'.
74 While Stavrakakis's analysis is an interesting macro-political provocation, it does not grapple with the detail of specific texts and their articulatory practices. PDT has tended to focus on the 'big' picture, a telescope that may be contrasted with the microscope offered in this book.
75 Machen, Davoudi, and Brooks, 'Climate Imaginaries and Their Mediums', 1.
76 Machen, Davoudi, and Brooks, 2.
77 Thompson and Ban, '"Turning to the Territory"'.
78 For an example of the former category, see Levy and Spicer, 'Contested Imaginaries and the Cultural Political Economy of Climate Change'; which draws on the deployment of Gramsci in Jessop, 'Cultural Political Economy and Critical Policy Studies'.

79 Behagel and Mert, 'The Political Nature of Fantasy and Political Fantasies of Nature', 84.
80 Behagel and Mert, 81. Michael Freeden has argued in more general terms that how fantasies work to structure and filter reality ought to be a 'major source of evidence and interpretation' for political discourse studies. Freeden, 'Discourse, Concepts, Ideologies', 58.
81 Chouliaraki and Fairclough, *Discourse in Late Modernity*, 14.
82 Jørgensen and Phillips, *Discourse Analysis as Theory and Method*, 15–16.
83 Fairclough, *Language and Power*, 32.
84 Blommaert, *Discourse*, 4.
85 Blommaert, 30.
86 Fairclough, *Discourse and Social Change*, 101.
87 Blommaert and Bulcaen, 'Critical Discourse Analysis'.
88 Agha, 'Voice, Footing, Enregisterment', 39.
89 Gal, 'Registers in Circulation', 4.
90 Eckert, 'Variation, Meaning and Social Change', 76.
91 See Tebaldi, 'Metapolitical Seduction'.
92 Eckert, 'Variation, Meaning and Social Change', 76.
93 Bakhtin, *The Dialogic Imagination: Four Essays*, 270–1; Milani and Jonsson, 'Who's Afraid of Rinkeby Swedish?'
94 Shotter and Billig, 'A Bakhtinian Psychology: From out of the Heads of Individuals and into the Dialogues between Them'.
95 Fairclough, *Analysing Discourse*, 24–6.
96 Fairclough, 145–50.
97 Leeuwen, *Discourse and Practice*, 32–47.
98 Machin and Mayr, *How to Do Critical Discourse Analysis*, Chapter 5.
99 Leeuwen, *Discourse and Practice*, 137–41.
100 Leeuwen, 'The Semiotics of Movement and Mobility', 101.
101 Collins, *Black Feminist Thought*, 78.
102 hooks, *Talking Back*, 42.
103 Collins, *Black Feminist Thought*, 294–5.
104 Fairclough, *Language and Power*, 38–9 emphasis in original.
105 Fairclough, *Analysing Discourse*, 41.
106 Dwyer and Jones III, 'White Socio-Spatial Epistemology', 213.
107 Milani and Richardson, 'Discourse and Affect'; Milani, 'Affect: Discourse, Politics, Intersectionality'.
108 The distinction between emotions and affect, while productive and interesting, is not directly relevant to my project here and I thus use these terms interchangeably.
109 Wetherell, *Affect and Emotion*, 19.
110 Wetherell, 24.
111 Kiesling, 'Masculine Stances and the Linguistics of Affect', 192.
112 Du Bois, 'The Stance Triangle', 163.

113 Kiesling, 'Masculine Stances and the Linguistics of Affect', 194.
114 Kiesling, 196.
115 Kiesling, 200.
116 Ahmed, *The Cultural Politics of Emotion*, 194.
117 Wetherell argues that Ahmed's work both reifies and mystifies emotions. For my purposes here, however, these differences are a little too arcane and granular to impact on the proposed approach to textual analysis. See Wetherell, 'Trends in the Turn to Affect'.
118 Kiesling, 'Masculine Stances and the Linguistics of Affect', 201.
119 Marttila, *Post-Foundational Discourse Analysis*, 37.
120 Fairclough, *Analysing Discourse*, 32.
121 Fairclough, 70–83.
122 Fairclough, 67.
123 Mayr, *Language and Power*, 103–9.
124 Mayr, 1–3.
125 Blommaert, *Discourse*, 218–21.
126 Bakhtin, *The Dialogic Imagination: Four Essays*, 243.
127 Bakhtin, 85.
128 see Laclau, *The Rhetorical Foundations of Society*.
129 Laclau, *New Reflections on the Revolution of Our Time*, 66.
130 Bakhtin, *The Dialogic Imagination: Four Essays*, 84.
131 Neely and Samura, 'Social Geographies of Race', 1939.
132 Neely and Samura, 1937.
133 Dwyer and Jones III, 'White Socio-Spatial Epistemology', 212–13.
134 Dwyer and Jones III, 'White Socio-Spatial Epistemology', 212.
135 Tuck and McKenzie, *Place in Research*.
136 Neely and Samura, 'Social Geographies of Race', 1938.
137 Feola et al., 'Collective Memories, Place-Framing and the Politics of Imaginary Futures in Sustainability Transitions and Transformation'.
138 Baldwin, 'Whiteness and Futurity', 174.
139 Baldwin, 175.
140 Bluwstein and Paolo De Rosa, 'Political Ecologies of the Future'.
141 Fairclough, *Analysing Discourse*, 151–2.
142 Gal, 'Scale-Making: Comparison and Perspective as Ideological Projects', 91.
143 Gal, 91.
144 Blommaert, *Discourse*, 223–4.
145 Blommaert, 222.
146 Kress, 'Against Arbitrariness', 174.
147 Fairclough, *Analysing Discourse*, 45.
148 Fairclough, 143.
149 Chouliaraki and Fairclough, *Discourse in Late Modernity*, 13; Fairclough, *Analysing Discourse*, 142.

150 Green, *Fragments from the History of Loss*, 15.
151 Fairclough, *Discourse and Social Change*, 138; Fairclough, *Analysing Discourse*, 108.
152 Fairclough, *Analysing Discourse*, 167–70.
153 Kress and Van Leeuwen, *Reading Images: The Grammar of Visual Design*, 120.
154 Fairclough, *Analysing Discourse*, 43.
155 Fairclough, 55.
156 van Leeuwen, *Discourse and Practice*, 106–19.
157 Fairclough, *Analysing Discourse*, 98–9.
158 See Martin, *Politics and Rhetoric.*
159 Ledin and Machin, *Doing Visual Analysis*, 47–61.
160 Kress and Leeuwen, 'Colour as a Semiotic Mode'.
161 See, for example, Markus and Cameron, *The Words between the Spaces*; Burnett and Walsh, 'Semiotic Citizenship and the Politics of Shack-Building'; Milani and Levon, 'Sexing Diversity'; Stroud, 'Turbulent Linguistic Landscapes and the Semiotics of Citizenship'.
162 Ledin and Machin, *Doing Visual Analysis*, 94–8.
163 Richardson and Wodak, 'Anti-Sorosism: Reviving the "Jewish World Conspiracy"'; Wodak, *The Politics of Fear (2nd Ed).*
164 Wodak, 'The Discourse-Historical Approach', 74.
165 Krzyżanowski, 'Discursive Shifts and the Normalisation of Racism'.
166 Lazar, 'Feminist Critical Discourse Analysis', 152.
167 Dardot and Laval, *The New Way of the World*, 3.
168 Available at https://x.com/RangaMberi/status/1773593707087831253, last accessed 25 September 2024.
169 Available at https://x.com/radical_carib/status/1774030003610533911, last accessed 25 September 2024.
170 Mautner, 'A Research Note on Corpora and Discourse', 2.
171 Billig, 'Critical Discourse Analysis and the Rhetoric of Critique', 36.
172 Lazar, 'Feminist Critical Discourse Analysis', 143. In highlighting a distinct politics of articulation within CDA, Lazar sought to make explicit what, if left implicit, could create the false impression that all CDA of gender/sexuality-related topics is informed by feminist theoretical precepts and emancipatory ends, which is clearly not the case. Her point is that, because feminist theory is not hardwired into CDA, attention both to its application in method and to research impact makes the analysis more robust.

## Chapter 3

1 The Starbucks Vision 2030 sustainability plan is available at https://www.starbucks.com/responsibility/planet/, last accessed 1 October 2024.

2 The Body Shop's Sustainability Commitments, which include the Refill Scheme, can be found at https://www.thebodyshop.com/en-au/about-us/brand-values/sustainability/sustainability-commitments/a/a00063, last accessed 1 October 2024.

3 The atmosphere of Prohibition-era Ontario is picked up in a number of details of the design of the hotel space, including the presentation of one of the lounges off the lobby, the Library Bar, as a 'famed cocktail den cloaked beyond a secret corner of the illustrious Fairmont Royal York hotel' – https://www.fairmont.com/royal-york-toronto/ last accessed 1 October 2024.

4 van Leeuwen, 'Towards a Semiotics of Typography', 148.

5 Ledin and Machin, *Doing Visual Analysis*, 94.

6 A quick online search suggests that I could buy RFID keycards in plastic for about $0.50 each and in wood for between $1 and $2.

7 Thurlow, 'Queering Critical Discourse Studies', 496.

8 See Deleuze and Guattari, *A Thousand Plateaus: Capitalism and Schizophrenia*, chapter 14.

9 Gough, 'Flows of Capitalism, Flows of Architecture'.

10 Tuana, *Racial Climates, Ecological Indifference.*

11 See Ballard, 'When in Rome'; and Dixon et al., 'Discourse and the Politics of Space in South Africa'.

12 Tweet by @POTUS on 23 January 2023 https://x.com/POTUS/status/1620138541937205250, last accessed 3 October 2024.

13 Laclau and Mouffe, *Hegemony and Socialist Strategy*, 95.

14 Green, *Fragments from the History of Loss*, 23.

15 Seele and Gatti, 'Greenwashing Revisited', 240–1.

16 Ormesher and Greenfield, 'Greenwashing Era Is over, Say Ad Agencies, as Regulators Get Tough'.

17 This is Olúfẹ́mi O.Táíwò's term for the accommodation of social and environmental justice struggles within existing capitalist, racist and colonial power structures. See Táíwò, *Elite Capture.*

18 ESG or 'sustainability' metrics are themselves arguably 'smokescreens' behind which corporations hide unethical practices. See Norman and MacDonald, 'Getting to the Bottom of "Triple Bottom Line"'.

19 Goldstein and Flitter, 'Cracking Down on a Wall Street Trend'. I was alerted to this article by the political scientist Thea Riofrancos's tweet: https://x.com/triofrancos/status/1571871066988990464, last accessed 3 October 2024.

20 These problems are in addition to the conceptual difficulty discourse analysts would face in proving wilful disinformation or deception on the part of the suspect entity, which has been made intrinsic to definitions of greenwashing. Our concern is to trace the socio-political *effects* of discursive acts, to which the evanescent and inaccessible *intentions* of agents are relevant only as far as they are discursively constructed post hoc.

21 Moore, *Capitalism in the Web of Life*, 17.
22 See, for example, Daly, *Beyond Growth*; and Raworth, *Doughnut Economics*. These figures are however still considered peripheral and their adherents positioned as 'renegades'.
23 Trucost, 'Natural Capital at Risk'.
24 Fraser, *Cannibal Capitalism*, 23.
25 Fraser, 23–4.
26 McCarthy and Prudham, 'Neoliberal Nature and the Nature of Neoliberalism', 277.
27 Moore, *Capitalism in the Web of Life*, 63.
28 Fraser, 'Climates of Capital', 111.
29 Burnett, *White Belongings*, Introduction; McCarthy and Prudham, 'Neoliberal Nature and the Nature of Neoliberalism', 277. Locke was heavily invested in slavery, but saw no contradiction in grounding his theory of property in a principle of self-ownership. See Mills, 'Racial Liberalism', 1382. This is the signal move of liberal philosophy: to invoke a universal human subject clearly modelled on and ultimately only intended for white European men.
30 Fraser, 'Climates of Capital', 112.
31 Malm, *Fossil Capital*.
32 Hornborg, 'Footprints in the Cotton Fields', 79.
33 Carrington, 'Revealed: Oil Sector's "Staggering" $3bn-a-Day Profits for Last 50 Years'.
34 Fraser, 'Climates of Capital', 116. See also Bhambra and Holmwood, 'Colonialism, Postcolonialism and the Liberal Welfare State'.
35 Mignolo, 'The Geopolitics of Knowledge and the Colonial Difference'.
36 Fressoz, *More and More and More*.
37 Locke, *Two Treatises of Government*, 307.
38 Bhandar, *Colonial Lives of Property*.
39 Du Bois, *Darkwater: Voices from within the Veil*, 18.
40 See Bhambra and Holmwood, 'Colonialism, Postcolonialism and the Liberal Welfare State'; and Roediger, *The Wages of Whiteness*.
41 Harris, 'Whiteness as Property'.
42 Moreton-Robinson, *The White Possessive*, 113.
43 McCarthy and Prudham, 'Neoliberal Nature and the Nature of Neoliberalism', 277.
44 For a discussion of the intellectual history of the term 'racial capitalism', see Levenson and Paret, 'The Three Dialectics of Racial Capitalism'. Racial capitalism has proved a productive analytic for a variety of contemporary phenomena. Writing about the global financial crisis in the late 2000s, for example, Paula Chakravarty and Denise Ferreira da Silva ask: 'How could the predatory targeting of economically dispossessed communities and the subsequent bailout of the nation's largest investment banks, instantly and volubly, be recast as a problem caused by the racial other ("illegal immigrants" and "state-dependent minorities")?' Chakravartty and da Silva, 'Accumulation, Dispossession, and Debt', 364.

45 Narayan, ‘The Wages of Whiteness in the Absence of Wages’, 2487. In *The History of the Russian Revolution,* Leon Trotsky defined the comprador in relation to the history of Portuguese imperial agents in China, from whence the term ‘comprador’ (‘buyer’) originates. Trotsky writes: ‘The Chinese compradors are the classic type of the national bourgeoisie, a kind of mediating agency between foreign finance capital and the economy of their own country. In the world hierarchy of the powers, Russia occupied before the war a considerably higher position than China. What position she would have occupied after the war, if there had been no revolution, is a different question. But the Russian autocracy on the one hand, the Russian bourgeoisie on the other, contained features of compradorism, ever more and more clearly expressed. They lived and nourished themselves upon their connections with foreign imperialism, served it, and without their support could not have survived. To be sure, they did not survive in the long run even with its support. The semi-comprador Russian bourgeoisie had world imperialistic interests in the same sense in which an agent working on percentages lives by the interests of his employer.’ Trotsky, *The History of the Russian Revolution*.

46 Elamin, ‘“When Mud Comes between Us”’.

47 Green, *Fragments from the History of Loss*, 9.

48 Wolf, ‘Market Forces Are Not Enough to Halt Climate Change’.

49 Vanclay, ‘Principles to Gain a Social Licence to Operate for Green Initiatives and Biodiversity Projects’.

50 Taylor, ‘Just an Energy Transition?’, 9.

51 Murray, Lenzen, and Murray, ‘Forest Carbon – Questions of Indigenous Rights and Market Forces’.

52 Galer, ‘Tree-Planting Schemes Are Just Creating Tree Cemeteries’.

53 ‘The Green Energy Scandal Exposed’.

54 Harlan, ‘How One Company’s Plan to Help the Planet Went off the Rails’. Africa is ground zero for this kind of thing; the former CEO of C-Quest Capital, which is based in Washington, DC, Ken Newcombe. Formerly of Goldman Sachs and the World Bank. Selling millions of carbon credits to international corporations, with little benefit to the continent.

55 ‘The Not-So-Benevolent Billionaire’.

56 See ‘Surviving the Broligarchy’, https://broligarchy.substack.com/, last accessed 7 May 2025.

57 Hultman, ‘Natures of Masculinities’; see also Hultman, ‘The Making of an Environmental Hero’.

58 Kendrick and Nagel, ‘The Cowboy Scientist Saves the Planet’.

59 Brooks and Wingard, ‘Evolutions in Hegemonic Discourses of Climate Change’, 313.

60 Rushkoff, ‘How Tech’s Richest Plan to Save Themselves after the Apocalypse’.

61 Gebru and Torres, ‘The TESCREAL Bundle’. I explore TESCREAL discourses in more detail in the Conclusion.

62 Tuana, *Racial Climates, Ecological Indifference*, 142–3.
63 McCarthy and Prudham, 'Neoliberal Nature and the Nature of Neoliberalism', 277.
64 Huber, '"Green Capitalism" Is a Lie'.
65 Riofrancos, 'The Security–Sustainability Nexus'; Riofrancos, *Resource Radicals*.
66 Ajl, *A People's Green New Deal*, no page.
67 Perry and Sealey-Huggins, 'Racial Capitalism and Climate Justice', 1.
68 Lebdioui, *Survival of the Greenest*.
69 Molek-Kozakowska, 'The Hybrid Discourse of the "European Green Deal"', 196.
70 Pulido, 'Geographies of Race and Ethnicity II'.
71 Heron, 'Capitalist Catastrophism and Eco-Apartheid'.
72 Deberdt and Le Billon, 'Green Transition's Necropolitics'; Franz and McNelly, 'The "Finance-Extraction-Transitions Nexus"'.
73 Mcallister, 'U.S. Climate Envoy Kerry Calls on African Nations to Help Curb Emissions'.
74 Haag, 'Old Colonial Power in New Green Financing Instruments. Approaching Financial Subordination from the Perspective of Racial Capitalism in Renewable Energy Finance in Senegal'.
75 Fischer and Knuth, 'Greenwashing in der afrikanischen Savanne'.
76 See the Survival International petition at https://act.survivalinternational.org/page/123852/action/1, last accessed 10 October 2024.
77 Per capita $CO_2$ emissions in Ghana in 2022 were 0,6t/annum while in Switzerland the figure was 4t/annum Data extracted from Hannah Ritchie, Pablo Rosado and Max Roser (2023) – 'Per capita, national, historical: how do countries compare on CO2 metrics?' https://ourworldindata.org/co2-emissions-metrics, last accessed 10 October 2024.
78 Goodman et al., 'Spectacular Environmentalisms'.
79 Büscher, 'Nature 2.0'; Büscher, Koot, and Nelson, 'Introduction. Nature 2.0'; Büscher, *The Truth about Nature*. At the same time, it is possible to overstate the ideological stranglehold that the media has on subjectivity. Kai Bosworth reads in the sardonic, irony-laden and 'bad environmentalism' of 'nature is healing' memes shared in the context of the coronavirus pandemic the potential for counterhegemonic affective responses to the naturalization of exploitative social orders. See Bosworth, 'The Bad Environmentalism of "Nature Is Healing" Memes'.
80 Agha, 'Meet Mediatization'; Chouliaraki and Fairclough, *Discourse in Late Modernity*, 10.
81 See https://app.wedonthavetime.org/, last accessed 10 October 2024.
82 Video of sales pitch to investors previously accessible at https://www.wedonthavetime.org/invest
83 Buller, *The Value of a Whale*, 5.
84 As Christian Chun observes, part of the common-sense of capitalism is that it is the only system that 'works'. Chun, *The Discourses of Capitalism*, 79–88.

85 Buller, *The Value of a Whale*, 13.
86 Buller, 8.
87 Méndez, '"The River Told Me"'.
88 Park, Liu, and Kaye, 'Analysis of Ageism, Sexism, and Ableism in User Comments on YouTube Videos about Climate Activist Greta Thunberg'.
89 Niceforo, 'Hate Speech and Environmental Activist Discourse', 15.
90 'The Not-So-Benevolent Billionaire'.
91 Ideland, *The Eco-Certified Child*.
92 Ideland, *Den KRAV-märkta människan*, 72.
93 Burnett, '"Without White People, the Animals Will Go!"'
94 Walenta, 'The Making of the Corporate Carbon Footprint'.
95 Wittmer, 'Dirty Work in the Clean City'.
96 Ledin and Machin, *Doing Visual Analysis*, 46.
97 In a fascinating study, the Holocaust historian Anne Berg has shown how Nazi fastidiousness about recycling started in the early 1930s. Their vast and expensive system of transporting waste around the Reich endured well into the war, when what was collected could not be processed. Goebbels apparently prized the practice as allowing all Germans to feel as if they were part of the war effort, and thus an effective propaganda tool for maintaining morale. See Berg, *Empire of Rags and Bones*.
98 Collard and Dempsey, 'Future Eco-Perfect', 1550.
99 Foucault, 'Of Other Spaces', 24. For a compelling deployment of heterotopias in the CDS tradition, see Milani and Levon, 'Israel as Homotopia'.
100 Alkon and McCullen, 'Whiteness and Farmers Markets'; Ballard and Jones, 'Natural Neighbors'; Hickcox, 'Green Belt, White City'; Quinton, Nesbitt, and Sax, 'How Well Do We Know Green Gentrification?'; Raidoo, 'The Unruly in the Anodyne: Nature in Gated Communities'.
101 I make this argument in greater detail in Burnett, '"Without White People, the Animals Will Go!"'
102 Schwartz, 'Two Climate Activists, 40 Years Apart in Age, on the Movement's Future'.
103 Mapping on to Tzvetan Todorov's classic equilibrium theory of narrative. See Todorov, 'The 2 Principles of Narrative'.
104 Dobruszkes, Mattioli, and Gozzoli, 'The Elephant in the Room', 2.
105 Gössling and Humpe, 'The Global Scale, Distribution and Growth of Aviation'.
106 See especially Brown, *In the Ruins of Neoliberalism*; and Dardot and Laval, *The New Way of the World*.
107 Mouffe, *For a Left Populism*, 4.
108 Wilton, 'Colouring Special Needs'. NIMBY is an acronym that stands for 'not in my backyard!'.
109 Graeber, *Debt*; Hickel, *The Divide*.
110 Parenti, *Against Empire*, 97.

111 Saito, *Marx in the Anthropocene.*
112 Oswald et al., 'Global Redistribution of Income and Household Energy Footprints'.
113 Hickel, 'The Anti-Colonial Politics of Degrowth'.
114 More popular accounts, see Schmelzer, Vetter, and Vansintjan, *The Future Is Degrowth*; Hickel, Klu, and Read, *Less Is More.*
115 Huber, 'Reflections on Climate, Class, and Strategy'; Huber, *Climate Change as Class War.*
116 Levien, 'White Energy Workers of the North, Unite?'
117 Ajl, *A People's Green New Deal.*
118 The Red Nation, *The Red Deal.*
119 Heron, 'The Great Unfettering'.

## Chapter 4

1 *Borrell at Launch of European Diplomatic Academy.*
2 Liboreiro, 'Borrell Apologises for Controversial Metaphor but Defends Speech'.
3 Borrell, 'On Metaphors and Geo-Politics'.
4 Borrell.
5 Borrell.
6 Koller, 'Cognitive Linguistics and Ideology', 241.
7 Kipling's Jungle Law is not however the dog-eat-dog world Borrell envisages; his relative subtlety and sympathy in constructing his characters seems however to have been lost in later definitions of the 'law of the jungle'. See Hotchkiss, 'The Jungle of Eden: Kipling, Wolf Boys, and the Colonial Imagination'.
8 'Jungle'.
9 Frenkel, 'Geography, Empire, and Environmental Determinism'.
10 de Montesquieu, *The Spirit of the Laws*, 196. I found this quote in a journal article, published in 2020 by political scientists at the University of Groningen in the Netherlands, who treat this quote as an empirically testable hypothesis rather than a fragment of racist ideology, thereby perpetuating the 'scientific' basis for racism. See Oumer, Maseland, and Garretsen, 'Was de Montesquieu (Only Half) Right?'. In her *Stanford Encyclopedia of Philosophy* entry on Montesquieu, Hilary Bok argues that he is not strictly speaking a determinist, as he believes that climate creates cultural tendencies and preconditions that the law should take into account. See Bok, 'Baron de Montesquieu, Charles-Louis de Secondat'.
11 Byron, *Lord Byron: The Major Works.*
12 Neil and Cerruti, 'South(s) of the South(s)'.
13 J.M. Coetzee has analysed how this idea plays out in accounts of the Indigenous Khoe people at the Cape written by early colonists. See Coetzee, *White Writing.*
14 For the classic account of racial liberalism see Mills, 'Racial Liberalism'.

15 Fanon, *Black Skin, White Masks*, 118.
16 Dictionary.com, 'LAW OF THE JUNGLE', accessed at https://www.dictionary.com/browse/law-of-the-jungle on 15 October 2024.
17 Borrell, 'On Metaphors and Geo-Politics'.
18 *Borrell at Launch of European Diplomatic Academy.*
19 Taylor, 'Metaphors of Migration over Time'.
20 Nguyen and McCallum, 'Drowning in Our Own Home'.
21 Crenshaw, 'Mapping the Margins', 1252.
22 Murray, 'The Idea of Gardening: Plants, Bewilderment, and Indigenous Identity in South Africa'. In Jonathan Glazer's 2023 film *The Zone of Interest* Hedwig Höss, the wife of Auschwitz commandant Rudolf Höss, is depicted as having planted an elaborately pretty garden against the wall of the concentration camp through which screams and gunshots can frequently be heard. The garden comes to stand in for the family's utter indifference to the most depraved forms of human suffering.
23 'Colony'.
24 *Borrell at Launch of European Diplomatic Academy.*
25 Sachseder, Stachowitsch, and Binder, 'Gender, Race, and Crisis-Driven Institutional Growth'.
26 Barkham, 'The Real David Attenborough'.
27 Gray, 'David Attenborough – Humans Are Plague on Earth'.
28 Bhattacharjee et al., 'Global Fertility in 204 Countries and Territories, 1950–2021, with Forecasts to 2100'. For the United Nations definition of replacement-level fertility, see https://www.un.org/esa/sustdev/natlinfo/indicators/methodology_sheets/demographics/total_fertility_rate.pdf last accessed 22 October 2024.
29 Soloway, *Demography and Degeneration.*
30 Bhattacharjee et al., 'Global Fertility in 204 Countries and Territories, 1950–2021, with Forecasts to 2100'.
31 Weldemichel, 'Inventing Hell'. For updated information on food security in Ethiopia, see https://fews.net/east-africa/ethiopia, last accessed 2 December 2024.
32 Tran, 'David Attenborough: Trying to Tackle Famine with Bags of Flour Is "Barmy"'.
33 O'Boyle, 'Classical Economics and the Great Irish Famine', 23.
34 O'Boyle, 24.
35 Tran, 'David Attenborough: Trying to Tackle Famine with Bags of Flour Is "Barmy"'.
36 *Securing a Sustainable Future for the Amazon.*
37 Lentin, *Why Race Still Matters*, chapter 2.
38 Amoah, 'Reflections'.
39 Farage, 'Britain's Population Surge'.
40 Farage.
41 Burnett and Richardson, '"Breeders for Race and Nation"', 351.

42 Farage, 'Britain's Population Surge'.
43 Baldwin, 'Premediation and White Affect', 78.
44 Baldwin, 81.
45 Baldwin, 81.
46 Baldwin, 82.
47 Gosine, 'Dying Planet, Deadly People', 72.
48 Gergan, Smith, and Vasudevan, 'Earth beyond Repair'; see also Whyte, 'Indigenous Science (Fiction) for the Anthropocene'.
49 Gergan, Smith, and Vasudevan, 'Earth beyond Repair', 92.
50 Gergan, Smith, and Vasudevan, 102.
51 Watts, *Postracial Fantasies and Zombies*. In the video-game-turned-television-series *The Last of Us* (HBO, 2023) for example, a global pandemic caused by a *Cordyceps* fungus turns swathes of humanity into hollowed-out zombie-like parts of interconnected mycelial structures. Groups of human survivors are pitted both against each other, and against a shadowy fascist state which eliminates anyone it suspects of being infected. Much of the drama of this genre is staged in moments where humans must kill their former friends and comrades as they become infected, and thus threats to the survival of others.
52 Puwar, *Space Invaders*.
53 Schultz, 'Demographic Futurity'.
54 Baldwin, 'Whiteness and Futurity', 181.
55 Malthus, *An Essay on the Principle of Population*, 13.
56 McNally, 'Political Economy to the Fore: Burke, Malthus and the Whig Response to Popular Radicalism in the Age of the French Revolution'.
57 McNally, 437–8.
58 McNally, 436.
59 Ehrlich, *The Population Bomb*.
60 See https://www.heinzawards.org/pages/paul-anne-ehrlich last accessed 24 October 2024.
61 Hardin, 'The Tragedy of the Commons'.
62 Mildenberger, 'The Tragedy of "The Tragedy of the Commons"'.
63 For example, Reynolds, Blackmore and Smith, *Environmental Responsibility Reader*.
64 Hardin, 'The Tragedy of the Commons', 1248.
65 Hardin, 1244.
66 Tuana, *Racial Climates, Ecological Indifference*, 134–5.
67 Rao, 'An Imagined Reality'.
68 As shown extensively in Ostrom, *Governing the Commons*.
69 Hartmann, 'Population Control in the New World Order'.
70 Kaplan, 'The Coming Anarchy'.
71 Dalby, 'The Environment as Geopolitical Threat', 474.
72 Homer-Dixon, *Environment, Scarcity, and Violence*, 18.

73 Peluso and Watts, 'Violent Environments', 3.
74 Dalby, 'The Environment as Geopolitical Threat', 483. Peluso and Watts make a similar point, arguing that the very idea of environmental security assumes 'an ineluctable connection between environmental degradation, population growth, alleged resource scarcity, and the proliferation of "small wars"'. Peluso and Watts, 'Violent Environments', 7.
75 Livingstone, 'The Climate of War'.
76 Austin et al., 'Critical Perspectives on the Study of Climate Security'.
77 Selby, Daoust, and Hoffmann, *Divided Environments*, 6.
78 Hartmann, 'Rethinking Climate Refugees and Climate Conflict'.
79 Duffield, 'Total War as Environmental Terror'.
80 Bettini, 'Climate Barbarians at the Gate?'
81 Tuana, 'Populationism and Climate Change: Controlling Images, Ecointersectional Alternatives', no page.
82 This phenomenon has been described by Sophia Siddiqui within European contexts where pro-natal initiatives target white 'native' women while people marked as 'demographic threats' face mounting obstacles in accessing sexual health and rights. See Siddiqui, 'Racing the Nation'.
83 Fletcher, Breitling, and Puleo, 'Barbarian Hordes', 1196.
84 Buckley, 'Earth Now Has 8 Billion Humans. This Man Wishes There Were None'.
85 Millward-Hopkins, 'Inequality Can Double the Energy Required to Secure Universal Decent Living'.
86 Kamara, 'Reframing Population'.
87 Martin, 'The West, Natural Resources and Population Control Policies in Africa in Historical Perspective'.
88 Lomborg, *The Skeptical Environmentalist*, 49.
89 Appropriately enough, the US Defense Advanced Research Projects Agency (DARPA) programme to convert plastics to food, which some astrobiologists are convinced might turn mined materials from asteroids into sustenance, is called 'cornucopia'. See Pilles, Nicklin, and Pearce, 'How We Can Mine Asteroids for Space Food'.
90 For a recent rehearsal of these ideas, see Klein and Thompson, *Abundance*.
91 Torres, 'The New Optimist Movement Has a Numbers Problem'. The figures Torres critiques here are the celebrity Canadian psycholinguist Steven Pinker and Max Roser, who runs the website 'Our World in Data'. See also Torres, 'Do We Live in Hell?'.
92 Rajaram, 'Refugees as Surplus Population'.
93 Billig, 'Rhetorical Uses of Precise Numbers and Semi-Magical Round Numbers'.
94 To return to Josep Borrell's speech at the launch of the European Diplomatic Academy.
95 Anderson, 'Servants and Slaves: Europe's Domestic Workers'.
96 Dalby, 'The Environment as Geopolitical Threat', 477.

97 Leeuwen, 'The Visual Representation of Social Actors'.

98 Fanon, *Black Skin, White Masks*, 43. For an application of the concept of the imago in critical discourse studies, see Burnett and Milani, 'Fatal Masculinities'.

99 Oreskes, 'Eight Billion People in the World Is a Crisis, Not an Achievement'.

100 Vidal, 'It Should Not Be Controversial to Say a Population of 8 Billion Will Have a Grave Impact on the Climate'.

101 If a bottom-decile poor person consumes *x* units of energy, then a top decile person consumes 20*x*. If the rich person cuts their consumption in half, freeing up 10*x*, then there is enough 'extra' energy for the members of the bottom decile to increase up to tenfold before the total energy used increases.

102 This shallow instrumentalization is typical of the discourse of femonationalism, which Sara Farris has identified as a key theme in right-wing politics in Europe and elsewhere. However, as Simon Bauer and colleagues have argued in the Swedish context, femonationalist discourses are pervasive outside of right-wing politics in the mundane practice of civic integration classes in social democratic welfare states. See Farris, *In the Name of Women's Rights*; Bauer et al., 'Gender Equality in the Name of the State'.

103 Dickie, 'Climate Justice Gets Harder as World Population Passes 8 Billion'.

104 See Táíwò, *Reconsidering Reparations.*

105 Norval, *Deconstructing Apartheid Discourse*, 106–9.

106 Steyn, *Whiteness Just Isn't What It Used to Be*, xxvi–xxvii.

107 Steyn, 162–4.

108 Dickie, 'Climate Justice Gets Harder as World Population Passes 8 Billion'.

109 Davies, '"Use Your £11bn Climate Fund to Pay for Family Planning," UK Told'.

110 For the classic treatment of the racial dynamics of representation, see Alcoff, 'The Problem of Speaking for Others'.

111 Yamada et al., 'Sustainability as a Moral Discourse'.

112 Oyěwùmí, 'The White Woman's Burden: African Women in Western Feminist Discourse'.

113 Chun, *The Discourses of Capitalism*, 143.

114 Tilley and Ajl, 'Eco-Socialism Will Be Anti-Eugenic or It Will Be Nothing'.

## Chapter 5

1 Brockington, *Fortress Conservation*; Brockington, Duffy, and Igoe, *Nature Unbound*; Büscher and Fletcher, *The Conservation Revolution*; Duffy, 'Waging a War to Save Biodiversity'; Fletcher, *Failing Forward.*

2 See https://unitedforwildlife.org/about-us/, last accessed 9 January 2025.

3 This is my transcription of William's speech, which was posted by Sky News on Twitter. See https://x.com/SkyNews/status/1853848235384656272, last accessed 9

January 2025. For excerpts of his speech, see the United for Wildlife press release at https://unitedforwildlife.org/news/game-changing-package-of-support-for-rangers-and-nature-protectors-across-africa/, last accessed 9 January 2025.

4 Chiapello and Fairclough, 'Understanding the New Management Ideology', 207.

5 Management discourse has its origins in military discourse. See Cummings, 'Shifting Foundations'.

6 Machin and Mayr, *How to Do Critical Discourse Analysis*, 40; see Dijk, 'Ideological Discourse Analysis'.

7 Galliers et al., 'Conservation Casualties'.

8 Mitchell et al., '30×30: A Guide to Inclusive, Equitable and Effective Implementation of Target 3 of the Kunming-Montreal Global Biodiversity Framework'.

9 See the United for Wildlife press release in note 3 above.

10 Sène, 'Against Wildlife Republics'.

11 Curtin, *The Image of Africa*.

12 Adams and McShane, *The Myth of Wild Africa*.

13 Neumann, *Imposing Wilderness*, 18.

14 Cloete, 'Women, Home and Personal Narrative in Kenya', 84.

15 A US reader of this manuscript commented that they doubted that whites feel any real resonance with Africa as a home. As the scholarship I build on in this section attests, however, this idea does seem to be quite prevalent in settler colonial contexts on the continent (such as South Africa, Kenya and Namibia) and their historical colonial metropoles. I am thus intentional here in naming Europe and not the 'West'.

16 I make this argument at greater length in Burnett, 'Constructing White Autochthony in South Africa's "Soul Country"'; and Burnett, *White Belongings*.

17 Neumann, *Imposing Wilderness*, 1.

18 Neumann, 15.

19 The distinction was introduced in Frykman and Löfgren, *Culture Builders*, 51.

20 Neumann, *Imposing Wilderness*, 24.

21 Carruthers, *The Kruger National Park*; Neumann, *Imposing Wilderness*, 32.

22 Neumann, *Imposing Wilderness*, 50.

23 Green, *Fragments from the History of Loss*, 126.

24 Green, 128.

25 Green, 137.

26 Burnett, *White Belongings*, chapter 2.

27 Green, *Fragments from the History of Loss*, 130–1. This idea will be explored in more depth in the next chapter in the discussion of the trope of the 'ecological Indian'.

28 Hughes, 'Third Nature', 160–1.

29 Neumann, 'Moral and Discursive Geographies in the War for Biodiversity in Africa'.

30 Burnett, '"Without White People, the Animals Will Go!"'
31 Green, *Fragments from the History of Loss*, 130.
32 Hughes, 'Third Nature', 161.
33 See Gressier, *At Home in the Okavango* for a case study of Botswana; Hughes, *Whiteness in Zimbabwe*; and Suzuki, *The Nature of Whiteness* for Zimbabwe; and McIntosh, 'Land, Belonging and Structural Oblivion among Contemporary White Kenyans' for Kenya.
34 Guha, 'Radical American Environmentalism and Wilderness Preservation'.
35 See Lane and Clark, 'The Solitary Walker in the Political World' for an excellent account of the philosophical issues at stake.
36 Dwyer and Jones III, 'White Socio-Spatial Epistemology', Hanes, *White Man's Game*.
37 Ratcliffe, 'Prince William Accused of "White Saviour" Mentality in Africa Wildlife Film'.
38 Staff Reporter, 'Prince Harry Visits Rwanda on a Wildlife Conservation Trip'.
39 Abidin et al., 'The Tropes of Celebrity Environmentalism'; Koot, 'Enjoying Extinction'.
40 Laville, 'Twelve Billionaires' Climate Emissions Outpollute 2.1m Homes, Analysis Finds'; Mancini, 'It Costs $25 Million a Year to Keep Jeff Bezos' $500 Million Luxury Superyacht Afloat – But He Earns That in Just 3 Hours'.
41 Koot, 'Enjoying Extinction'.
42 Mohamed, 'No Thanks, KP, the Proteas Don't Need You'.
43 Burnett and Milani, 'Fatal Masculinities' offers an account of the scale of the crisis at the time.
44 'Kevin Pietersen: Beast of Man'.
45 'Umganu Lodge' https://umganulodge.com/. Accessed 14 January 2025.
46 In the first episode of the podcast the title is explained as relating to a 'famous saying' recited by Pietersen's tracker friend Anton Lategan at the fireside one evening: 'What is man without the beasts? For if all the beasts were gone, man would die of a great loneliness of spirit. For whatever happens to the beast will also happen to man.' This is a slightly paraphrased version of a statement attributed to the Native American Chief Seattle, leader of the Suquamish and Duwamish nations, referring to the colonial slaughter of North American wildlife.
47 Büscher and Ramutsindela, 'Green Violence'.
48 Büscher and Ramutsindela, 12.
49 Neumann, 'Moral and Discursive Geographies in the War for Biodiversity in Africa'.
50 Lunstrum, 'Feed Them to the Lions'. As Tiza Mfuni pointed out in his comments to me on this chapter, the discursive construction of killable people cleaves not only to white-Black racial difference, but also within African societies along urban/rural and class divides.

51 Büscher, 'Rhino Poaching Is out of Control!', 992.
52 Burnett, '"Without White People, the Animals Will Go!"'
53 Lindsey et al., 'Conserving Africa's Wildlife and Wildlands through the COVID-19 Crisis and Beyond', 1302.
54 Lindsey et al., 'Shepherding Sub-Saharan Africa's Wildlife through Peak Anthropogenic Pressure toward a Green Anthropocene'.
55 Hughes et al., 'Smaller Human Populations Are Neither a Necessary nor Sufficient Condition for Biodiversity Conservation', 2.
56 Neumann, *Imposing Wilderness*, 37.
57 Lindsey et al., 'Conserving Africa's Wildlife and Wildlands through the COVID-19 Crisis and Beyond'.
58 Dr Dilys Roe, chair of the International Union for the Conservation of Nature's sustainable use and livelihoods specialist group, thus stated that 'hunting can mitigate the threat of poaching.' Source: https://x.com/FACEforHunters/status/1577344333262602254. As Frowin Becker observes in the Namibian context, trophy hunting relies on deeply racist and colonial social relations and landholding patterns. Becker, 'Trophy's Choice'.
59 Brockington, *Fortress Conservation*.
60 Lunstrum, 'Green Militarization'.
61 Duffy, 'Waging a War to Save Biodiversity'; Hughes, 'Third Nature'; Humphreys and Smith, 'The "Rhinofication" of South African Security'; Lenggenhager, 'Nature, War and Development'; Ramutsindela, Matose, and Mushonga, *The Violence of Conservation in Africa*.
62 Neumann, *Imposing Wilderness*, 6.
63 Duffy and Brockington, 'Political Ecology of Security', 32.
64 Abrams, '"We Just Saw the Fence"'.
65 Burnett and Milani, 'Fatal Masculinities'.
66 Thakholi, 'Conservation Labour Geographies'.
67 Massé, Givá, and Lunstrum, 'A Feminist Political Ecology of Wildlife Crime'.
68 Koot and Veenenbos, 'The Spectacle of Inclusive Female Anti-Poaching'.
69 Massé, 'Anti-Poaching's Politics of (In)Visibility'.
70 Bosworth, 'The Bad Environmentalism of "Nature Is Healing" Memes'.
71 Lindsey et al., 'Conserving Africa's Wildlife and Wildlands through the COVID-19 Crisis and Beyond'.
72 Lindsey et al., 1306.
73 Büscher and Fletcher, *The Conservation Revolution*, 20.
74 See Apostolopoulou et al., 'Reviewing 15 Years of Research on Neoliberal Conservation' for a comprehensive review of this term.
75 Conservation International, for example, advocates for the 3 per cent of global climate funding currently spent on biodiversity to be increased dramatically. See

https://www.conservation.org/priorities/stabilizing-our-climate-by-protecting-and-restoring-nature, last accessed 17 January 2025.

76 Fisher, *Capitalist Realism*.

77 Fraser, *Cannibal Capitalism*; Fressoz, *More and More and More*.

78 Fletcher, *Failing Forward*.

79 Fletcher, 11.

80 The boards of the Big 4 conservation NGOs have now been so stocked with people with links to the financial industry that they new form a majority. See: https://africanarguments.org/2023/08/revealed-big-conservation-ngos-majority-governed-finance-africa-carbon-markets/.

81 Fletcher, *Failing Forward*, 9.

82 Developed in Procházka and Blommaert, 'Ergoic Framing in New Right Online Groups'.

83 Igoe, Neves, and Brockington, 'A Spectacular Eco-Tour around the Historic Bloc'.

84 Brockington and Scholfield, 'The Conservationist Mode of Production and Conservation NGOs in Sub-Saharan Africa'; Büscher and Fletcher, 'Accumulation by Conservation'.

85 Thakholi, 'Conservation Labour Geographies'.

86 Alasow, 'Capitalism and Private Nature Reserves'.

87 Büscher, Koot, and Thakholi, 'Fossilized Conservation'; see also Cocks, 'The Pleasures of Degeneration'.

88 Tweet by @LongoFiore on 14 December 2022, https://x.com/LongoFiore/status/1603029491328106496 last accessed 20 January 2025. Transcription of the video by the author.

89 Locke, 'Nature Needs (at Least) Half'.

90 Dinerstein et al., 'An Ecoregion-Based Approach to Protecting Half the Terrestrial Realm'. See also the interactive of ecoregions available at https://ecoregions.appspot.com/, last accessed 20 January 2024.

91 Wilson, *Half-Earth: Our Planet's Fight for Life*.

92 Wilson, chapter 9 (e-book version).

93 Bontempi et al., 'Conflict and Conservation'.

94 Massarella et al., 'Transformation beyond Conservation'.

95 Thing, Jones, and Jones, 'The Politics of Conservation'.

96 Mitchell et al., '30×30: A Guide to Inclusive, Equitable and Effective Implementation of Target 3 of the Kunming-Montreal Global Biodiversity Framework'.

97 Tweet by @VMatiru, 5 October 2022, https://x.com/VMatiru/status/1577670899322523655, last accessed 20 January 2025.

98 Büscher and Fletcher, *The Conservation Revolution*, 117–19.

99 See the tweet shared by Fiore Longo (@LongoFiore) on 2 March 2022, https://x.com/LongoFiore/status/1498984291178532864, last accessed 17 January 2025. See also Survival International, 'The Big Green Lie' https://www.survivalinternational.org/campaigns/biggreenlie last accessed 17 January 2025.

100 No Deal for Nature, https://nodealfornature.wixsite.com/info, last accessed 20 January 2025.

101 Asia Indigenous Peoples Pact (AIPP) et al., 'Reconciling Conservation and Global Biodiversity Goals with Community Land Rights in Asia', 8.

102 Amnesty International, 'Amnesty International Recommendations to States Ahead of COP 15 (Convention On Biological Diversity) 30 X 30 Proposal'.

103 Ogada, 'Marseille 2021'.

104 Ogada, 'Stealthy, Surreptitious Second Coming of Western Colonialism to Africa'.

105 Survival International, 'A defeat and some successes with a bitter after taste', December 2022. https://www.survivalinternational.org/articles/30-per-cent-statement, last accessed 17 January 2025.

106 Büscher and Duffy, 'Biodiversity Treaty'.

107 Spence, *Dispossessing the Wilderness*, 23; See also DeLuca and Demo, 'Imagining Nature and Erasing Class and Race'. Muir also makes an appearance in a useful graphic representation of fortress conservation, by Gord Hill published in the non-profit news outlet Grist. Hill, 'How Protecting the Earth Became an Excuse for Murder'.

108 Spence, *Dispossessing the Wilderness*, 86.

109 Beinart and Coates, *Environment and History*; Carruthers, *The Kruger National Park*.

110 Koot, Büscher, and Thakholi, 'The New Green Apartheid?'

111 Murdock, 'Conserving Dispossession? A Genealogical Account of the Colonial Roots of Western Conservation'.

112 Agrawal and Redford, 'Conservation and Displacement'.

113 Bluwstein, Cavanagh, and Fletcher, 'Securing Conservation *Lebensraum*?'

114 Corry, 'Ngorongoro Nazi'.

115 To the extent that his own foundation published a statement distancing them from Wilson's views after his death. See https://eowilsonfoundation.org/eow/a-statement-on-e-o-wilson-and-the-j-philippe-rushton-correspondence/, last accessed 20 January 2024.

116 Sène, 'Against Wildlife Republics'.

117 Ndesanjo, 'Is Ngorongoro's Multiple Land-Use Model a Scam?'

118 Singo, 'When Maasaiphobia Became Policy'.

119 Mittal, 'Justice Denied'.

120 Chanzo Reporter, 'Tanzania Delists All Wards and Villages in the Contested Ngorongoro Area'.

121 McCrummen, 'This Will Finish Us'.

122 Holterman, 'Stopping the Slaughter'. The combination of conservation with mining concessions is unfortunately not an isolated phenomenon. See Widengård, 'The Mining-Conservation Nexus in Jamaica's Cockpit Country' for a similar case in Jamaica.

123 Neumann, *Imposing Wilderness*, 8.
124 Pouchet, 'Participation, Inequality, and Conversations about Conservation'.
125 Neumann, *Imposing Wilderness*, 40.
126 Bersaglio and Cleaver, 'Green Grab by Bricolage'.
127 Fox, 'The 2017 Shooting of Kuki Gallmann and the Politics of Conservation in Northern Kenya'.
128 German, Unks, and King, 'Green Appropriations through Shifting Contours of Authority and Property on a Pastoralist Commons'.
129 Pas, Watson, and Butt, 'Land Tenure Transformation'.
130 Mittal, Moloo, and Mousseau, 'Stealth Game'.
131 Mkutu Agade, '"We Will Not Watch Like Monkeys"', 203.
132 Hoeven, 'Guns and Conservation', 9.
133 Schetter, Mkutu Agade, and Müller-Koné, 'Frontier NGOs', 2.
134 Garnett et al., 'A Spatial Overview of the Global Importance of Indigenous Lands for Conservation'.
135 Neumann, *Imposing Wilderness*, 9.
136 Blanc, *The Invention of Green Colonialism*, 18.
137 Sène, 'Land Grabs and Conservation Propaganda'.
138 Gilmore, 'Fatal Couplings of Power and Difference'.
139 See Burnett, *White Belongings*, 45–6.
140 Kalof, Dietz, and Guagnano, 'Race, Gender and Environmentalism'; Kepe, 'Shaped by Race'.
141 Hughes, *Whiteness in Zimbabwe*, 36–7; Burnett, *White Belongings*, 77–9.
142 Connell, *Southern Theory*, 47.
143 Bluwstein, Cavanagh, and Fletcher, 'Securing Conservation *Lebensraum*?'
144 Green, *Fragments from the History of Loss*, 128.
145 Gettleman, 'Loss of Fertile Land Fuels "Looming Crisis" across Africa'.
146 The late Kenyan writer Binyavanga Wainaina satirically advised people writing about Africa to treat the continent 'as if it were one country'. See Wainaina, 'How to Write about Africa'.
147 Gettleman, 'Loss of Fertile Land Fuels "Looming Crisis" across Africa'.
148 Goyanes, 'This All-Female Anti-Poaching Unit in South Africa Is Saving Rhinos and Wildlife'.
149 Koot and Veenenbos, 'The Spectacle of Inclusive Female Anti-Poaching'.
150 Vettese and Pendergrass, *Half-Earth Socialism*.
151 Sène, 'Justice in Nature Conservation', 844.
152 Büscher and Fletcher, 'Towards Convivial Conservation'; Büscher and Fletcher, *The Conservation Revolution*.
153 Massarella et al., 'Transformation beyond Conservation'; Massarella et al., *Convivial Conservation: From Principles to Practice*.

# Chapter 6

1 Said, *Culture and Imperialism*, xiii.

2 Grand Calais Tourism and Culture, 'The Eastern Coast of Calais: Horizon XXL!' https://www.calaisxxl.com/en/inspirations/a-lest-de-calais-lhorizon-xxl/. Page last accessed 23 January 2025.

3 Hansen and Machin, 'Visually Branding the Environment', 792.

4 Eden 62, 'Nos missions', https://eden62.fr/qui-sommes-nous/nos-missions-2/. Page last accessed 23 January 2025.

5 Davies et al., 'Eco-Coloniality', 779.

6 As Lilie Chouliaraki and colleagues show, the 'crisis' as constructed in European media had the effect of constructing a migrant Other. While reportage was initially sympathetic, this changed in the wake of the November terror attacks in Paris. The voices of migrants themselves were seldom heard in the news, and connections were not drawn between conflicts and social distress in their home countries and their decision to migrate. See Chouliaraki et al., 'The European "Migration Crisis" and the Media: A Cross-European Press Content Analysis'.

7 Tyerman, 'Everyday Borders in Calais', 465.

8 Davies et al., 'Eco-Coloniality', 789.

9 Brito, '(Dis)Possessive Borders, (Dis)Possessed Bodies', 3; also cited in Davies et al., 'Eco-Coloniality', 779.

10 Davies et al., 'Eco-Coloniality', 783.

11 Davies et al., 786–9.

12 This is a quote from Jordan Bardella, a far-right French politician cited in Turner and Bailey, '"Ecobordering"', 112.

13 Turner and Bailey, 112.

14 Hultgren, 'Natural Exceptions to Green Sovereignty?'; see also Hultgren, *Border Walls Gone Green*.

15 The second half of this quote was translated slightly differently in Tyerman, 'Everyday Borders in Calais', 472, where I found the link to the report. In the original French, Le Pen says 'Il n'y a plus de loi qui vaille à Calais, plus d'État de droit, il n'y a plus que la jungle, la violence et la loi du plus fort. Bref, l'inverse de la République […] L'immigration est un véritable poison, dont Calais est la conséquence la plus visible et la plus choquante. Les clandestins sont des gens en situation irrégulière, des délinquants qui n'ont rien à faire sur notre territoire national'. The text is available at 'Marine Le Pen à Calais: « Il faut renvoyer les clandestins chez eux » (VIDÉO)' https://www.lavoixdunord.fr/art/region/marine-le-pen-a-calais-il-faut-renvoyer-les-ia33b0n2454355, last accessed 23 January 2025. The English text I present here combines the Google Translate (French to English) translation with an edited version of Tyerman's translation.

16 Krzyżanowski, 'Discursive Shifts and the Normalisation of Racism'. See also Mondon and Winter, *Reactionary Democracy*.
17 Milman, 'Climate Denial Is Waning on the Right. What's Replacing It Might Be Just as Scary'; also cited in Turner and Bailey, '"Ecobordering"', 120.
18 See Richardson, 'Racial Populism in British Fascist Discourse'; and Stoegner and Wodak, '"The Man Who Hated Britain" – the Discursive Construction of "National Unity" in the Daily Mail'.
19 Tyerman, 'Everyday Borders in Calais', 471.
20 Post on 17 August 2015, https://www.facebook.com/groups/chaseshelloutofthekaroo/permalink/980299551993609 last accessed 25 January 2025.
21 Raper, *New Dictionary of South African Place Names*, 120.
22 Adhikari, *The Anatomy of a South African Genocide*. The term 'genocidal moment' was first used in Moses, 'An Antipodean Genocide?'. See also Bam-Hutchison, 'Epistemicide in the Cape'.
23 Through forced sale or stock theft. See Laband, *The Land Wars*.
24 See Keegan, *Colonial South Africa and the Origins of the Racial Order*.
25 This is an argument I develop at length in Burnett, *White Belongings*. For analysis of a contemporary land justice movement in Graaff-Reinet, see Burnett and Walsh, 'Semiotic Citizenship and the Politics of Shack-Building'.
26 Burnett, 'Constructing White Autochthony in South Africa's "Soul Country"'.
27 Foster, *Washed with Sun*, 1–2.
28 Coetzee, *White Writing*, 5.
29 Coetzee, 10.
30 This is an argument I make at length in Burnett, 'Adamastor Unbound? Whiteness and Landscape in Post-1994 South Africa'.
31 Barren expanses and uniform rows of shoebox houses in Black townships on the urban periphery meanwhile reflected to their occupants that they were but sojourners whose 'real' homes were back in the Native reserves. Foster, 'From Socio-Nature to Spectral Presence', 192–7. Despite the best efforts of apartheid planners, however, the townships did become 'home'. See Dlamini, *Native Nostalgia*, 151–2.
32 Ballard and Jones, 'Natural Neighbors', 138. See also Warren, 'Nativeness and Nationhood'.
33 Cane, *Civilising Grass*.
34 Catriona Elder identifies a similar anxiety in Australian film, where settler belonging is metaphorized in indigenous and exotic flora and fauna arranged in moral hierarchies. See Elder, 'Narratives of National Belonging'.
35 Raidoo, 'The Unruly in the Anodyne: Nature in Gated Communities'.
36 Westhuizen, 'Afrikaners in Post-Apartheid South Africa'.

37 Smith, *Myths and Memories of the Nation*, 150. Anthony D. Smith's use of the word 'ethnoscape' differs from that of Arjun Appadurai, who defines it as 'the landscape of persons who constitute the shifting world in which we live: tourists, immigrants, refugees, guest workers and other moving groups and individuals'. Appadurai, *Modernity at Large*, 33.
38 Smith, *Myths and Memories of the Nation*, 69.
39 Geertsema, 'White Natives?', 108.
40 Burnett, 'Constructing White Autochthony in South Africa's "Soul Country"'.
41 This video is available on the YouTube channel Reelblack One, as 'Keep America Beautiful: The Crying Indian (1970)', https://www.youtube.com/watch?v=h0sxwGlTLWw, last accessed 25 January 2025.
42 Dunaway, 'The "Crying Indian" Ad That Fooled the Environmental Movement'.
43 Krech, 'Reflections on Conservation, Sustainability, and Environmentalism in Indigenous North America', 78; see also Krech, *The Ecological Indian*.
44 Krech, *The Ecological Indian*; Krech, 'Reflections on Conservation, Sustainability, and Environmentalism in Indigenous North America'.
45 Deloria, *Red Earth, White Lies*, 93.
46 Garrard, *Ecocriticism*, 162.
47 Reid, *Native American Racism in the Age of Donald Trump*.
48 Fanon, *Black Skin, White Masks*, 84.
49 As I argued in Burnett, *White Belongings*, 143.
50 Dunaway, 'The "Crying Indian" Ad That Fooled the Environmental Movement'.
51 Nadasdy, 'Transcending the Debate over the Ecologically Noble Indian', 292.
52 Nouala, 'Livestock Is a Form of Climate Justice in the Global South'.
53 Öhman, 'Embodied Vulnerability in Large-Scale Technical Systems'; Össbo and Lantto, 'Colonial Tutelage and Industrial Colonialism'; see also Lawrence and Larsen, 'The Politics of Planning'.
54 Kim, 'Makah Whaling and the (Non) Ecological Indian', 86.
55 Conklin and Graham, 'The Shifting Middle Ground'.
56 Tănăsescu, 'Rights of Nature, Legal Personality, and Indigenous Philosophies'.
57 Tuck and Yang, 'Decolonization Is Not a Metaphor', 26.
58 Powell, *Vanishing America*, 29.
59 Over 750,000 Palestinians were forced to flee their homes in conflicts leading up to the creation of the state of Israel in 1948, and have been denied the right of return. Around 150,000 stayed, however, and became citizens of Israel. Now numbering 20 per cent of the population (around 2 million people), Palestinian citizens of Israel are the target of policies of de-Arabization and denationalization, enjoying second-class citizenship. Those who left before 1948 now number 6 million, and live in and around refugee camps on the West Bank (incl. East Jerusalem), Gaza, and in Lebanon, Syria and Jordan. See 'Palestine Refugees', https://www.unrwa.org/palestine-refugees, last accessed 2 February 2025. For the status of Palestinian

citizens of Israel see Awayed-Bishara, 'Sumud Pedagogy as Linguistic Citizenship'; and Milani and Levon, 'Israel as Homotopia'.

60 Eghbariah, 'The Struggle for Za'atar and Akkoub'.

61 Gorney, 'Roots of Identity, Canopy of Collision', 329.

62 Gorney, 330–1.

63 George, 'Making the Desert Bloom'.

64 Gorney, 'Roots of Identity, Canopy of Collision', 331.

65 Braverman, *Planted Flags*.

66 Braverman, 105–9; Gorney, 'Roots of Identity, Canopy of Collision', 332.

67 Said, 'Palestine: Memory, Invention and Space', 19. Also cited in Eghbariah, 'The Struggle for Za'atar and Akkoub'.

68 Gorney, 'Roots of Identity, Canopy of Collision', 331.

69 Said, *Orientalism*, 99.

70 Gafter and Milani, 'Eco-Orientalism'. As they point out, this term was originally coined by the literary scholar Tori Bush in the US context. See Bush, 'Eco-Orientalism'.

71 Gafter and Milani, 'Eco-Orientalism', no page.

72 Braverman, *Settling Nature*, 4.

73 Braverman, 247.

74 Rinat, 'Tree Hugging Settlers Make a Green Case for Seizing Palestinian Land'. For an account of how organic farming by Zionist settlers cements claims to autochthony, see Handel, Monterescu, and Grosglik, 'Soulful Soil and Colonial Quality: Organic Farming in the West Bank'.

75 Sasa, 'Oppressive Pines'.

76 Feldman and McGonigle, 'Introduction', 3.

77 Awayed-Bishara, 'Sumud Pedagogy as Linguistic Citizenship', 5.

78 Milani and Levon, 'Israel as Homotopia'.

79 Eghbariah, 'The Struggle for Za'atar and Akkoub'.

80 'Foragers', https://www.jumanamanna.com/Foragers, last accessed 2 February 2025.

81 Valle, 'When Environmentalism Meets Xenophobia'.

82 Mills, 'Black Trash', 84. The word he uses is 'pleonastic' – i.e. using more words than necessary to express an idea.

83 Staudenmeier, *Ecology Contested*.

84 Conversi, 'The Ultimate Challenge'; Conversi and Posocco, 'Which Nationalism for the Anthropocene?'

85 Conversi and Posocco, 'Which Nationalism for the Anthropocene?'

86 Hau, 'From Local Concerns to Global Challenges'; Warren, 'Nativeness and Nationhood'.

87 Posocco and Watson, 'Nationalism and Environmentalism'.

88 Following Morgan Margulies. See Margulies, 'Eco-Nationalism'.

89 Hultgren, 'Natural Exceptions to Green Sovereignty?', 301.

90 Tutt and Heubaum, 'Nationalism and the Energy Transition: The Case of the SNP'.
91 Bettini and Casaglia, 'From Denial to Domestication'.
92 Margulies, 'Eco-Nationalism', 27–8.
93 Arantes, 'Towards a Green Nationalism with Chinese Characteristics?'; Koch, 'Green Nationalism from Above'.
94 Brain, 'Introduction', 612.
95 Turner and Bailey, '"Ecobordering"', 119.
96 Beirich, 'Nativists, Environmentalism & the Hypocrisy of Hate'; Mix, 'The Greening of White Separatism'.
97 Malm and The Zetkin Collective, *White Skin, Black Fuel*, chapter 5 (e-book version).
98 See the contributions in Brüggemeier, Cioc, and Zeller, *How Green Were the Nazis?*; See also Lekan, 'From Landscape to Lebensraum: Race and Environment under Nazism'; and Staudenmaier, 'Organic Farming in Nazi Germany'.
99 Biehl and Staudenmaier, *Ecofascism*; Forchtner and Kølvraa, 'Extreme Right Images of Radical Authenticity'.
100 Lawton, 'The Rise of Real Eco-Fascism'.
101 Macklin, 'The Extreme Right, Climate Change and Terrorism'.
102 Moore and Roberts, *The Rise of Ecofascism*, 11.
103 Moore and Roberts, 11.
104 Campion, 'Defining Ecofascism', 933.
105 Lubarda, 'Beyond Ecofascism?'
106 Olsen, *Nature and Nationalism*.
107 Lubarda, 'Beyond Ecofascism?', 723–5.
108 Lubarda, 725.
109 Lubarda, *Far-Right Ecologism*. See also Pál, 'Illiberal Environmentalism?'
110 There is an extensive and growing literature on FRE in Europe that pays close attention to local specificities. In Romania, for example, Orthodox spirituality and an idealized Romanian peasantry are invoked as the foundations of national self-sufficiency. See Mihai and Ungureanu, 'Far-Right Ecology and Geopolitical Resentment at Europe's Periphery'. Romanian forests and mountains are constructed by nationalists as sentient protectors of the borders of Europe. See Coţofană, 'Ingrained Ontologies: How Romania's Institutionalized Processes Teach Us to Think with Xenophobic Sentient Landscapes'. In Finland, the out-of-place littering migrant is constructed in opposition to the environmentally conscious 'True Finn'. See Pietiläinen, '"They Will Not Survive Here"'. In France, localism centres on locally produced goods and foods, while French culture is imagined as an endangered 'species'. See Bivar, 'The Patriot Ecology of the French Far Right'. In Germany, the Alternative für Deutschland (AfD) prioritizes the beautification of the Heimat (homeland) in a similar vein to historical Nazism. See Obertreis, 'Environmental and Climate Policies as the New Hobby Horse'. A thoroughgoing

anti-urbanism that problematizes capitalism, the state and financial centres as 'unnatural' is contrasted to 'natural' formations such as local, rural families. See Staudenmeier, *Ecology Contested*. In so-called '*völkisch*' (of the German *Volk*) settlements in northern Germany, meanwhile, as Matt Varco has documented, racially homogeneous communities deploy nature as a signifier that connects national protection from ethnic outsiders to the protection of the German ecosystem. See Varco, 'Volk Utopia'.

111 Richards, Jones, and Brinn, 'Eco-Fascism Online'.

112 The idea of a society born from the ashes of the old order is referred to as 'palingenesis' and is a common feature of extreme-right social visions. See Richardson, 'Fascist Discourse'.

113 Rao, 'Saffronisation of the Holy Cow'; Sharma, *Green and Saffron*. Animal protection is an important theme of historical and contemporary ecofascism. The Nazi Animal Protection Law of 1933 is often cited as an evidence of the distorted moral economy of the Nazis. In Sweden, the far-right *Sverigedemokraterna* argue that immigrants are more violent to animals, and that animal cruelty must be even more strongly punished. See Backlund and Jungar, 'Animal Advocacy and the Radical Right'. For an overview of this issue, see 'The Ambiguities of Animal Rights', the third chapter in Staudenmeier, *Ecology Contested*. There are also a number of neo-Nazi movements that practice veganism. See Forchtner and Tominc, 'Kalashnikov and Cooking-Spoon'.

114 Chaudhary, *The Exhausted of the Earth*, no page.

115 Singh, 'A New Regime of Dispossession in Neo-Liberal India?'

116 D'Souza, 'Environmental History of South Asia in the Time of Hindutva'.

117 Burnett and Richardson, '"Breeders for Race and Nation"'.

118 Daggett, 'Petro-Masculinity'; see Lubarda and Forchtner, 'The Far Right and the Environment: Past-Present-Future' for an overview.

119 Hultman, 'The Making of an Environmental Hero'; Hultman, 'Natures of Masculinities'. In a distinct but related phenomenon, Jeanne Hamming analyses how popular climate change novels enlist climate change as an occasion to dramatize the recovery of an imperilled national masculinity. See Hamming, 'Nationalism, Masculinity, and the Politics of Climate Change in the Novels of Kim Stanley Robinson and Michael Crichton'.

120 Di Chiro, 'Polluted Politics? Confronting Toxic Discourse, Sex Panic, and Eco-Normativity', 201.

121 Di Chiro, 202.

122 Di Chiro, 203–4.

123 Chen, *Animacies*, 169.

124 Burnett, 'Healthy White Nationalists'; Tebaldi and Burnett, 'The Science of Desire'.

125 Boast, 'Theorizing the Gay Frog'. The hypothesis that homosexuality and trans identity are causally linked to prenatal exposure to environmental toxins is without any scientific bases. See Zabalegui, 'The Tale of EDCs and Trans Identities'.

126 Cited in Moore and Roberts, *The Rise of Ecofascism*, 81.

127 Perret, '"Transgender Frogs Turn Your Son Gay"'.

128 As Des Fitzgerald argues, during the Covid-19 pandemic green nationalism morphed into 'viral nationalism' in the UK border regime. Fitzgerald, 'COVID-19, Border Control, and Viral Nationalism'.

129 Varco, 'Volk Utopia'.

130 Anson, 'No One Is a Virus'.

131 Bigger and Nelson, 'Fighting Fire and Fascism in the American West'.

132 Hartman, 'Beyond Climate Denial'.

133 Rahman, Anbarci, and Ulubaşoğlu, '"Storm Autocracies"'.

134 Shanaah, Fritsche, and Osmundsen, 'The Effect of Climate Change Threat on Public Attitudes towards Ethnic and Religious Minorities and Climate Refugees'; Uhl et al., 'Undesirable Effects of Threatening Climate Change Information'.

135 Babcock, 'The Current Role of the Environment in Reinforcing Acts of Domestic Terrorism'; Bergman, 'Rising Sea Levels and the Right Wave'.

136 Anson et al., *Against the Ecofascist Creep*.

137 Bahrainwala, 'Shithole Rhetorics'.

138 As discussed above. See Gilmore, *Golden Gulag*.

139 Smith, *Myths and Memories of the Nation*.

140 Forchtner and Olsen, 'Against the Promethean'.

141 Bauman, *Retrotopia*.

142 Party statistics taken from the Wikipedia page 'Green Party (Sweden)', https://en.wikipedia.org/wiki/Green_Party_(Sweden) last accessed 4 February 2024.

143 Wetterstrand, Maria, 18 November 2009. 'Wetterstrand: De gröna ett naturligt hem för socialliberaler', https://web.archive.org/web/20110728040457/http://www.newsmill.se/artikel/2009/11/17/miljopartiet-de-grona-ett-naturligt-hem-socialliberaler, last accessed 4 February 2024.

144 Translation by the author from the original Swedish: 'solidaritet med djur, natur och det ekologiska systemet; solidaritet med kommande generationer; solidaritet med världens alla människor'. 'Grön ideologi – solidaritet i handling', https://www.mp.se/om/partiprogram/gron-ideologi/, last accessed 4 February 2025.

145 'gränslösa solidaritet'. Ibid.

146 Blomster, 'In a Red Little Cottage'.

147 'Det unika pigmentet', https://falurodfarg.com/om-falu-rodfarg/det-unika-pigmentet/ last accessed 4 February 2025.

148 Karlbom, 'Sweden's Iron Ore Exports to Germany, 1933–1944'.

149 Ojala and Nordin, 'Mining Sápmi'.

150 Or a river, or the sea, though the romance of the *dopp* does tend to focus on lakes.

151 Ljunggren, 'Nation-Building, Primitivism and Manliness – the Issue of Gymnastics in Sweden around 1800'.

152 Thurlow, Pellanda, and Wohlgemuth, 'The Discursive Chronotopes of Waste'.

153 The entire log of this mission is available at http://www.captain-charles-moore.org/chile-2016/, last accessed 5 February 2025.

154 Though Moore's project updates are no longer available due to ResearchGate discontinuing the functionality, his findings were reported on at the time in the ResearchGate blog. See ResearchGate, 'Scientists Announce the Existence of Another Ocean Garbage Patch'.

155 de Guzman, 'Second Garbage Patch the Size of Texas Discovered in Pacific Ocean'; *Ocean Garbage Patch Could Be Bigger Than Mexico*; Kjellberg, 'Ny "skräp-ö" hittad i Stilla havet – större än Mexiko'.

156 Hale, 'A Garbage Patch the Size of Mexico Was Just Discovered in the Pacific Ocean'.

157 IFLScience stands for 'I Fucking Love Science' and was started by the English blogger and science communicator Elise Andrew. See https://en.wikipedia.org/wiki/Elise_Andrew, last accessed 5 February 2025.

158 Ideland, *Den KRAV-märkta människan*, 116–17. Translation from the Swedish by the author. Original text: "Smuts' och 'förorening' är i läromedlen ord som klibbar fast vid färgade kroppar […] 'Några vita människor ser vi inte bada i smuts. Tvärtom är de världsmästare i pantning och sorterar med lust och glädje sina sopor'.

159 Strick, 'The Alternative Right, Masculinities, and Ordinary Affect', 225.

160 See Van Valkenburgh, 'Digesting the Red Pill'. In far-right online parlance, to be 'red-pilled' is to be conscientized about the true nature of society, which is understood as organized to discriminate against white men specifically. To be blue-pilled is to be sanguine about society and believe what the 'mainstream' tells you; blue-pilled men are tolerant of feminism, liberal values and multiculturalism.

161 Strick, 'The Alternative Right, Masculinities, and Ordinary Affect', 223.

162 For an extended discussion of race and gender in RPGs, see Josefsen and Burnett, 'Visions of Degeneracy'.

163 Patel, Prachi, 'Stemming the Plastic Tide: 10 Rivers Contribute Most of the Plastic in the Oceans', https://www.scientificamerican.com/article/stemming-the-plastic-tide-10-rivers-contribute-most-of-the-plastic-in-the-oceans/ last accessed 6 February 2025.

164 Majavu, 'The White International'.

165 Though 'white man bad' has been used in memes, its origin is obscure. A cursory online search suggests that it is a phrase often used in historical texts written by white Americans that put words into the mouths of Native American and African American people.

166 Bettini and Casaglia, 'From Denial to Domestication'.

167 Axelsson, Nasiritousi, and Miljand, 'Stöd för och attityder till klimatpolitik i Sverige'.

168 Westberg and Årman, 'Family Politics in Contemporary Fascist Propaganda'.

169 Copsey and Worley, 'Introduction'.

170 This bio is taken from Orwoll's profile on the website of the publisher of a range of fascist literature, Arktos Media. See https://arktos.com/author/orwoll/, last accessed 6 February 2025.

171 'Return to the Land', https://returntotheland.org/#/home, last accessed 6 February 2025.

172 A white subculture in evidence in many locations across the world, which relies on myths of racial superiority to articulate a 'resilient' lifestyle outside of degenerate and racially mixed modern societies. Orwoll names preppers explicitly later in the video, saying that RTTL is 'great for preppers, identitarians and people who just want to save money'. For analysis of the prepper movement in the United States, see Ford, '"They Will Be Like a Swarm of Locusts"'. For a similar phenomenon in South Africa, see Burnett, 'Right as Rain'.

173 Du Mez, *Jesus and John Wayne*; see also Bjork-James, *The Divine Institution*.

174 Simpson, *As We Have Always Done*, 22.

175 Simpson, 23.

176 Pugliese, 'More-Than-Human Lifeworlds, Settler Modalities of Geno-Ecocide and Border Questions'.

## Conclusion

1 Brown, 'Revaluing Critique', 470.

2 Cohen and Riofrancos, 'Biden Left Us with a "Prius Economy." It's Time for Something Different'.

3 'Wealthier homeowners nab billions in tax credits for energy efficiency', https://www.eenews.net/articles/wealthy-homeowners-nab-billions-in-tax-credits-for-energy-efficiency/, last accessed 8 February 2025.

4 Shahid, 'The Left Didn't Sink Kamala Harris. Here's What Did'.

5 This image is available in many places online. The link to Musk's original tweet is https://x.com/elonmusk/status/1599671964582391808

6 Taylor, '"F*ck Earth"'.

7 Chun, *The Discourses of Capitalism*, 143.

8 When I started my doctoral research on the anti-fracking movement in South Africa in 2015, Musk was often celebrated as an environmental hero, such as when a planned Tesla investment in PowerWall battery production in Cape Town was reported. See https://www.facebook.com/groups/chaseshelloutofthekaroo/permalink/1058165687540328/, last accessed 8 February 2025.

9 Gramsci, *Selections from the Prison Notebooks of Antonio Gramsci*, 418.
10 Solnit, 'Why Climate Despair Is a Luxury'.
11 Solnit.
12 Aguon, 'Julian Aguon'.
13 Whyte, 'Too Late for Indigenous Climate Justice', 2.
14 Paperson, 'A Ghetto Land Pedagogy'.

# Works cited


Abidin, Crystal, Dan Brockington, Michael K. Goodman, Mary Mostafanezhad, and Lisa Ann Richey. 'The Tropes of Celebrity Environmentalism'. *Annual Review of Environment and Resources* 45, no. 1 (17 October 2020): 387–410. https://doi.org/10.1146/annurev-environ-012320-081703.

Abrams, Amber. '"We Just Saw the Fence": Infrastructural Violence, Fencing and the Legacy of South Africa's Bantustan'. In *The Violence of Conservation in Africa: State, Militarization and Alternatives*, edited by Maano Ramutsindela, Frank Matose, and Tafadzwa Mushonga, 113–33. Cheltenham: Edward Elgar Publishing, 2022.

Adams, Jonathan S., and Thomas O. McShane. *The Myth of Wild Africa: Conservation without Illusion*. Oakland: University of California Press, 1996.

Adhikari, Mohamed. *The Anatomy of a South African Genocide: The Extermination of the Cape San Peoples*. Athens, OH: Ohio University Press, 2011.

Agha, Asif. 'Meet Mediatization'. *Language & Communication* 31, no. 3 (July 2011): 163–70. https://doi.org/10.1016/j.langcom.2011.03.006.

Agha, Asif. 'Voice, Footing, Enregisterment'. *Journal of Linguistic Anthropology* 15, no. 1 (2005): 38–59. https://doi.org/10.1525/jlin.2005.15.1.38.

Agrawal, Arun, and Kent Redford. 'Conservation and Displacement: An Overview'. *Conservation and Society* 7, no. 1 (March 2009): 1. https://doi.org/10.4103/0972-4923.54790.

Aguon, Julian. 'No Country for Eight-Spot Butterflies'. *Progressive International* (blog), 12 June 2020. https://progressive.international/wire/2020-06-12-julian-aguon-no-country-for-eight-spot-butterflies/en.

Ahmed, Sara. *The Cultural Politics of Emotion*. 2nd edition. New York: Routledge, 2014.

Ahmed, Yunana. 'Political Discourse Analysis: A Decolonial Approach'. *Critical Discourse Studies* 18, no. 1 (2 January 2021): 139–55. https://doi.org/10.1080/17405904.2020.1755707.

Ajl, Max. *A People's Green New Deal*. London: Pluto Press, 2021.

Alasow, Khadra Ghedi. 'Capitalism and Private Nature Reserves: The Taming of Mala Mala Land Claim'. *Master of Science, University of Cape Town*, 2020. https://open.uct.ac.za/bitstream/handle/11427/32641/thesis_sci_2020_alasow%20khadra%20ghedi.pdf?sequence=1&isAllowed=y.

Alcoff, Linda. 'The Problem of Speaking for Others'. *Cultural Critique*, no. 20 (Winter 1991–1992): 5. https://doi.org/10.2307/1354221.

Alexander, Richard. *Framing Discourse on the Environment: A Critical Discourse Approach*. New York and Abingdon: Routledge, 2009.

Alexander, Richard, and Arran Stibbe. 'From the Analysis of Ecological Discourse to the Ecological Analysis of Discourse'. *Language Sciences*, Ecolinguistics: the Ecology of Language and the Ecology of Science 41 (2014): 104–10. https://doi.org/10.1016/j.langsci.2013.08.011.

Alkon, Alison Hope, and Christie Grace McCullen. 'Whiteness and Farmers Markets: Performances, Perpetuations … Contestations?' *Antipode* 43, no. 4 (September 2011): 937–59. https://doi.org/10.1111/j.1467-8330.2010.00818.x.

Allen, Amy. *The End of Progress: Decolonizing the Normative Foundations of Critical Theory*. New York: Columbia University Press, 2016.

Althusser, Louis. 'Ideology and Ideological State Apparatuses (Notes towards an Investigation)'. In *Lenin and Philosophy, and Other Essays*, 121–73. London: New Left Books, 1971.

Altvater, Elmar, Eileen C. Crist, Donna J. Haraway, Daniel Hartley, Christian Parenti, and Justin McBrien. *Anthropocene or Capitalocene?: Nature, History, and the Crisis of Capitalism*. Edited by Jason W. Moore. 1st edition. Oakland: PM Press, 2016.

Alvarado, Genie. 'Index of Winds'. *University of Hawai'i Library*, September 2005. https://www2.hawaii.edu/~hawnpac/libguides_docs/indexofwinds.pdf.

Amnesty International. 'Amnesty International Recommendations to States Ahead of COP 15 (Convention on Biological Diversity) 30 X 30 Proposal'. *Amnesty International*, 2022. https://www.amnesty.org/en/documents/ior40/6162/2022/en/.

Amoah, Susuana. 'Reflections: Overpopulation & the Unbearable Whiteness of Green'. *O N C A* (blog), 24 July 2019. https://onca.org.uk/overpopulation-the-unbearable-whiteness-of-green/.

Anderson, Bridget. 'Servants and Slaves: Europe's Domestic Workers'. *Race & Class* 39, no. 1 (July 1997): 37–49. https://doi.org/10.1177/030639689703900104.

Anker, Peder. *Imperial Ecology: Environmental Order in the British Empire, 1895–1945*. Cambridge and London: Harvard University Press, 2001.

Anson, April. 'No One Is a Virus: On American Ecofascism'. *Environmental History Now* (blog), 6 September 2023. https://envhistnow.com/2023/09/06/no-one-is-a-virus-on-american-ecofascism/.

Anson, April, Cassie Galentine, Shane Hall, Alex Menrisky, Bruno Seraphin, Melody Kennan, and Anti Creep Climate Initiative. *Against the Ecofascist Creep*. San Diego: The Anti-Creep Climate Initiative, 2022.

Apostolopoulou, Elia, Anastasia Chatzimentor, Sara Maestre-Andrés, Marina Requena-i-Mora, Alejandra Pizarro, and Dimitrios Bormpoudakis. 'Reviewing 15 Years of Research on Neoliberal Conservation: Towards a Decolonial, Interdisciplinary, Intersectional and Community-Engaged Research Agenda'. *Geoforum* 124 (2021): 236–56. https://doi.org/10.1016/j.geoforum.2021.05.006.

Appadurai, Arjun. *Modernity at Large: Cultural Dimensions of Globalization*. Minneapolis: University of Minnesota Press, 1996. http://ebookcentral.proquest.com/lib/gu/detail.action?docID=310379.

Arantes, Virginie. 'Towards a Green Nationalism with Chinese Characteristics?' *Journal of Contemporary China* 33, no. 145 (January 2024): 65–82. https://doi.org/10.1080/10670564.2023.2196514.

Argyrou, Vassos. *The Logic of Environmentalism: Anthropology, Ecology and Postcoloniality*. Oxford and Brooklyn: Berghahn Books, 2005.

Arvin, Maile Renee. *Possessing Polynesians: The Science of Settler Colonial Whiteness in Hawai`i and Oceania*. Durham and London: Duke University Press Books, 2019.

Asia Indigenous Peoples Pact, Badan Registrasi Wilayah Adat, Cambodian Indigenous Peoples Alliance, Cambodia Indigenous Peoples Organization, Centre for Orang Asli Concerns, Center for Indigenous Peoples' Research and Development, Federation of Community Forestry Users Nepal, Indigenous Media Network, Indigenous Peoples Foundation for Education and Environment, Indigenous Peoples Partnership, Indonesian Institute for Forest and Environment, Inter Mountain Peoples Education and Culture in Thailand Association, Jaringan Kerja Pemetaan Partisipatif, Land Conflict Watch, Nepal Federation of Indigenous Nationalities, Network of Indigenous Peoples in Thailand, Non-Timber Forest Products Exchange Programme, Partners of Community Organizations in Sabah Trust, Promotion of Indigenous and Nature Together, Rights and Resources Initiative, and Working Group ICCAs Indonesia. 'Reconciling Conservation and Global Biodiversity Goals with Community Land Rights in Asia'. Washington, DC: Rights and Resources Initiative, 7 February 2022. https://doi.org/10.53892/HEUK4095.

Augé, Anaïs. *Metaphor and Argumentation in Climate Crisis Discourse*. New York: Routledge, 2023. https://doi.org/10.4324/9781003342908.

Austin, Jonathan Luke, David Chandler, Marwa Daoudy, Rita Floyd, Delf Rothe, and Dahlia Simangan. 'Forum: At the Crossroads Critical Perspectives on the Study of Climate Security'. *Geoforum* 155 (2024): 104101. https://doi.org/10.1016/j.geoforum.2024.104101.

Awayed-Bishara, Muzna. 'Sumud Pedagogy as Linguistic Citizenship: Palestinian Youth in Israel against Imposed Subjectivities'. *Language in Society*, First View, 1–23. https://doi.org/10.1017/S0047404523000891.

Axelsson, Sofia, Naghmeh Nasiritousi, and Matilda Miljand. 'Stöd för och attityder till klimatpolitik i Sverige'. Stockholm: Klimatpolitiska Rådet, March 2024. https://www.klimatpolitiskaradet.se/wp-content/uploads/2024/03/stodforochattitydertillklimatpolitikisverige.pdf.

Babcock, Hope M. 'The Current Role of the Environment in Reinforcing Acts of Domestic Terrorism'. *Virginia Environmental Law Journal* 38, no. 2 (2020): 207–31.

Backlund, Anders, and Ann-Cathrine Jungar. 'Animal Advocacy and the Radical Right: The Case of Sweden'. *Journal of Political Ideologies* 29, no. 3 (2024): 614–33. https://doi.org/10.1080/13569317.2022.2138292.

Bahrainwala, Lamiyah. 'Shithole Rhetorics'. *Journal of International and Intercultural Communication* 14, no. 3 (3 July 2021): 185–201. https://doi.org/10.1080/17513057.2020.1795224.

Bakhtin, Mikhail M. *The Dialogic Imagination: Four Essays*. Edited by M. Holquist. Translated by C. Emerson and M. Holquist. Austin: University of Texas Press, 1981. http://site.ebrary.com/id/10439450.

Baldwin, Andrew. 'Premediation and White Affect: Climate Change and Migration in Critical Perspective'. *Transactions of the Institute of British Geographers* 41, no. 1 (1 January 2016): 78–90. https://doi.org/10.1111/tran.12106.

Baldwin, Andrew. 'Whiteness and Futurity: Towards a Research Agenda'. *Progress in Human Geography* 36, no. 2 (1 April 2012): 172–87. https://doi.org/10.1177/0309132511414603.

Ballard, Richard. 'When in Rome: Claiming the Right to Define Neighbourhood Character in South Africa's Suburbs'. *Transformation: Critical Perspectives on Southern Africa* 57, no. 1 (2005): 64–87.

Ballard, Richard, and Gareth A. Jones. 'Natural Neighbors: Indigenous Landscapes and Eco-Estates in Durban, South Africa'. *Annals of the Association of American Geographers* 101, no. 1 (21 January 2011): 131–48. https://doi.org/10.1080/00045608.2010.520224.

Bam-Hutchison, June. 'Epistemicide in the Cape: Symbolic and Restorative Justice in South Africa'. In *Gender, Transitional Justice and Memorial Arts: Global Perspectives on Commemoration and Mobilization*, edited by Jelke Boesten and Helen Scanlon, 117–35. London: Routledge, 2021. https://www.taylorfrancis.com/chapters/edit/10.4324/9781003174462-9/epistemicide-cape-june-bam-hutchison.

Barkham, Patrick. 'The Real David Attenborough'. *The Guardian*, 22 October 2019, sec. Television & radio. https://www.theguardian.com/tv-and-radio/2019/oct/22/david-attenborough-climate-change-bbc.

Bartlett, Tom, and Nicolina Montesano Montessori. 'Towards Webs of Equivalence and the Political Nomad in Agonistic Debate: Contributions from CDA and Scales Theory'. *Journal of Language and Politics* 20, no. 1 (21 January 2021): 129–44. https://doi.org/10.1075/jlp.20046.bar.

Bauer, Simon, Tommaso M. Milani, Kerstin von Brömssen, and Andrea Spehar. 'Gender Equality in the Name of the State: State Feminism or Femonationalism in Civic Orientation for Newly Arrived Migrants in Sweden?' *Critical Discourse Studies* 21, no. 5 (2 September 2024): 591–609. https://doi.org/10.1080/17405904.2023.2228933.

Bauman, Zygmunt. *Retrotopia*. Cambridge: Polity Press, 2017.

Bearak, Max. 'It's Not Just Willow: Oil and Gas Projects Are Back in a Big Way'. *New York Times*, 6 April 2023, sec. Climate. https://www.nytimes.com/2023/04/06/climate/oil-gas-drilling-investment-worldwide-willow.html.

Becker, Frowin. 'Trophy's Choice: How the Link between Trophy Hunting and White Land Ownership Challenges Notions of Community Empowerment in Namibia'. *Current Conservation* (blog), 4 February 2022. https://www.currentconservation.org/trophys-choice-how-the-link-between-trophy-hunting-and-white-land-ownership-challenges-notions-of-community-empowerment-in-namibia/.

Behagel, Jelle Hendrik, and Ayşem Mert. 'The Political Nature of Fantasy and Political Fantasies of Nature'. *Journal of Language and Politics* 20, no. 1 (21 January 2021): 79–94. https://doi.org/10.1075/jlp.20049.beh.

Beinart, William, and Peter Coates. *Environment and History: The Taming of Nature in the USA and South Africa*. London and New York: Routledge, 1995.

Beirich, Heidi. 'Greenwash: Nativists, Environmentalism & the Hypocrisy of Hate'. Montgomery, AL: Southern Poverty Law Center, July 2010. https://www.splcenter.org/resources/reports/greenwash-nativists-environmentalism-and-hypocrisy-hate/.

Bennett, Jane, and William Chaloupka. *In the Nature of Things: Language, Politics, and the Environment*. Minneapolis, MN: University of Minnesota Press, 1993. https://ebookcentral.proquest.com/lib/pensu/detail.action?docID=310252.

Berg, Anne. *Empire of Rags and Bones: Waste and War in Nazi Germany*. New York, NY: Oxford University Press, 2024.

Bergman, Harriet M. 'Rising Sea Levels and the Right Wave: An Analysis of the Climate Change Communication That Enables "the Fascist Creep"'. *Krisis* 41, no. 2 (2021): 2–18. https://doi.org/10.21827/krisis.41.2.37164.

Bersaglio, Brock, and Frances Cleaver. 'Green Grab by Bricolage – The Institutional Workings of Community Conservancies in Kenya'. *Conservation and Society* 16, no. 4 (December 2018): 467. https://doi.org/10.4103/cs.cs_16_144.

Bettini, Giovanni. 'Climate Barbarians at the Gate? A Critique of Apocalyptic Narratives on "Climate Refugees"'. *Geoforum* 45 (March 2013): 63–72. https://doi.org/10.1016/j.geoforum.2012.09.009.

Bettini, Giovanni, and Anna Casaglia. 'From Denial to Domestication: Unpacking Italy's Right-Wing Approach to Climate Migration and Security'. *Geoforum* 155 (1 October 2024): 104079. https://doi.org/10.1016/j.geoforum.2024.104079.

Bettini, Giovanni, and Lazaros Karaliotas. 'Exploring the Limits of Peak Oil: Naturalising the Political, de-Politicising Energy: Exploring the Limits of Peak Oil'. *The Geographical Journal* 179, no. 4 (December 2013): 331–41. https://doi.org/10.1111/geoj.12024.

Bhambra, Gurminder K. 'Decolonizing Critical Theory?: Epistemological Justice, Progress, Reparations'. *Critical Times* 4, no. 1 (1 April 2021): 73–89. https://doi.org/10.1215/26410478-8855227.

Bhambra, Gurminder K., and John Holmwood. 'Colonialism, Postcolonialism and the Liberal Welfare State'. *New Political Economy* 23, no. 5 (3 September 2018): 574–87. https://doi.org/10.1080/13563467.2017.1417369.

Bhambra, Gurminder K., and Peter Newell. 'More than a Metaphor: "Climate Colonialism" in Perspective'. *Global Social Challenges Journal* 2, no. 2 (1 December 2023): 179–87. https://doi.org/10.1332/EIEM6688.

Bhandar, Brenna. *Colonial Lives of Property: Law, Land, and Racial Regimes of Ownership*. Durham, NC: Duke University Press, 2018.

Bhaskar, Roy. *Enlightened Common Sense: The Philosophy of Critical Realism*. London: Routledge, 2016. https://doi.org/10.4324/9781315542942.

GBD 2021 Fertility and Forecasting Collaborators. 'Global Fertility in 204 Countries and Territories, 1950–2021, with Forecasts to 2100: A Comprehensive Demographic Analysis for the Global Burden of Disease Study 2021'. *The Lancet* 403, no. 10440 (18 May 2024): 2057–99. https://doi.org/10.1016/S0140-6736(24)00550-6.

Biehl, Janet, and Peter Staudenmaier. *Ecofascism: Lessons from the German Experience*. 1st edition. Edinburgh: AK Press, 1995.

Bigger, Patrick, and Sara Nelson. 'Fighting Fire and Fascism in the American West'. *Dissent Magazine*, Spring 2023. https://www.dissentmagazine.org/article/fighting-fire-and-fascism-in-the-american-west/.
Billig, Michael. 'Critical Discourse Analysis and the Rhetoric of Critique'. In *Critical Discourse Analysis: Theory and Interdisciplinarity*, edited by Gilbert Weiss and Ruth Wodak, 35–46. London: Palgrave Macmillan, 2003. https://doi.org/10.1057/9780230514560_2.
Billig, Michael. 'Rhetorical Uses of Precise Numbers and Semi-Magical Round Numbers in Political Discourse about COVID-19: Examples from the Government of the United Kingdom'. *Discourse & Society* 32, no. 5 (2 May 2021): 542–58. https://doi.org/10.1177/09579265211013115.
Bivar, Venus. 'The Patriot Ecology of the French Far Right'. *Environmental History* 27, no. 4 (October 2022): 618–24. https://doi.org/10.1086/721413.
Bjork-James, Sophie. *The Divine Institution: White Evangelicalism's Politics of the Family*. New Brunswick: Rutgers University Press, 2021.
Blanc, Guillaume. *The Invention of Green Colonialism*. Translated by Helen Morrison. Cambridge: Polity, 2022.
Blommaert, Jan. *Discourse: A Critical Introduction*. Cambridge: Cambridge University Press, 2005.
Blommaert, Jan, and Chris Bulcaen. 'Critical Discourse Analysis'. *Annual Review of Anthropology* 29, no. 1 (2000): 447–66. https://doi.org/10.1146/annurev.anthro.29.1.447.
Blomster, Anna Jenny Katarina. 'In a Red Little Cottage: Icons of Identity and Nation in Sweden'. Ph.D., University of California, 2016. https://escholarship.org/uc/item/0j25f2jn.
Bluwstein, Jevgeniy, and Salvatore Paolo De Rosa. 'Political Ecologies of the Future: Introduction to the Special Issue'. *Geoforum* 153 (1 July 2024): 104023. https://doi.org/10.1016/j.geoforum.2024.104023.
Bluwstein, Jevgeniy, Connor Cavanagh, and Robert Fletcher. 'Securing Conservation *Lebensraum*? The Geo-, Bio-, and Ontopolitics of Global Conservation Futures'. *Geoforum* 153 (1 July 2024): 103752. https://doi.org/10.1016/j.geoforum.2023.103752.
Boast, Hannah. 'Theorizing the Gay Frog'. *Environmental Humanities* 14, no. 3 (1 November 2022): 661–79. https://doi.org/10.1215/22011919-9962959.
Bok, Hilary. 'Baron de Montesquieu, Charles-Louis de Secondat'. In *The Stanford Encyclopedia of Philosophy*, edited by Edward N. Zalta. Stanford, CA: Metaphysics Research Lab, Stanford University, 2018. https://plato.stanford.edu/archives/win2018/entries/montesquieu/.
Bonilla, Yarimar. 'The Coloniality of Disaster: Race, Empire, and the Temporal Logics of Emergency in Puerto Rico, USA'. *Political Geography* 78 (1 April 2020): 102181. https://doi.org/10.1016/j.polgeo.2020.102181.
Bontempi, Antonio, Pietro Venturi, Daniela Del Bene, Arnim Scheidel, Quim Zaldo-Aubanell, and Roser Maneja Zaragoza. 'Conflict and Conservation: On the Role

of Protected Areas for Environmental Justice'. *Global Environmental Change* 82 (1 September 2023): 102740. https://doi.org/10.1016/j.gloenvcha.2023.102740.

*Borrell at Launch of European Diplomatic Academy Called Europe 'Garden', Outside World 'Jungle'.* EU Debates, 2022. https://www.youtube.com/watch?v=cmNALPfGq-A.

Borrell, Josep. 'On Metaphors and Geo-Politics'. *EEAS European Union External Action* (blog), 18 October 2022. https://www.eeas.europa.eu/eeas/metaphors-and-geo-politics_en.

Bosworth, Kai. 'The Bad Environmentalism of "Nature Is Healing" Memes'. *Cultural Geographies*, 22 April 2021, 14744740211012007. https://doi.org/10.1177/14744740211012007.

Brain, Stephen. 'Introduction: Nature and the New Right'. *Environmental History* 27, no. 4 (1 October 2022): 610–17. https://doi.org/10.1086/721416.

Braverman, Irus. *Planted Flags: Trees, Land, and Law in Israel/Palestine.* New York: Cambridge University Press, 2009.

Braverman, Irus. *Settling Nature: The Conservation Regime in Palestine-Israel.* Minneapolis, MN: University of Minnesota Press, 2023.

Bresnihan, Patrick, and Naomi Millner. 'Decolonising Environmental Politics'. In *Handbook of Critical Environmental Politics*, edited by Luigi Pellizzoni, Emanuele Leonardi, and Viviana Asara, 521–39. Cheltenham and Northampton: Edward Elgar Publishing, 2022.

Brito, Tarsis. '(Dis)Possessive Borders, (Dis)Possessed Bodies: Race and Property at the Postcolonial European Borders'. *International Political Sociology* 17, no. 2 (20 March 2023): olad009. https://doi.org/10.1093/ips/olad009.

Brockington, Dan, International African Institute. *Fortress Conservation: The Preservation of the Mkomazi Game Reserve, Tanzania*. Oxford: James Currey, 2002.

Brockington, Dan, and Katherine Scholfield. 'The Conservationist Mode of Production and Conservation NGOs in Sub-Saharan Africa'. *Antipode* 42, no. 3 (2010): 551–75. https://doi.org/10.1111/j.1467-8330.2010.00763.x.

Brockington, Dan, Rosaleen Duffy, and Jim Igoe. *Nature Unbound: Conservation, Capitalism and the Future of Protected Areas.* 1st edition. London: Routledge, 2008.

Brooks, James, and Leah Wingard. 'Evolutions in Hegemonic Discourses of Climate Change: An Ecomodern Enactment of Implicatory Denial'. *Discourse & Society* 35, no. 3 (1 May 2024): 312–28. https://doi.org/10.1177/09579265231211487.

Brotherton, Chloe. '"We Have the Best Gays, Folks": Homonationalism and Islamophobia in a Critical Discourse Analysis of a pro-Trump Social Media Forum'. *Gender and Language* 17, no. 1 (15 May 2023): 39–53. https://doi.org/10.1558/genl.18550.

Brown, Wendy. *In the Ruins of Neoliberalism: The Rise of Antidemocratic Politics in the West.* New York: Columbia University Press, 2019.

Brown, Wendy. 'Revaluing Critique: A Response to Kenneth Baynes'. *Political Theory* 28, no. 4 (1 August 2000): 469–79. https://doi.org/10.1177/0090591700028004002.

Brown, Wendy. *Undoing the Demos: Neoliberalism's Stealth Revolution*. New York: Zone Books, 2017.

Brüggemeier, Franz-Josef, Mark Cioc, and Thomas Zeller, eds. *How Green Were the Nazis? Nature, Environment, and Nation in the Third Reich*. Athens: Ohio University Press, 2005. http://site.ebrary.com/id/10141108.

Bryant, Bunyan I., and Paul Mohai. *Race and the Incidence of Environmental Hazards: A Time for Discourse*. Boulder, CO: Westview Press, 1992.

Buckingham, Susan, and Rakibe Kulcur. 'Gendered Geographies of Environmental Injustice'. *Antipode* 41, no. 4 (1 September 2009): 659–83. https://doi.org/10.1111/j.1467-8330.2009.00693.x.

Buckley, Cara. 'Earth Now Has 8 Billion Humans. This Man Wishes There Were None.' *New York Times*, 23 November 2022, sec. Climate. https://www.nytimes.com/2022/11/23/climate/voluntary-human-extinction.html.

Buell, Lawrence. *The Environmental Imagination: Thoreau, Nature Writing, and the Formation of American Culture*. Harvard, MA: Harvard University Press, 1996.

Bullard, Robert D., Paul Mohai, Robin Saha, and Beverly Wright. 'Toxic Wastes and Race at Twenty: 1987–2007'. Cleveland: United Church of Christ Witness Ministries, 2007. http://citeseerx.ist.psu.edu/viewdoc/download?doi=10.1.1.366.4145&rep=rep1&type=pdf.

Buller, Adrienne. *The Value of a Whale: On the Illusions of Green Capitalism*. Manchester: Manchester University Press, 2022.

Burnett, Scott. 'Adamastor Unbound? Whiteness and Landscape in Post-1994 South Africa'. In *Sentient Ecologies: Xenophobic Imaginaries of Landscape*, edited by Alexandra Coţofană and Hikmet Kuran, 23–42. New York: Berghahn Books, 2022. https://doi.org/10.3167/9781800736627.

Burnett, Scott. 'Constructing White Autochthony in South Africa's "Soul Country": Intersections of Race and Land'. *Discourse, Context & Media*, no. 30 (4 January 2019): 1–11. https://doi.org/10.1016/j.dcm.2018.12.002.

Burnett, Scott. 'Healthy White Nationalists: Far-Right Selbstbilder in a Digital Age'. In *Masculinities and Discourses of Men's Health*, edited by Gavin Brookes and Małgorzata Chałupnik, 161–88. Palgrave Studies in Language, Gender and Sexuality. Cham: Springer International Publishing, 2023. https://doi.org/10.1007/978-3-031-38407-3_7.

Burnett, Scott. 'Right as Rain: Affective Publics and the Changing Visual Rhetoric of the Far Right in South Africa'. In *Visualising Far-Right Environments: Communication and the Politics of Nature*, edited by Bernhard Forchtner, 24–42. Manchester: Manchester University Press, 2023. https://manchesteruniversitypress.co.uk/9781526165381/.

Burnett, Scott. *White Belongings: Race, Land, and Property in Post-Apartheid South Africa*. Lanham, MD: Lexington Books, 2022. https://rowman.com/ISBN/9781793654946/White-Belongings-Race-Land-and-Property-in-Post-Apartheid-South-Africa.

Burnett, Scott. '"Without White People, the Animals Will Go!": COVID-19 and the Struggle for the Future of South African Conservation'. *Environment and Planning E: Nature and Space* (11 January 2022): 25148486211069884. https://doi.org/10.1177/25148486211069884.

Burnett, Scott, and Aylwyn M. Walsh. 'Semiotic Citizenship and the Politics of Shack-Building in Graaff-Reinet, South Africa'. In *The Routledge Handbook of Architecture, Urban Space and Politics, Volume II*, 339–409. Abingdon: Routledge, 2024.

Burnett, Scott, and John E. Richardson. '"Breeders for Race and Nation": Gender, Sexuality and Fecundity in Post-War British Fascist Discourse'. *Patterns of Prejudice* 55, no. 4 (27 April 2022): 331–56. https://doi.org/10.1080/0031322X.2021.2011088.

Burnett, Scott, and Tommaso M. Milani. 'Fatal Masculinities: A Queer Look at Green Violence'. *ACME: An International Journal for Critical Geographies* 16, no. 3 (2017): 548–75.

Burns, Robert. 'To a Mouse, On Turning Her Up in Her Nest with the Plough'. *Scottish Poetry Library* (blog), November 1785. https://www.scottishpoetrylibrary.org.uk/poem/mouse/.

Büscher, Bram. 'Nature 2.0'. *Geoforum*, Global Production Networks, Labour and Development 44 (1 January 2013): 1–3. https://doi.org/10.1016/j.geoforum.2012.08.004.

Büscher, Bram. '"Rhino Poaching Is out of Control!" Violence, Race and the Politics of Hysteria in Online Conservation'. *Environment and Planning A* 48, no. 5 (1 May 2016): 979–98. https://doi.org/10.1177/0308518X16630988.

Büscher, Bram. *The Truth about Nature: Environmentalism in the Era of Post-Truth Politics and Platform Capitalism*. Oakland, CA: University of California Press, 2020.

Büscher, Bram, and Maano Ramutsindela. 'Green Violence: Rhino Poaching and the War to Save Southern Africa's Peace Parks'. *African Affairs* 115, no. 458 (1 January 2016): 1–22. https://doi.org/10.1093/afraf/adv058.

Büscher, Bram, and Robert Fletcher. 'Accumulation by Conservation'. *New Political Economy* 20, no. 2 (4 March 2015): 273–98. https://doi.org/10.1080/13563467.2014.923824.

Büscher, Bram, and Robert Fletcher. *The Conservation Revolution: Radical Ideas for Saving Nature beyond the Anthropocene*. London: Verso, 2020.

Büscher, Bram, and Robert Fletcher. 'Towards Convivial Conservation'. *Conservation and Society* 17, no. 3 (1 July 2019): 283. https://doi.org/10.4103/cs.cs_19_75.

Büscher, Bram, and Rosaleen Duffy. 'Biodiversity Treaty: UN Deal Fails to Address the Root Causes of Nature's Destruction'. *The Conversation*, 21 December 2022. http://theconversation.com/biodiversity-treaty-un-deal-fails-to-address-the-root-causes-of-natures-destruction-196905.

Büscher, Bram, Stasja Koot, and Ingrid L. Nelson. 'Introduction. Nature 2.0: New Media, Online Activism and the Cyberpolitics of Environmental Conservation'. *Geoforum* 79 (February 2017): 111–13. https://doi.org/10.1016/j.geoforum.2016.12.001.

Büscher, Bram, Stasja Koot, and Lerato Thakholi. 'Fossilized Conservation, or the Unsustainability of Saving Nature in South Africa'. *Environment and Planning E: Nature and Space* 7, no. 1 (1 February 2024): 50–68. https://doi.org/10.1177/25148486211062002.

Bush, Tori. 'Eco-Orientalism: Power Discourses on Isle de Jean Charles'. *ISLE: Interdisciplinary Studies in Literature and Environment* 29, no. 2 (8 July 2022): 248–59. https://doi.org/10.1093/isle/isac011.

Butler, Judith. 'What Is Critique? An Essay on Foucault's Virtue'. *Transversal Texts*, May 2001. https://transversal.at/transversal/0806/butler/en.

Byron, George Gordon Lord. *Lord Byron: The Major Works*. Edited by Jerome J. McGann. Oxford: Oxford University Press, 2008.

Cameron, Deborah. *Verbal Hygiene*. Abingdon: Routledge, 1995.

Campion, Kristy. 'Defining Ecofascism: Historical Foundations and Contemporary Interpretations in the Extreme Right'. *Terrorism and Political Violence* 35, no. 4 (19 May 2023): 926–44. https://doi.org/10.1080/09546553.2021.1987895.

Cane, Jonathan. *Civilising Grass: The Art of the Lawn on the South African Highveld*. Johannesburg: Wits University Press, 2019.

Carrington, Damian. 'Revealed: Oil Sector's "Staggering" $3bn-a-Day Profits for Last 50 Years'. *The Guardian*, 21 July 2022, sec. Environment. https://www.theguardian.com/environment/2022/jul/21/revealed-oil-sectors-staggering-profits-last-50-years.

Carruthers, Jane. *The Kruger National Park: A Social and Political History*. Pietermaritzburg: University of Natal Press, 1995.

Carson, Rachel. *Silent Spring*. Boston, MA: Houghton Mifflin, 1962.

Chakravartty, Paula, and Denise Ferreira da Silva. 'Accumulation, Dispossession, and Debt: The Racial Logic of Global Capitalism – An Introduction'. *American Quarterly* 64, no. 3 (2012): 361–85. https://doi.org/10.1353/aq.2012.0033.

Chanzo Reporter. 'Tanzania Delists All Wards and Villages in the Contested Ngorongoro Area, Stakeholders Warn the Plan Is Unconstitutional'. *The Chanzo*, 20 August 2024, sec. News. https://thechanzo.com/2024/08/20/tanzania-delists-all-wards-and-villages-in-the-contested-ngorongoro-area-stakeholders-warn-the-plan-is-unconstitutional/.

Chaudhary, Ajay Singh. *The Exhausted of the Earth: Politics in a Burning World*. London: Repeater, 2024.

Chen, Mel Y. *Animacies: Biopolitics, Racial Mattering, and Queer Affect*. Durham: Duke University Press, 2012.

Chiapello, Eve, and Norman Fairclough. 'Understanding the New Management Ideology: A Transdisciplinary Contribution from Critical Discourse Analysis and New Sociology of Capitalism'. *Discourse & Society* 13, no. 2 (March 2002): 185–208. https://doi.org/10.1177/0957926502013002406.

Chilton, Paul, and Ruth Wodak. 'Preface'. In *New Agenda in (Critical) Discourse Analysis: Theory, Methodology and Interdisciplinarity*, edited by Ruth Wodak and

Paul Chilton, xi–xviii. Amsterdam and Philadelphia, PA: John Benjamins, 2005. http://ebookcentral.proquest.com/lib/pensu/detail.action?docID=622905.

Chouliaraki, Lilie, and Norman Fairclough. *Discourse in Late Modernity: Rethinking Critical Discourse Analysis*. 1st edition. Edinburgh: Edinburgh University Press, 1999.

Chouliaraki, Lilie, Myria Georgiou, Rafal Zaborowski, and Wouter Oomen. 'The European "Migration Crisis" and the Media: A Cross-European Press Content Analysis'. London: Department of Media and Communications London School of Economics and Political Science, 2017.

Chun, Christian W. *The Discourses of Capitalism: Everyday Economists and the Production of Common Sense*. 1st edition. London; New York: Routledge, 2017.

Cloete, Elsie. 'Women, Home and Personal Narrative in Kenya'. *Journal of Literary Studies* 24, no. 1 (March 2008): 82–99. https://doi.org/10.1080/02564710701789057.

Cocks, Catherine. 'The Pleasures of Degeneration: Climate, Race, and the Origins of the Global Tourist South in the Americas'. *Discourse* 29, no. 2 (2007): 215–35.

Coetzee, J. M. *White Writing: On the Culture of Letters in South Africa*. New Haven, CT: Yale University Press, 1988.

Cohen, Daniel Aldana, and Thea Riofrancos. 'Biden Left Us with a "Prius Economy." It's Time for Something Different.' *New York Times*, 7 January 2025, sec. Opinion. https://www.nytimes.com/2025/01/07/opinion/electric-vehicles-tax-credits.html.

Cohen, Erik. 'Tourism and Land Grab in the Aftermath of the Indian Ocean Tsunami'. *Scandinavian Journal of Hospitality and Tourism* 11, no. 3 (1 October 2011): 224–36. https://doi.org/10.1080/15022250.2011.593359.

Collard, Rosemary-Claire, and Jessica Dempsey. 'Future Eco-Perfect: Temporal Fixes of Liberal Environmentalism'. *Antipode* 54, no. 5 (2022): 1545–65. https://doi.org/10.1111/anti.12849.

Collins, Patricia Hill. *Black Feminist Thought*. Routledge Classics. New York and Abingdon: Routledge Classics, 2009. https://futuresinitiative.org/para/wp-content/uploads/sites/196/2017/08/Collins-Patricia.-Black-Feminist-Thought-copy.pdf?x83816.

Collins, Patricia Hill. 'Intersectionality's Definitional Dilemmas'. *Annual Review of Sociology* 41, no. 1 (14 August 2015): 1–20. https://doi.org/10.1146/annurev-soc-073014-112142.

'Colony'. In *Merriam-Webster.Com Dictionary*. Merriam-Webster, no date. https://www.merriam-webster.com/dictionary/colony.

Conklin, Beth A., and Laura R. Graham. 'The Shifting Middle Ground: Amazonian Indians and Eco-Politics'. *American Anthropologist* 97, no. 4 (1995): 695–710. https://doi.org/10.1525/aa.1995.97.4.02a00120.

Connell, Raewyn. *Southern Theory: The Global Dynamics of Knowledge in Social Science*. Cambridge; Malden: Polity, 2007.

Conversi, Daniele. 'The Ultimate Challenge: Nationalism and Climate Change'. *Nationalities Papers* 48, no. 4 (July 2020): 625–36. https://doi.org/10.1017/nps.2020.18.

Conversi, Daniele, and Lorenzo Posocco. 'Which Nationalism for the Anthropocene? A Comparative Study of Exemplary Green Nation-States'. *Frontiers in Political Science* 4 (2022): 1–16. https://doi.org/10.3389/fpos.2022.857597.

Copsey, Nigel, and Matthew Worley. 'Introduction'. In *Tomorrow Belongs to Us: The British Far Right since 1967*, edited by Nigel Copsey and Matthew Worley, i–xx. Abingdon and New York: Routledge, 2018.

Corry, Stephen. 'Ngorongoro Nazi', 18 April 2022. https://www.theelephant.info/analysis/2022/04/18/ngorongoro-nazi/.

Coțofană, Alexandra. 'Ingrained Ontologies: How Romania's Institutionalized Processes Teach Us to Think with Xenophobic Sentient Landscapes'. In *Sentient Ecologies: Xenophobic Imaginaries of Landscape*, edited by Alexandra Coțofană and Hikmet Kuran, 1st edition, 67–97. Environmental Anthropology and Ethnobiology 31. New York: Berghahn Books, 2022.

Cox, Robert. *Environmental Communication and the Public Sphere*. Thousand Oaks: SAGE, 2013.

Cox, Robert. 'Nature's "Crisis Disciplines": Does Environmental Communication Have an Ethical Duty?' *Environmental Communication* 1, no. 1 (1 May 2007): 5–20. https://doi.org/10.1080/17524030701333948.

Cox, Robert, and Stephen Depoe. 'Environment, Communication and Environmental Communication: Emergence and Development of a Field'. In *The Routledge Handbook of Environment and Communication*, edited by Anders Hansen and Robert Cox, 2nd edition, 13–32. Abingdon and New York: Routledge, 2023.

Crenshaw, Kimberlé W. 'Demarginalizing the Intersection of Race and Sex: A Black Feminist Critique of Antidiscrimination Doctrine, Feminist Theory and Antiracist Politics'. *University of Chicago Legal Forum* 1989, no. 1 (1989): 139–67. https://doi.org/10.12691/education-3-7-4.

Crenshaw, Kimberlé W. 'Mapping the Margins: Intersectionality, Identity Politics, and Violence against Women of Color'. *Stanford Law Review* 43, no. 6 (July 1991): 1241–99. https://doi.org/10.2307/1229039.

Cretney, Raven, and Sylvia Nissen. 'Emergent Spaces of Emergency Claims: Possibilities and Contestation in a National Climate Emergency Declaration'. *Antipode* 54, no. 5 (2022): 1566–84. https://doi.org/10.1111/anti.12843.

Cummings, Stephen. 'Shifting Foundations: Redrawing Strategic Management's Military Heritage'. *Critical Perspectives on International Business* 3, no. 1 (6 February 2007): 41–62. https://doi.org/10.1108/17422040710722551.

Curtin, Philip D. *The Image of Africa: British Ideas and Action, 1780–1850*. Madison, WI: University of Wisconsin Press, 1964.

Daggett, Cara. 'Petro-Masculinity: Fossil Fuels and Authoritarian Desire'. *Millennium: Journal of International Studies* 47, no. 1 (September 2018): 25–44. https://doi.org/10.1177/0305829818775817.

Dalby, Simon. 'The Environment as Geopolitical Threat: Reading Robert Kaplan's "Coming Anarchy"'. *Cultural Geographies* 3, no. 4 (1 October 1996): 472–96. https://doi.org/10.1177/147447409600300406.

Daly, Herman E. *Beyond Growth: The Economics of Sustainable Development*. New edition. Boston, MA: Beacon Press, 1997.

D'Angelo, Paul. 'Framing: Media Frames'. In *The International Encyclopedia of Media Effects*, edited by Patrick Rössler, Cynthia A. Hoffner, and Liesbet Zoonen, 1st edition, 1–10. Wiley, 2017. https://doi.org/10.1002/9781118783764.wbieme0048.

Dardot, Pierre, and Christian Laval. *The New Way of the World: On Neo-Liberal Society*. Translated by Gregory Elliot. London: Verso, 2017.

Darrah-Okike, Jennifer. 'Theorizing Race in Hawai'i: Centering Place, Indigeneity, and Settler Colonialism'. *Sociology Compass* 14, no. 7 (2020): e12791. https://doi.org/10.1111/soc4.12791.

Davies, Lizzy. '"Use Your £11bn Climate Fund to Pay for Family Planning," UK Told'. *The Guardian*, 26 August 2021, sec. Global development. https://www.theguardian.com/global-development/2021/aug/26/use-your-11bn-climate-fund-to-pay-for-family-planning-uk-told.

Davies, Thom, Tesfalem H. Yemane, Joe Turner, Lucy Mayblin, and Arshad Isakjee. 'Eco-Coloniality and the Violent Environmentalism of the UK–France Border'. *Environment and Planning D: Society and Space* 42, no. 5–6 (December 2024): 776–802. https://doi.org/10.1177/02637758241264415.

Davis, Heather, and Zoe Todd. 'On the Importance of a Date, or, Decolonizing the Anthropocene'. *ACME: An International Journal for Critical Geographies* 16, no. 4 (20 December 2017): 761–80. https://doi.org/10.14288/acme.v16i4.1539.

De Cleen, Benjamin, Jana Goyvaerts, Nico Carpentier, Jason Glynos, and Yannis Stavrakakis. 'Moving Discourse Theory Forward: A Five-Track Proposal for Future Research'. *Journal of Language and Politics* 20, no. 1 (21 January 2021): 22–46. https://doi.org/10.1075/jlp.20076.dec.

Deberdt, Raphael, and Philippe Le Billon. 'Green Transition's Necropolitics: Inequalities, Climate Extractivism, and Carbon Classes'. *Antipode* 56, no. 4 (2024): 1264–88. https://doi.org/10.1111/anti.13032.

Deleuze, Gilles, and Félix Guattari. *A Thousand Plateaus: Capitalism and Schizophrenia*. Translated by Brian Massumi. 1988. Reprint, London and New York: Continuum, 2004.

Deloria, Vine. *Red Earth, White Lies: Native Americans and the Myth of Scientific Fact*. Golden, CO: Fulcrum Publishing, 1997.

DeLuca, Kevin Michael, and Anne Teresa Demo. 'Imagining Nature and Erasing Class and Race: Carleton Watkins, John Muir, and the Construction of Wilderness'. *Environmental History* 6, no. 4 (1 October 2001): 541–60. https://doi.org/10.2307/3985254.

Descola, Philippe. *Beyond Nature and Culture*. Chicago, IL: University of Chicago Press, 2013.

Di Chiro, Giovanna. 'Polluted Politics? Confronting Toxic Discourse, Sex Panic, and Eco-Normativity'. In *Queer Ecologies: Sex, Nature, Politics, Desire*, edited by Catriona Mortimer-Sandilands and Bruce Erickson, 190–230. Bloomington, IN: Indiana University Press, 2010.

Dickie, Gloria. 'Climate Justice Gets Harder as World Population Passes 8 Billion'. *Daily Maverick*, 15 November 2022. https://www.dailymaverick.co.za/article/2022-11-15-climate-justice-gets-harder-as-world-population-passes-8-billion/.

Dijk, Teun A. van. 'Analyzing Frame Analysis: A Critical Review of Framing Studies in Social Movement Research'. *Discourse Studies* 25, no. 2 (1 April 2023): 153–78. https://doi.org/10.1177/14614456231155080.

Dijk, Teun A. van. 'Ideological Discourse Analysis'. *New Courant* 4, no. 1 (1995): 135–61.

Dinerstein, Eric, David Olson, Anup Joshi, Carly Vynne, Neil D. Burgess, Eric Wikramanayake, Nathan Hahn, et al. 'An Ecoregion-Based Approach to Protecting Half the Terrestrial Realm'. *BioScience* 67, no. 6 (1 June 2017): 534–45. https://doi.org/10.1093/biosci/bix014.

Dixon, John A., Don H. Foster, Kevin Durrheim, and Lindy Wilbraham. 'Discourse and the Politics of Space in South Africa: The "Squatter Crisis"'. *Discourse & Society* 5, no. 3 (1 July 1994): 277–96. https://doi.org/10.1177/0957926594005003002.

Dlamini, Jacob S. T. *Native Nostalgia*. Auckland Park: Jacana, 2009.

Dobruszkes, Frédéric, Giulio Mattioli, and Enzo Gozzoli. 'The Elephant in the Room: Long-Haul Air Services and Climate Change'. *Journal of Transport Geography* 121 (1 December 2024): 104022. https://doi.org/10.1016/j.jtrangeo.2024.104022.

Dryzek, John S. *The Politics of the Earth: Environmental Discourses*. 4th edition. Oxford: Oxford University Press, 2022.

D'Souza, Rohan. 'Environmental History of South Asia in the Time of Hindutva'. *Environmental History* 27, no. 4 (October 2022): 625–33. https://doi.org/10.1086/721410.

Du Bois, John W. 'The Stance Triangle'. In *Stancetaking in Discourse: Subjectivity, Evaluation, Interaction*, edited by R. Englebretson, 139–82. Amsterdam and Philadelphia, PA: John Benjamins Publishing, 2007.

Du Bois, W. E. B. *Darkwater: Voices from within the Veil*. London: Verso, 2016. https://public.ebookcentral.proquest.com/choice/publicfullrecord.aspx?p=5069597.

Du Mez, Kristin Kobes. *Jesus and John Wayne: How White Evangelicals Corrupted a Faith and Fractured a Nation*. New York, NY: Liveright, 2020.

Duffield, Mark. 'Total War as Environmental Terror: Linking Liberalism, Resilience, and the Bunker'. *South Atlantic Quarterly* 110, no. 3 (1 July 2011): 757–69. https://doi.org/10.1215/00382876-1275779.

Duffy, Rosaleen V., and Dan Brockington. 'Political Ecology of Security: Tackling the Illegal Wildlife Trade'. *Journal of Political Ecology* 29, no. 1 (20 January 2022): 21–35.

Duffy, Rosaleen. 'Waging a War to Save Biodiversity: The Rise of Militarized Conservation'. *International Affairs* 90, no. 4 (July 2014): 819–34. https://doi.org/10.1111/1468-2346.12142.

Dunaway, Finis. 'The "Crying Indian" Ad That Fooled the Environmental Movement'. *Chicago Tribune*, 21 November 2017.

Dunlap, Alexander. 'Decolonizing Energy Justice: Identity Politics, Attachment to Land, and Warfare'. *Progress in Human Geography* 47, no. 2 (1 April 2023): 348–50. https://doi.org/10.1177/03091325221149553.

Dwyer, Owen J., and John Paul Jones III. 'White Socio-Spatial Epistemology'. *Social & Cultural Geography* 1, no. 2 (2000): 209–22. https://doi.org/10.1080/14649360020010211.

Eckert, Penelope. 'Variation, Meaning and Social Change'. In *Sociolinguistics: Theoretical Debates*, edited by Nikolas Coupland, 68–85. Cambridge: Cambridge University Press, 2016. https://doi.org/10.1017/CBO9781107449787.004.

Eddins, J. J. Board of Land and Natural Resources v. Crabtree, West's Hawai'i Reports and Pacific Reporter (The Supreme Court of the State of Hawai'i 2024).

Eder, Klaus. 'The Institutionalisation of Environmentalism: Ecological Discourse and the Second Transformation of the Public Sphere'. In *Risk, Environment and Modernity: Towards a New Ecology*, edited by Scott Lash, Bronislaw Szerszynski & Brian Wynne, 204–23. London: SAGE, 1996. https://doi.org/10.4135/9781446221983.

Eghbariah, Rabea. 'The Struggle for Za'atar and Akkoub: Israeli Nature Protection Laws and the Criminalization of Palestinian Herb-Picking Culture'. Oxford: Oxford Food Symposium, 2020.

Ehrlich, Paul R. *The Population Bomb*. New York: Ballantine Books, 1968.

Elamin, Nisrin Omer. '"When Mud Comes between Us": Contemporary Land Enclosures and the Politics of Claim-Making in Central Sudan'. Ph.D., Stanford University, 2019. https://www.proquest.com/docview/2734703755/abstract/C1D57AA143E54DC5PQ/1.

Elder, Catriona. 'Narratives of National Belonging: The Domestic, the Wild, and the Mutant in Australian Film.' *Social Identities* 22, no. 5 (1 September 2016): 521–36. https://doi.org/10.1080/13504630.2016.1145586.

Ellis, Erle. 'Why I Resigned from the Anthropocene Working Group'. *Anthroecology Lab* (blog), 13 July 2023. https://anthroecology.org/why-i-resigned-from-the-anthropocene-working-group/.

euronews green with Reuters. 'Climate Activists Stop Berlin Traffic by Gluing Themselves to Roads'. *Euronews*, 25 April 2023. https://www.euronews.com/green/2023/04/25/berlin-activists-glue-themselves-to-roads-causing-massive-disruption-across-the-city.

Fairclough, Norman. 'A Dialectical-Relational Approach to Critical Discourse Analysis in Social Research'. In *Methods for Critical Discourse Analysis*, edited by Ruth Wodak and Michael Meyer, 2nd edition, 162–86. London: SAGE, 2009.

Fairclough, Norman. *Analysing Discourse: Textual Analysis for Social Research*. London: Routledge, 2003.

Fairclough, Norman. *Discourse and Social Change*. Cambridge: Polity Press, 1992.

Fairclough, Norman. *Language and Power*. 2nd edition. London; New York: Longman, 2001.

Fairclough, Norman, and Ruth Wodak. 'Critical Discourse Analysis'. In *Discourse Studies: A Multidisciplinary Introduction – Volume 2: Discourse as Social Interaction*, edited by Teun A. van Dijk, 258–84. London: SAGE, 1997.

Fanon, Frantz. *Black Skin, White Masks*. 1952. Reprint, London: Pluto Books, 2008.

Farage, Nigel. 'It's Official … Britain's Population Surge Is TOTALLY out of Control'. *The Express*, 30 October 2015, sec. Comment. https://www.express.co.uk/comment/expresscomment/615678/Nigel-Farage-Ukip-Britain-immigration-figures-soar-uncontrolled.

Farris, Sara R. *In the Name of Women's Rights: The Rise of Femonationalism*. Durham: Duke University Press, 2017.

Feldman, Rachel Z., and Ian McGonigle. 'Introduction'. In *Settler-Indigeneity in the West Bank*, edited by Rachel Z. Feldman and Ian McGonigle, 3–44. Montreal: McGill-Queen's University Press, 2023.

Feola, Giuseppe, Michael K. Goodman, Jaime Suzunaga, and Jenny Soler. 'Collective Memories, Place-Framing and the Politics of Imaginary Futures in Sustainability Transitions and Transformation'. *Geoforum* 138 (1 January 2023): 103668. https://doi.org/10.1016/j.geoforum.2022.103668.

Fill, Alwin, and Hermine Penz, eds. *The Routledge Handbook of Ecolinguistics*. Routledge Handbooks in Linguistics. New York: Routledge, 2018.

Fill, Alwin, and Peter Mühlhäusler. *Ecolinguistics Reader: Language, Ecology and Environment*. London: Continuum, 2001.

Fischer, Tin, and Hannah Knuth. '$CO_2$-Zertifikate: Greenwashing in der afrikanischen Savanne'. *Die Zeit*, 17 March 2023. https://www.zeit.de/wirtschaft/2023-03/co2-zertifikate-netflix-luxus-kritik.

Fisher, Mark. *Capitalist Realism: Is There No Alternative?* Winchester: Zero Books, 2010.

Fitzgerald, Des. 'Normal Island: COVID-19, Border Control, and Viral Nationalism in UK Public Health Discourse'. *Sociological Research Online* 28, no. 2 (June 2023): 596–606. https://doi.org/10.1177/13607804211049464.

Flavelle, Christopher, and Manuela Andreoni. 'How Climate Change Turned Lush Hawaii into a Tinderbox'. *The New York Times*, 10 August 2023, sec. Climate. https://www.nytimes.com/2023/08/10/climate/hawaii-fires-climate-change.html.

Fletcher, Robert. *Failing Forward: The Rise and Fall of Neoliberal Conservation*. Oakland, CA: University of California Press, 2023.

Fletcher, Robert, Jan Breitling, and Valerie Puleo. 'Barbarian Hordes: The Overpopulation Scapegoat in International Development Discourse'. *Third World Quarterly* 35, no. 7 (9 August 2014): 1195–215. https://doi.org/10.1080/01436597.2014.926110.

Flowerdew, John. 'Critical Discourse Analysis and Strategies of Resistance'. In *Advances in Discourse Studies*, edited by Vijay Kumar Bhatia, John Flowerdew, and Rodney H. Jones, 195–210. London: Routledge, 2008.

Forchtner, Bernhard, and Ana Tominc. 'Kalashnikov and Cooking-Spoon: Neo-Nazism, Veganism and a Lifestyle Cooking Show on YouTube'. *Food, Culture & Society* 20, no. 3 (September 2017): 415–41. https://doi.org/10.1080/15528014.2017.1337388.

Forchtner, Bernhard, and Christoffer Kølvraa. 'Extreme Right Images of Radical Authenticity: Multimodal Aesthetics of History, Nature, and Gender Roles in Social Media'. *European Journal of Cultural and Political Sociology* 4, no. 3 (3 July 2017): 252–81. https://doi.org/10.1080/23254823.2017.1322910.

Forchtner, Bernhard, and Jonathan Olsen. 'Against the Promethean: Energy Throughput and the Far-Right Politics of Degrowth'. *Environment and Planning E: Nature and Space* (23 July 2024) 25148486241266794. https://doi.org/10.1177/25148486241266794.

Ford, Allison E. '"They Will Be Like a Swarm of Locusts": Race, Rurality, and Settler Colonialism in American Prepping Culture'. *Rural Sociology* 86, no. 3 (2021): 469–93. https://doi.org/10.1111/ruso.12382.

Foster, Jeremy A. 'From Socio-Nature to Spectral Presence: Re-Imagining the Once and Future Landscape of Johannesburg'. *Safundi* 10, no. 2 (1 April 2009): 175–213. https://doi.org/10.1080/17533170902793117.

Foster, Jeremy A. *Washed with Sun: Landscape and the Making of White South Africa.* Pittsburgh, PA: University of Pittsburgh Press, 2008.

Foucault, Michel. 'Of Other Spaces'. Translated by Jay Miskowiec. *Diacritics* 16, no. 1 (Spring 1986): 22–7. https://doi.org/10.4324/9780429399671-24.

Foucault, Michel. *Power/Knowledge: Selected Interviews and Other Writings: 1972–1977.* Edited by Colin Gordon. New York: Pantheon Books, 1980.

Fox, Graham R. 'The 2017 Shooting of Kuki Gallmann and the Politics of Conservation in Northern Kenya'. *African Studies Review* 61, no. 2 (2018): 210–36. https://doi.org/10.1017/asr.2017.130.

Franz, Tobias, and Angus McNelly. 'The "Finance-Extraction-Transitions Nexus": Geographies of the Green Transition in the 21st Century'. *Antipode* n/a, no. n/a. Accessed 16 May 2024. https://doi.org/10.1111/anti.13049.

Fraser, Nancy. *Cannibal Capitalism: How Our System Is Devouring Democracy, Care, and the Planetand What We Can Do about It.* London: Verso, 2022.

Fraser, Nancy. 'Climates of Capital'. *New Left Review*, no. 127 (25 February 2021): 94–127.

Freeden, Michael. 'Discourse, Concepts, Ideologies: Pausing for Thought'. *Journal of Language and Politics* 20, no. 1 (21 January 2021): 47–61. https://doi.org/10.1075/jlp.20051.fre.

Frenkel, Stephen. 'Geography, Empire, and Environmental Determinism'. *Geographical Review* 82, no. 2 (April 1992): 143. https://doi.org/10.2307/215428.

Fressoz, Jean-Baptiste. *More and More and More: An All-Consuming History of Energy.* London: Allen Lane, 2024.

Frykman, Jonas, and Orvar Löfgren. *Culture Builders: A Historical Anthropology of Middle-Class Life.* New Brunswick, NJ: Rutgers University Press, 1987.

Gafter, Roey J., and Tommaso M. Milani. 'Eco-Orientalism: The Discursive Dark Side of Wild-Life Conservation in Israel'. In *Subverting Space and Race: Towards a Geopolitically Conscious and Antiracist Sociology of Language*, edited by Mingdan Wu, Sibo Rugwiza Kanobana, and Jaspal Naveel Singh. Berlin: DeGruyter, Forthcoming.

Gal, Susan. 'Gender and the Discursive Authority of Far Right Politics'. *Gender and Language* 15, no. 1 (25 March 2021): 96–103. https://doi.org/10.1558/genl.19526.

Gal, Susan. 'Registers in Circulation: The Social Organization of Interdiscursivity'. *Signs and Society* 6, no. 1 (1 January 2018): 1–24. https://doi.org/10.1086/694551.

Gal, Susan. 'Scale-Making: Comparison and Perspective as Ideological Projects'. In *Scale: Discourse and Dimensions of Social Life*, edited by E. Summerson Carr and Michael Lempert, 91–111. Oakland, CA: University of California Press, 2016.

Galer, Sophia Smith. '"Greenwashing": Tree-Planting Schemes Are Just Creating Tree Cemeteries'. *VICE*, 1 September 2022. https://www.vice.com/en/article/tree-planting-schemes-england/.

Galliers, Chris, Roger Cole, Rohit Singh, Jeff Ohlfs, Hamera Aisha, Amon Benoit Koutoua, Carlien Roodt, and Mónica Álvarez Malvido. 'Conservation Casualties: An Analysis of on-Duty Ranger Fatalities (2006–2021)'. *Parks* 28, no. 1 (2022): 39–50.

Garnett, Stephen T., Neil D. Burgess, Julia E. Fa, Álvaro Fernández-Llamazares, Zsolt Molnár, Cathy J. Robinson, James E. M. Watson, Kerstin K. Zander, Beau Austin, Eduardo S. Brondizio, Neil French Collier, Tom Duncan, Erle Ellis, Hayley Geyle, Micha V. Jackson, Harry Jonas, Pernilla Malmer, Ben McGowan, Amphone Sivongxay, and Ian Leiper. 'A Spatial Overview of the Global Importance of Indigenous Lands for Conservation'. *Nature Sustainability* 1, no. 7 (July 2018): 369–74. https://doi.org/10.1038/s41893-018-0100-6.

Garrard, Greg. *Ecocriticism*. 3rd edition. Abingdon; New York: Routledge, 2023.

Gebru, Timnit, and Émile P. Torres. 'The TESCREAL Bundle: Eugenics and the Promise of Utopia through Artificial General Intelligence'. *First Monday*, 14 April 2024. https://doi.org/10.5210/fm.v29i4.13636.

Geertsema, Johan. 'White Natives? Dan Roodt, Afrikaner Identity and the Politics of the Sublime'. *The Journal of Commonwealth Literature* 41, no. 3 (1 September 2006): 103–20. https://doi.org/10.1177/0021989406068737.

George, Alan. '"Making the Desert Bloom" A Myth Examined'. *Journal of Palestine Studies* 8, no. 2 (Winter 1979): 88–100. https://doi.org/10.2307/2536511.

Gergan, Mabel, Sara Smith, and Pavithra Vasudevan. 'Earth beyond Repair: Race and Apocalypse in Collective Imagination'. *Environment and Planning D: Society and Space* (7 February 2018): 026377581875607. https://doi.org/10.1177/0263775818756079.

German, Laura A., Ryan Unks, and Elizabeth King. 'Green Appropriations through Shifting Contours of Authority and Property on a Pastoralist Commons'. *The Journal of Peasant Studies* 44, no. 3 (4 May 2017): 631–57. https://doi.org/10.1080/03066150.2016.1176562.

Gettleman, Jeffrey. 'Loss of Fertile Land Fuels "Looming Crisis" across Africa'. *New York Times*, 29 July 2017, sec. World. https://www.nytimes.com/2017/07/29/world/africa/africa-climate-change-kenya-land-disputes.html.

Ghosh, Amitav. *The Nutmeg's Curse: Parables for a Planet in Crisis*. 1st edition. Chicago, IL: University of Chicago Press, 2022.

Gilmore, Ruth Wilson. 'Fatal Couplings of Power and Difference: Notes on Racism and Geography'. *Professional Geographer* 54, no. 1 (February 2002): 15.

Gilmore, Ruth Wilson. *Golden Gulag: Prisons, Surplus, Crisis, and Opposition in Globalizing California*. American Crossroads 21. Berkeley, CA: University of California Press, 2007.

Goldstein, Matthew, and Emily Flitter. 'Cracking Down on a Wall Street Trend: E.S.G. Makeovers'. *New York Times*, 17 September 2022, sec. Business. https://www.nytimes.com/2022/09/17/business/dealbook/esg-wall-street.html.

Goodman, Michael K., Jo Littler, Dan Brockington, and Maxwell Boykoff. 'Spectacular Environmentalisms: Media, Knowledge and the Framing of Ecological Politics'. *Environmental Communication* 10, no. 6 (November 2016): 677–88. https://doi.org/10.1080/17524032.2016.1219489.

Gorney, Edna. 'Roots of Identity, Canopy of Collision: Re-Visioning Trees as an Evolving National Symbol within the Israeli-Palestinian Conflict'. In *Environmental History in the Making*, edited by Cristina Joanaz de Melo, Estelita Vaz, and Lígia M Costa Pinto, II: Acting: 327–44. Cham: Springer International Publishing, 2017. https://doi.org/10.1007/978-3-319-41139-2_18.

Gosine, Andil. 'Dying Planet, Deadly People: "Race"-Sex Anxieties and Alternative Globalizations'. *Social Justice* 32, no. 4 (2005): 69–86.

Gosine, Andil. 'Non-White Reproduction and Same-Sex Eroticism: Queer Acts against Nature'. In *Queer Ecologies: Sex, Nature, Politics, Desire*, edited by Catriona Mortimer-Sandilands and Bruce Erickson, 149–72. Bloomington, IN: Indiana University Press, 2010.

Gössling, Stefan, and Andreas Humpe. 'The Global Scale, Distribution and Growth of Aviation: Implications for Climate Change'. *Global Environmental Change* 65 (1 November 2020): 102194. https://doi.org/10.1016/j.gloenvcha.2020.102194.

Gough, Tim. 'Flows of Capitalism, Flows of Architecture'. *Ardeth*, MONEY: The Economies of Architecture II, no. 3 (30 November 2018): 96–114.

Goyanes, Cristina. 'This All-Female Anti-Poaching Unit in South Africa Is Saving Rhinos and Wildlife'. *National Geographic Online*, 12 October 2017. https://www.nationalgeographic.com/adventure/article/black-mambas-anti-poaching-wildlife-rhino-team.

Graeber, David. *Debt: The First 5,000 Years,Updated and Expanded*. Updated, Expanded edition. New York: Melville House, 2014.

Gramsci, Antonio. *Selections from the Prison Notebooks of Antonio Gramsci*. Edited and translated by Quintin Hoare and Geoffrey Nowell-Smith. London: Lawrence and Wishart, 1971.

Gray, Louise. 'David Attenborough – Humans Are Plague on Earth'. *The Telegraph*, 22 January 2013. http://www.telegraph.co.uk/news/earth/earthnews/9815862/Humans-are-plague-on-Earth-Attenborough.html.

Green, Louise. *Fragments from the History of Loss: The Nature Industry and the Postcolony*. University Park, PA: The Pennsylvania State University Press, 2021.

Gressier, Catie. *At Home in the Okavango: White Batswana Narratives of Emplacement and Belonging*. New York and Oxford: Berghahn Books, 2015.

Grove, Richard H. *Green Imperialism: Colonial Expansion, Tropical Island Edens and the Origins of Environmentalism, 1600–1860*. Edition Unstated. Cambridge: Cambridge University Press, 1996.

Guha, Ramachandra. 'Radical American Environmentalism and Wilderness Preservation: A Third World Critique'. *Environmental Ethics* 11, no. 1 (Spring 1989): 71–83.

Guzman, Dianne de. 'Second Garbage Patch the Size of Texas Discovered in Pacific Ocean'. News. SFGATE, 3 August 2017. https://www.sfgate.com/news/article/garbage-patch-south-pacific-texas-size-discovered-11730144.php.

Haag, Steffen. 'Old Colonial Power in New Green Financing Instruments. Approaching Financial Subordination from the Perspective of Racial Capitalism in Renewable Energy Finance in Senegal'. *Geoforum*, 21 October 2022. https://doi.org/10.1016/j.geoforum.2022.09.018.

Halberstam, Jack. *Female Masculinity*. 1st edition. Durham, NC: Duke University Press Books, 1998.

Hale, Tom. 'A Garbage Patch the Size of Mexico Was Just Discovered in the Pacific Ocean'. IFLScience, 31 July 2017. https://www.iflscience.com/yet-another-huge-plastic-garbage-patch-has-been-found-in-the-pacific-43017.

Hall, Stuart, Chas Critcher, Tony Jefferson, John Clarke, and Brian Roberts. *Policing the Crisis: Mugging, the State, and Law and Order*. Critical Social Studies. London and Basingstoke: Macmillan Press, 1978.

Halliday, Michael A. K. 'New Ways of Meaning: The Challenge to Applied Linguistics'. In *Thirty Years of Linguistic Evolution: Studies in Honour of René Dirven on the Occasion of His Sixtieth Birthday*, edited by Martin Pütz, 59–95. Amsterdam and Philadelphia, PA: John Benjamins Publishing, 1992.

Hamming, Jeanne. 'Nationalism, Masculinity, and the Politics of Climate Change in the Novels of Kim Stanley Robinson and Michael Crichton'. *Extrapolation* 54, no. 1 (2013): 21–45.

Handel, Ariel, Daniel Monterescu, and Rafi Grosglik. 'Soulful Soil and Colonial Quality: Organic Farming in the West Bank'. In *Settler-Indigeneity in the West Bank*, edited by Rachel Z. Feldman and Ian McGonigle, 68–97. Montreal, Quebec: McGill-Queen's University Press, 2023. https://doi.org/10.2307/jj.8085217.7.

Hanes, Stephanie. *White Man's Game: Saving Animals, Rebuilding Eden, and Other Myths of Conservation in Africa*. New York: Henry Holt, 2017.

Hannigan, John. *Environmental Sociology*. 4th edition. London: Routledge, 2022. https://doi.org/10.4324/9781003193777.

Hansen, Anders, and David Machin. 'Visually Branding the Environment: Climate Change as a Marketing Opportunity'. *Discourse Studies* 10, no. 6 (December 2008): 777–94. https://doi.org/10.1177/1461445608098200.

Haraway, Donna J. 'Anthropocene, Capitalocene, Plantationocene, Chthulucene: Making Kin'. *Environmental Humanities* 6, no. 1 (1 May 2015): 159–65. https://doi.org/10.1215/22011919-3615934.

Hardin, Garrett. 'The Tragedy of the Commons'. *Science* 162, no. 3859 (13 December 1968): 1243–8. https://doi.org/10.1126/science.162.3859.1243.

Harding, Sandra. *Whose Science? Whose Knowledge? By Sandra G. Harding | Paperback*. Ithaca, NY: Cornell University Press, 1991. https://www.cornellpress.cornell.edu/book/9780801497469/whose-science-whose-knowledge/.

Harlan, Chico. 'How One Company's Plan to Help the Planet Went Off the Rails'. *The Washington Post*, 1 September 2024, sec. Climate. https://www.washingtonpost.com/climate-environment/2024/08/24/carbon-credits-cook-stoves-africa/.

Harris, Cheryl I. 'Whiteness as Property'. *Harvard Law Review* 106, no. 8 (1993): 1707–91.

Hartman, Matthew R. 'Beyond Climate Denial: White Supremacy and the Growth of Nationalist Environmentalism on the Right'. *Journal for the Study of Religion, Nature and Culture* 18, no. 3 (30 March 2024): 376–92. https://doi.org/10.1558/jsrnc.23634.

Hartmann, Betsy. 'Population Control in the New World Order'. *Development in Practice* 2, no. 3 (1992): 210–15.

Hartmann, Betsy. 'Rethinking Climate Refugees and Climate Conflict: Rhetoric, Reality and the Politics of Policy Discourse'. *Journal of International Development* 22, no. 2 (1 March 2010): 233–46. https://doi.org/10.1002/jid.1676.

Hau, Mark F. 'From Local Concerns to Global Challenges: Continuity and Change in Sub-State "Green Nationalism"'. *Frontiers in Political Science* 3 (2021). https://doi.org/10.3389/fpos.2021.764939.

Hawai'i Alliance for Progressive Action (HAPA). 'Stop Land Grabs in Lāhainā and Commit to a Just Recovery'. Action Network, no date. https://actionnetwork.org/petitions/stopmauilandgrabs.

'Hawaii Governor Calls on People to Visit West Maui When It Reopens in October: "Helping Our People Heal"'. *CBS Mornings*. CBS, 19 September 2023. https://www.cbsnews.com/news/hawaii-governor-josh-green-visitors-west-maui-lahaina-reopens-october-8/.

Hayasaki, Erika. 'A Year Ago, He Ran into the Flames. Today, the Scars Remain'. *Men's Health*, 8 August 2024. https://www.menshealth.com/health/a61765941/maui-wildfire-survivors-mental-health/.

Heise, Ursula K. 'Introduction: Planet, Species, Justice – and the Stories We Tell about Them'. In *The Routledge Companion to the Environmental Humanities*, edited by Ursula K. Heise, Jon Christensen, and Michelle Niemann, 1–10. Abingdon and New York: Taylor & Francis, 2017.

Heron, Kai. 'Capitalist Catastrophism and Eco-Apartheid'. *Geoforum* (13 October 2023): 103874. https://doi.org/10.1016/j.geoforum.2023.103874.

Heron, Kai. 'The Great Unfettering'. *New Left Review*, September 2022. https://newleftreview.org/sidecar/posts/the-great-unfettering.

Hickcox, Abby. 'Green Belt, White City: Race and the Natural Landscape in Boulder, Colorado'. *Discourse* 29, no. 2 (2007): 236–59.
Hickel, Jason. 'The Anti-Colonial Politics of Degrowth'. *Political Geography* (April 2021): 102404. https://doi.org/10.1016/j.polgeo.2021.102404.
Hickel, Jason. *The Divide: Global Inequality from Conquest to Free Markets*. New York: W. W. Norton, 2018.
Hickel, Jason, Kofi Klu, and Rupert Read. *Less Is More: How Degrowth Will Save the World*. Ely: Windmill Books, 2021.
Hill, Gord. 'How Protecting the Earth Became an Excuse for Murder'. News. *Grist* (blog), 12 April 2023. https://grist.org/indigenous/fortress-conservation-legacy-violence-comic-indigenous-land-30×30/.
Hiraishi, Ku'uwehi. 'Lahaina Fires Reveal Ongoing Power Struggle for West Maui Water Rights'. *Hawai'i Public Radio*, 17 August 2023, sec. Local News. https://www.hawaiipublicradio.org/local-news/2023-08-17/lahaina-fires-reveal-ongoing-power-struggle-for-west-maui-water-rights.
Hoeven, Sara van der. 'Guns and Conservation: Protecting Wildlife and Ensuring "Peace and Security" in Northern Kenya'. *Mambo!*, 4 February 2021, https://mambo.hypotheses.org/3043.
Holterman, Devin. 'Stopping the Slaughter: The Economic Rationale to Anti-Poaching in Tanzania'. *Geoforum* 135 (October 2022): 106–17. https://doi.org/10.1016/j.geoforum.2022.08.003.
Homer-Dixon, Thomas F. *Environment, Scarcity, and Violence*. Princeton, NJ: Princeton University Press, 1999.
Hook, Derek. 'Genealogy, Discourse, "Effective History": Foucault and the Work of Critique'. *Qualitative Research in Psychology* 2, no. 1 (1 January 2005): 3–31. https://doi.org/10.1191/1478088705qp025oa.
Hooks, bell. *Talking Back: Thinking Feminist, Thinking Black*. Boston, MA: South End Press, 1989.
Hornborg, Alf. 'Footprints in the Cotton Fields: The Industrial Revolution as Time–Space Appropriation and Environmental Load Displacement'. *Ecological Economics* 59, no. 1 (5 August 2006): 74–81. https://doi.org/10.1016/j.ecolecon.2005.10.009.
Hotchkiss, Jane. 'The Jungle of Eden: Kipling, Wolf Boys, and the Colonial Imagination'. *Victorian Literature and Culture* 29, no. 2 (September 2001): 435–49. https://doi.org/10.1017/S1060150301002108.
Houghton, John T., G. J. Jenkins, and J. J. Ephraums, eds. *Climate Change: The Scientific Assessment*. Cambridge, New York and Melbourne: Cambridge University Press, 1990. https://archive.ipcc.ch/publications_and_data/publications_ipcc_first_assessment_1990_wg1.shtml.
Huber, Matt. '"Green Capitalism" Is a Lie'. *Jacobin*, 19 October 2022. https://jacobin.com/2022/10/green-capitalism-blackrock-climate-finance.
Huber, Matthew T. *Climate Change as Class War: Building Socialism on a Warming Planet*. London and Brooklyn, NY: Verso, 2022.
Huber, Matthew T. 'Reflections on Climate, Class, and Strategy'. *Critical Sociology* 49, no. 7–8 (1 November 2023): 1343–9. https://doi.org/10.1177/08969205231189452.

Hughes, Alice C., Kévin Tougeron, Dominic A. Martin, Filippo Menga, Bruno H. P. Rosado, Sebastian Villasante, Shweta Madgulkar, Fernando Gonçalves, Davide Geneletti, Luisa Maria Diele-Viegas, Sebastian Berger, Sheila R. Colla, Vitor de Andrade Kamimura, Holly Caggiano, Felipe Melo, Marcelo Guilherme de Oliveira Dias, Elke Kellner, and Edivando Vitor do Couto. 'Smaller Human Populations Are Neither a Necessary nor Sufficient Condition for Biodiversity Conservation'. *Biological Conservation* 277 (1 January 2023): 109841. https://doi.org/10.1016/j.biocon.2022.109841.

Hughes, David McDermott. 'Third Nature: Making Space and Time in the Great Limpopo Conservation Area'. *Cultural Anthropology* 20, no. 2 (May 2005): 157–84.

Hughes, David McDermott. *Whiteness in Zimbabwe: Race, Landscape, and the Problem of Belonging*. 1st edition. New York: Palgrave Macmillan, 2010.

Hultgren, John. *Border Walls Gone Green: Nature and Anti-Immigrant Politics in America*. Minneapolis, MN: University of Minnesota Press, 2015. http://ebookcentral.proquest.com/lib/pensu/detail.action?docID=4391835.

Hultgren, John. 'Natural Exceptions to Green Sovereignty? American Environmentalism and the "Immigration Problem"'. *Alternatives: Global, Local, Political* 37, no. 4 (2012): 300–16.

Hultman, Martin. 'Natures of Masculinities: Conceptualising Industrial, Ecomodern and Ecological Masculinities'. In *Understanding Climate Change through Gender Relations*, edited by Susan Buckingham and Virginie Le Masson, 87–103. London and New York: Routledge, 2017.

Hultman, Martin. 'The Making of an Environmental Hero: A History of Ecomodern Masculinity, Fuel Cells and Arnold Schwarzenegger'. *Environmental Humanities* 2, no. 1 (1 January 2013): 79–99. https://doi.org/10.1215/22011919-3610360.

Humphreys, Jasper, and M. L. R. Smith. 'The "Rhinofication" of South African Security'. *International Affairs* 90, no. 4 (July 2014): 795–818. https://doi.org/10.1111/1468-2346.12141.

Ideland, Malin. *Den KRAV-märkta människan*. Lund: Celanders förlag, 2016. https://www.bokus.com/bok/9789187393433/den-krav-markta-manniskan/.

Ideland, Malin. *The Eco-Certified Child*. Cham: Springer, 2019.

Igoe, Jim, Katja Neves, and Dan Brockington. 'A Spectacular Eco-Tour around the Historic Bloc: Theorising the Convergence of Biodiversity Conservation and Capitalist Expansion'. *Antipode* 42, no. 3 (June 2010): 486–512. https://doi.org/10.1111/j.1467-8330.2010.00761.x.

Ing, Kaniela. 'The Climate Crisis and Colonialism Destroyed My Maui Home'. *TIME*, 17 August 2023. https://time.com/6305817/maui-wildfires-climate-change-colonialism-essay/.

Jenkins, Mary. 'The Dystopian World of Blade Runner: An Ecofeminist Perspective'. *Trumpeter* 14, no. 4 (1999): 2–13.

Jessop, Bob. 'Cultural Political Economy and Critical Policy Studies'. *Critical Policy Studies* 3, no. 3–4 (28 April 2010): 336–56. https://doi.org/10.1080/19460171003619741.

Joosse, Sofie, Stina Powell, Hanna Bergeå, Steffen Böhm, Camilo Calderón, Elvira Caselunghe, Anke Fischer, Ann Grubbström, Lars Hallgren, Sara Holmgren,

Annette Löf, Helena Nordström Källström, Kaisa Raitio, Susan Senecah, Christoffer Söderlund Kanarp, Erica von Essen, Lotten Westberg, and Martin Westin. 'Critical, Engaged and Change-Oriented Scholarship in Environmental Communication. Six Methodological Dilemmas to Think With'. *Environmental Communication* 14, no. 6 (17 August 2020): 758–71. https://doi.org/10.1080/17524032.2020.1725588.

Jørgensen, Marianne, and Louise J. Phillips. *Discourse Analysis as Theory and Method.* London: SAGE, 2002. http://srmo.sagepub.com/view/discourse-analysis-as-theory-and-method/SAGE.xml.

Josefsen, Niels-Martin Ström, and Scott Burnett. 'Visions of Degeneracy: Race, Gender and Sexuality in "Exalted" Sourcebooks'. *Gender and Language* 18, no. 2 (23 August 2024): 150–74. https://doi.org/10.1558/genl.25912.

'Jungle'. In *Merriam-Webster.Com*. Merriam-Webster, no date. https://www.merriam-webster.com/dictionary/jungle.

Kalof, Linda, Thomas Dietz, and Gregory Guagnano. 'Race, Gender and Environmentalism: The Atypical Values and Beliefs of White Men'. *Race, Gender & Class* 9, no. 2 (30 April 2002). http://0-search.proquest.com.innopac.wits.ac.za/docview/218826193/D4E1A0AAAAB4424FPQ/8?accountid=15083.

Kamara, Yarri. 'Reframing Population: Other Perspectives on Africa's Population Growth'. *The Republic*, 17 August 2022. https://republic.com.ng/august-september-2022/reframing-population/.

Kaplan, Robert D. 'The Coming Anarchy'. *The Atlantic*, 1 February 1994. https://www.theatlantic.com/magazine/archive/1994/02/the-coming-anarchy/304670/.

Karlbom, Rolf. 'Sweden's Iron Ore Exports to Germany, 1933–1944'. *Scandinavian Economic History Review* 13, no. 1 (1 January 1965): 65–93. https://doi.org/10.1080/03585522.1965.10414365.

Katy Tur Reports [@KatyOnMSNBC]. '"Lahaina Is Not for Sale." Maui Resident Tiare Lawrence Talks to @KatyTurNBC about Real Estate Investors Reaching Out to Residents after the Destructive Fires on Maui. Https://T.Co/vDXEJIb8J1'. Tweet. *Twitter*, 14 August 2023. https://x.com/KatyOnMSNBC/status/1691172937100075008.

Keegan, Timothy J. *Colonial South Africa and the Origins of the Racial Order*. London: Leicester University Press, 1997.

Kendall, Gavin. 'What Is Critical Discourse Analysis?' *Forum Qualitative Sozialforschung* 2, no. 8, Article 29 (31 May 2007).

Kendrick, Sam, and Joane Nagel. 'The Cowboy Scientist Saves the Planet: Hegemonic Masculinity in Cli-Fi Films'. *Masculinities: A Journal of Identity and Culture*, no. 14 (15 November 2020): 5–34.

Kepe, Thembela. 'Shaped by Race: Why "Race" Still Matters in the Challenges Facing Biodiversity Conservation in Africa'. *Local Environment* 14, no. 9 (October 2009): 871–8. https://doi.org/10.1080/13549830903164185.

Kevany, Sophie. 'Non-Native Grass Species Blamed for Ferocity of Hawaii Wildfires'. *The Guardian*, 16 August 2023, sec. Environment. https://www.theguardian.com/global-development/2023/aug/16/non-native-grass-species-blamed-for-ferocity-of-hawaii-wildfires.

'Kevin Pietersen: Beast of Man'. BBC Radio 5, 2 May 2019. https://www.bbc.co.uk/programmes/articles/8mb5BLnX0qMpTrPjQJ9zhP/the-people-saving-the-rhino.

Khan, Farieda. 'Environmentalism in South Africa: A Sociopolitical Perspective'. *Macalester International* 9, no. 1 (2000): 11.

Kiesling, Scott F. 'Masculine Stances and the Linguistics of Affect: On Masculine Ease'. *NORMA* 13, no. 3–4 (2 October 2018): 191–212. https://doi.org/10.1080/18902138.2018.1431756.

Kim, Claire Jean. 'Makah Whaling and the (Non) Ecological Indian'. In *Colonialism and Animality: Anti-Colonial Perspectives in Critical Animal Studies*, edited by Kelly Struthers Montford and Chloë Taylor, 50–103. London: Routledge, 2021.

Kinefuchi, Etsuko. 'Critical Discourse Analysis and the Ecological Turn in Intercultural Communication'. *Review of Communication* 18, no. 3 (3 July 2018): 212–30. https://doi.org/10.1080/15358593.2018.1479882.

Kippen, Colin. 'Mai Hilahila: Uncomfortable Conversations Are Our Kuleana'. *Ka Wai Ola*, 1 October 2023. https://kawaiola.news/oha/mai-hilahila-uncomfortable-conversations-are-our-kuleana/.

Kjellberg, Tobias. 'Ny "skräp-ö" hittad i Stilla havet – större än Mexiko'. News. Göteborgs-Posten, 16 August 2017. https://www.gp.se/nyheter/varlden/ny-skrap-o-hittad-i-stilla-havet-storre-an-mexiko.1503ef10-fd4a-414e-b3c6-eff9b6fda4ec.

Klein, Ezra, and Derek Thompson. *Abundance*. New York: Avid Reader Press (Simon & Schuster), 2025.

Klein, Naomi. 'The Rise of Disaster Capitalism', 14 April 2005. https://www.thenation.com/article/archive/rise-disaster-capitalism/.

Klein, Naomi, and Kapuaʻala Sproat. 'Why Was There No Water to Fight the Fire in Maui?' *The Guardian*, 17 August 2023, sec. Opinion. https://www.theguardian.com/commentisfree/2023/aug/17/hawaii-fires-maui-water-rights-disaster-capitalism.

Koch, Natalie. 'Green Nationalism from Above: Authoritarian State Power and the Greening of UAE Nationalism'. *Nations and Nationalism*, 2024. https://doi.org/10.1111/nana.13042.

Kolden, Crystal A., John T. Abatzoglou, Matthew W. Jones, and Piyush Jain. 'Wildfires in 2023'. *Nature Reviews Earth & Environment* 5, no. 4 (April 2024): 238–40. https://doi.org/10.1038/s43017-024-00544-y.

Koller, Veronika. 'Cognitive Linguistics and Ideology'. In *The Bloomsbury Companion to Cognitive Linguistics*, edited by Jeannette Littlemore and John R. Taylor, 234–52. Bloomsbury Companions. London; New York: Bloomsbury, 2014.

Koot, Stasja. 'Enjoying Extinction: Philanthrocapitalism, Jouissance, and "Excessive Environmentourism" in the South African Rhino Poaching Crisis'. *Journal of Political Ecology* 28, no. 1 (17 August 2021). https://doi.org/10.2458/jpe.2984.

Koot, Stasja, Bram Büscher, and Lerato Thakholi. 'The New Green Apartheid? Race, Capital and Logics of Enclosure in South Africa's Wildlife Economy'. *Environment and Planning E: Nature and Space* (28 June 2022): 25148486221110438. https://doi.org/10.1177/25148486221110438.

Koot, Stasja, and Frederiek Veenenbos. 'The Spectacle of Inclusive Female Anti-Poaching: Heroines, Green Militarization and Invisible Violence'. *Geoforum* 144 (1 August 2023): 103806. https://doi.org/10.1016/j.geoforum.2023.103806.

Krech, Shepard. 'Reflections on Conservation, Sustainability, and Environmentalism in Indigenous North America'. *American Anthropologist* 107, no. 1 (March 2005): 78–86. https://doi.org/10.1525/aa.2005.107.1.078.

Krech, Shepard. *The Ecological Indian: Myth and History*. New York: W.W. Norton, 1999.

Kress, Gunther. 'Against Arbitrariness: The Social Production of the Sign as a Foundational Issue in Critical Discourse Analysis'. *Discourse & Society* 4, no. 2 (1 April 1993): 169–91. https://doi.org/10.1177/0957926593004002003.

Kress, Gunther, and Theo Van Leeuwen. 'Colour as a Semiotic Mode: Notes for a Grammar of Colour'. *Visual Communication* 1, no. 3 (1 October 2002): 343–68. https://doi.org/10.1177/147035720200100306.

Kress, Gunther, and Theo Van Leeuwen. *Reading Images: The Grammar of Visual Design*. Abingdon: Routledge, 2006.

Krzyżanowski, Michał. 'Discursive Shifts and the Normalisation of Racism: Imaginaries of Immigration, Moral Panics and the Discourse of Contemporary Right-Wing Populism'. *Social Semiotics* 30, no. 4 (7 August 2020): 503–27. https://doi.org/10.1080/10350330.2020.1766199.

Krzyżanowski, Michał, and Bernhard Forchtner. 'Theories and Concepts in Critical Discourse Studies: Facing Challenges, Moving beyond Foundations'. *Discourse & Society* 27, no. 3 (2016): 253–61.

Kumar, Ankit. 'Energy Geographies in/of the Anthropocene: Where Now?' *Geography Compass* 16, no. 10 (2022): e12659. https://doi.org/10.1111/gec3.12659.

Kurtz, Hilda E. 'Reflections on the Iconography of Environmental Justice Activism'. *Area* 37, no. 1 (March 2005): 79–88.

Laband, John. *The Land Wars: The Dispossession of the Khoisan and AmaXhosa in the Cape Colony*. Cape Town: Penguin Random House, 2020.

Laclau, Ernesto. 'Identity and Hegemony: The Role of Universality in the Constitution of Political Logics'. In *Contingency, Hegemony, Universality: Contemporary Dialogues on the Left*, edited by Judith Butler, Ernesto Laclau, and Slavoj Žižek, 44–89. London: Verso, 2000.

Laclau, Ernesto. *New Reflections on the Revolution of Our Time*. Phronesis. London: Verso, 1990.

Laclau, Ernesto. *The Rhetorical Foundations of Society*. London: Verso, 2014.

Laclau, Ernesto, and Chantal Mouffe. *Hegemony and Socialist Strategy: Towards a Radical Democratic Politics*. 2nd edition. 1985. Reprint, London and Brooklyn, NY: Verso, 2014.

Lamb, William F., Giulio Mattioli, Sebastian Levi, J. Timmons Roberts, Stuart Capstick, Felix Creutzig, Jan C. Minx, Finn Müller-Hansen, Trevor Culhane, and Julia K. Steinberger. 'Discourses of Climate Delay'. *Global Sustainability* 3 (January 2020): e17. https://doi.org/10.1017/sus.2020.13.

Lane, Joseph H., and Rebecca R. Clark. 'The Solitary Walker in the Political World: The Paradoxes of Rousseau and Deep Ecology'. *Political Theory* 34, no. 1 (1 February 2006): 62–94. https://doi.org/10.1177/0090591705279076.

Laville, Sandra. 'Twelve Billionaires' Climate Emissions Outpollute 2.1m Homes, Analysis Finds'. *The Guardian*, 20 November 2023, sec. Environment. https://www.theguardian.com/environment/2023/nov/20/twelve-billionaires-climate-emissions-jeff-bezos-bill-gates-elon-musk-carbon-divide.

Lawrence, Rebecca, and Rasmus Kløcker Larsen. 'The Politics of Planning: Assessing the Impacts of Mining on Sami Lands'. *Third World Quarterly* 38, no. 5 (4 May 2017): 1164–80. https://doi.org/10.1080/01436597.2016.1257909.

Lawton, Graham. 'The Rise of Real Eco-Fascism'. *New Scientist* 243, no. 3243 (17 August 2019): 24. https://doi.org/10.1016/S0262-4079(19)31529-5.

Lazar, Michelle M. 'Feminist Critical Discourse Analysis: Articulating a Feminist Discourse Praxis'. *Critical Discourse Studies* 4, no. 2 (1 August 2007): 141–64. https://doi.org/10.1080/17405900701464816.

Lebdioui, Amir. *Survival of the Greenest: Economic Transformation in a Climate-Conscious World*. Cambridge University Press, 2024. https://doi.org/10.1017/9781009339414.

Ledin, Per, and David Machin. *Doing Visual Analysis: From Theory to Practice*. Los Angeles, CA: SAGE, 2018.

Lee, Stephanie M. 'Big Oil Helped Shape Stanford's Latest Climate-Research Focus'. *The Chronicle of Higher Education*, 4 May 2023, sec. News. https://www.chronicle.com/article/big-oil-helped-shape-stanfords-latest-climate-research-focus.

Leeuwen, Theo van. *Discourse and Practice: New Tools for Critical Discourse Analysis*. New York: Oxford University Press, 2008.

Leeuwen, Theo van. 'The Semiotics of Movement and Mobility'. *Multimodality & Society* 1, no. 1 (1 March 2021): 97–118. https://doi.org/10.1177/2634979521992733.

Leeuwen, Theo van. 'The Visual Representation of Social Actors'. In *Discourse and Practice: New Tools for Critical Analysis*, 136–48. Oxford Studies in Sociolinguistics. New York: Oxford University Press, 2008. https://doi.org/10.1093/acprof:oso/9780195323306.003.0008.

Leeuwen, Theo van. 'Towards a Semiotics of Typography'. *Information Design Journal* 14, no. 2 (1 January 2006): 139–55. https://doi.org/10.1075/idj.14.2.06lee.

Legum, Judd, Tesnim Zekeria, and Rebecca Crosby. '1500 Environmental Lobbyists Are Double-Dealing with the Fossil Fuel Industry'. *Popular Information* (blog), 6 July 2023. https://popular.info/p/1500-environmental-lobbyists-are?utm_medium=email.

Lekan, Thomas M. 'From Landscape to Lebensraum: Race and Environment under Nazism'. In *Imagining the Nation in Nature*, 153–203. Landscape Preservation and German Identity, 1885–1945. Harvard: Harvard University Press, 2004. https://doi.org/10.2307/j.ctv1q3z2q5.7.

Lenggenhager, Luregn. 'Nature, War and Development: South Africa's Caprivi Strip, 1960–1980'. *Journal of Southern African Studies* 41, no. 3 (June 2015): 467–83. https://doi.org/10.1080/03057070.2015.1025337.

Lentin, Alana. *Why Race Still Matters*. Cambridge: Polity, 2020.

Leurs, Rob. 'The "Chain of Equivalence": Cultural Studies and Laclau & Mouffe's Discourse Theory'. *Politics and Culture* 2009, no. 4 (9 November 2009). https://politicsandculture.org/2009/11/09/the-chain-of-equivalence-cultural-studies-and-laclau-mouffes-discourse-theory/.

Levenson, Zachary, and Marcel Paret. 'The Three Dialectics of Racial Capitalism: From South Africa to the U.S. and Back Again'. *Du Bois Review: Social Science Research on Race* 20, no. 2 (October 2023): 333–51. https://doi.org/10.1017/S1742058X22000212.

Levien, Michael. 'White Energy Workers of the North, Unite? A Review of Huber's Climate Change as Class War'. *Historical Materialism* (blog), 23 February 2023. https://www.historicalmaterialism.org/white-energy-workers-of-the-north-unite-a-review-of-hubers-climate-change-as-class-war/.

Levitz, Eric. 'Blaming "Capitalism" Is Not an Alternative to Solving Problems'. *New York Magazine Intelligencer*, 10 April 2023. https://nymag.com/intelligencer/2023/04/blaming-capitalism-is-not-an-alternative-to-solving-problems.html.

Levy, David L., and André Spicer. 'Contested Imaginaries and the Cultural Political Economy of Climate Change'. *Organization* 20, no. 5 (1 September 2013): 659–78. https://doi.org/10.1177/1350508413489816.

Lewis, Desiree, Vasu Reddy, and Lynn Mafofo. 'Transnational Perspectives on Food, Ecology and the Anthropocene'. *Agenda* 37, no. 1 (2 January 2023): 3–11. https://doi.org/10.1080/10130950.2023.2205707.

Liboiron, Max. *Pollution Is Colonialism*. Durham, NC; London: Duke University Press, 2021.

Liboreiro, Jorge. 'Borrell Apologises for Controversial Metaphor but Defends Speech'. *Euronews*, 19 October 2022. https://www.euronews.com/my-europe/2022/10/19/josep-borrell-apologises-for-controversial-garden-vs-jungle-metaphor-but-stands-his-ground.

Lindsey, P. A., S. H. Anderson, A. Dickman, P. Gandiwa, S. Harper, A. B. Morakinyo, N. Nyambe, M. O'Brien-Onyeka, C. Packer, A. H. Parker, A. S. Robson, Alice Ruhweza, E. A. Sogbohossou, K. W. Steiner, and P. N. Tumenta. 'Shepherding Sub-Saharan Africa's Wildlife through Peak Anthropogenic Pressure toward a Green Anthropocene'. *Annual Review of Environment and Resources* 47 (17 October 2022): 91–121. https://doi.org/10.1146/annurev-environ-120920-125536.

Lindsey, Peter, James Allan, Peadar Brehony, Amy Dickman, Ashley Robson, Colleen Begg, Hasita Bhammar, Lisa Blanken, Thomas Breuer, Kathleen Fitzgerald, Michael Flyman, Patience Gandiwa, Nicia Giva, Dickson Kaelo, Simon Nampindo, Nyambe Nyambe, Kurt Steiner, Andrew Parker, Dilys Roe, Paul Thomson, Morgan Trimble, Alexandre Caron, and Peter Tyrrell. 'Conserving Africa's Wildlife and Wildlands through the COVID-19 Crisis and Beyond'. *Nature Ecology & Evolution* 4, no. 10 (October 2020): 1300–10. https://doi.org/10.1038/s41559-020-1275-6.

Livingstone, David N. 'The Climate of War: Violence, Warfare, and Climatic Reductionism'. *Wiley Interdisciplinary Reviews: Climate Change* 6, no. 5 (2015): 437–44. https://doi.org/10.1002/wcc.352.

Ljunggren, Jens. 'Nation-Building, Primitivism and Manliness – the Issue of Gymnastics in Sweden around 1800'. *Scandinavian Journal of History* 21, no. 2 (1996): 101–20. https://doi.org/10.1080/03468759608579319.

Locke, Harvey. 'Nature Needs (at Least) Half: A Necessary New Agenda for Protected Areas'. In *Protecting the Wild: Parks and Wilderness, the Foundation for Conservation*, edited by George Wuerthner, Eileen Crist, and Tom Butler, 3–15. Washington, DC: Foundations for Deep Ecology, 2015.

Locke, John. *Two Treatises of Government*. Edited by Peter Laslett. 1690. Reprint, Cambridge: Cambridge University Press, 1960.

Lomborg, Bjørn. *The Skeptical Environmentalist: Measuring the Real State of the World*. Cambridge and New York: Cambridge University Press, 2001.

Lorde, Audre. *Sister Outsider: Essays and Speeches*. Revised. 1984. Reprint, Berkeley, CA: Crossing Press, 2007.

Lousley, Cheryl. 'Ecocriticism'. In *Oxford Research Encyclopedia of Literature*, 2020. https://doi.org/10.1093/acrefore/9780190201098.013.974.

Lubarda, BalŠa. 'Beyond Ecofascism? Far-Right Ecologism (FRE) as a Framework for Future Inquiries'. *Environmental Values* 29, no. 6 (1 December 2020): 713–32. https://doi.org/10.3197/096327120X15752810323922.

Lubarda, Balša. *Far-Right Ecologism: Environmental Politics and the Far Right in Hungary and Poland*. 1st edition. Routledge Studies in Fascism and the Far Right. New York: Routledge, 2024. https://doi.org/10.4324/9781003306108.

Lubarda, Balša, and Bernhard Forchtner. 'The Far Right and the Environment: Past-Present-Future'. In *Populism and Far-Right: Trends in Europe*, edited by Valerio Alfonso Bruno, 85–112. Polidemos 4. Milano: EDUCatt, 2022.

Lunstrum, Elizabeth. 'Feed Them to the Lions: Conservation Violence Goes Online'. *Geoforum* 79 (1 February 2017): 134–43. https://doi.org/10.1016/j.geoforum.2016.04.009.

Lunstrum, Elizabeth. 'Green Militarization: Anti-Poaching Efforts and the Spatial Contours of Kruger National Park'. *Annals of the Association of American Geographers* 104, no. 4 (4 July 2014): 816–32. https://doi.org/10.1080/00045608.2014.912545.

Lyte, Brittany. 'Lahaina Wildfire Death Toll Now at 102'. *Honolulu Civil Beat*, 24 June 2024. https://www.civilbeat.org/2024/06/lahaina-wildfire-death-toll-now-at-102/.

Macgilchrist, Felicitas. 'Fissures in the Discourse-Scape: Critique, Rationality and Validity in Post-Foundational Approaches to CDS'. *Discourse & Society* 27, no. 3 (2016): 262–77.

Macgilchrist, Felicitas. 'Media Discourse and De/Coloniality: A Post-Foundational Approach'. In *Contemporary Critical Discourse Studies*, edited by Piotr Cap and Christopher Hart, 387–407. London: Continuum, 2014. https://www.torrossa.com/en/resources/an/5202523.

Machen, Ruth, Simin Davoudi, and Elizabeth Brooks. 'Climate Imaginaries and Their Mediums'. *Geoforum* 138 (1 January 2023): 103672. https://doi.org/10.1016/j.geoforum.2022.103672.

Machin, David, and Andrea Mayr. *How to Do Critical Discourse Analysis: A Multimodal Introduction*. London: SAGE, 2012. http://ebookcentral.proquest.com/lib/gu/detail.action?docID=880803.

Macklin, Graham. 'The Extreme Right, Climate Change and Terrorism'. *Terrorism and Political Violence* 34, no. 5 (July 2022): 979–96. https://doi.org/10.1080/09546553.2022.2069928.

Majavu, Mandisi. 'The White International: "The Cause of the White Man on the Pacific Coast"'. In *The Routledge Handbook on Africana Criminologies* edited by Biko Agozino, Viviane Saleh-Hanna, Emmanuel C. Onyeozili, and Nontyatyambo Pearl Dastile, chapter 15. Abingdon: Routledge, 2020.

Malm, Andreas. *Fossil Capital: The Rise of Steam Power and the Roots of Global Warming*. Illustrated edition. London: Verso, 2016.

Malm, Andreas, and The Zetkin Collective. *White Skin, Black Fuel: On the Danger of Fossil Fascism*. London and New York: Verso, 2021.

Malthus, Thomas. *An Essay on the Principle of Population*. Edited by Geoffrey Gilbert. 1798. Reprint, Oxford: Oxford University Press, 1999.

Mancini, Jeannine. 'It Costs $25 Million a Year to Keep Jeff Bezos' $500 Million Luxury Superyacht Afloat — But He Earns That in Just 3 Hours'. News. Yahoo Finance, 26 June 2024. https://finance.yahoo.com/news/costs-25-million-keep-jeff-133527432.html.

Maran, Timo. *Ecosemiotics: The Study of Signs in Changing Ecologies*. Cambridge: Cambridge University Press, 2020. https://www.cambridge.org/core/elements/ecosemiotics/D29658F0C2E12040454C776F82627253.

Margulies, Morgan. 'Eco-Nationalism: A Historical Evaluation of Nationalist Praxes in Environmentalist and Ecologist Movements'. *Consilience*, no. 23 (2021): 22–9.

Markus, Thomas A., and Deborah Cameron. *The Words between the Spaces: Buildings and Language*. London and New York: Routledge, 2002.

Martin, Guy. 'The West, Natural Resources and Population Control Policies in Africa in Historical Perspective'. *Journal of Third World Studies* 22, no. 1 (2005): 69–107.

Martin, James R. 'Grace: The Logogenesis of Freedom'. *Discourse Studies* 1, no. 1 (1999): 29–56.

Martin, James. *Politics and Rhetoric: A Critical Introduction*. Oxford and New York: Routledge, 2014.

Martin, James R. 'Positive Discourse Analysis: Solidarity and Change'. *Revista Canaria de Estudios Ingleses*, no. 49 (2004): 179–200.

Marttila, Tomas. *Post-Foundational Discourse Analysis: From Political Difference to Empirical Research*. New York: Palgrave Macmillan, 2016. http://search.ebscohost.com/login.aspx?direct=true&scope=site&db=nlebk&db=nlabk&AN=1091345.

Marx, Karl, and Friedrich Engels. *The German Ideology*. Edited by C. J. Arthur. 1846. Reprint, London: Lawrence & Wishart, 1970.

Massarella, Kate, Judith E. Krauss, Wilhelm A. Kiwango, and Robert Fletcher, eds. *Convivial Conservation: From Principles to Practice*. MayFly Books, 2023.

Massarella, Kate, Anja Nygren, Robert Fletcher, Bram Büscher, Wilhelm A. Kiwango, Sanna Komi, Judith E. Krauss, Mathew B. Mabele, Alex McInturff, Laila T. Sandroni, Peter S. Alagona, Dan Brockington, Robert Coates, Rosaleen Duffy, Katia M. P. M. B. Ferraz, Stasja Koot, Silvio Marchini, and Alexandre R. Percequillo. 'Transformation beyond Conservation: How Critical Social Science Can Contribute to a Radical New Agenda in Biodiversity Conservation'. *Current Opinion in Environmental Sustainability* 49 (1 April 2021): 79–87. https://doi.org/10.1016/j.cosust.2021.03.005.

Massé, Francis. 'Anti-Poaching's Politics of (in)Visibility: Representing Nature and Conservation amidst a Poaching Crisis'. *Geoforum* 98 (1 January 2019): 1–14. https://doi.org/10.1016/j.geoforum.2018.09.011.

Massé, Francis, Nicia Givá, and Elizabeth Lunstrum. 'A Feminist Political Ecology of Wildlife Crime: The Gendered Dimensions of a Poaching Economy and Its Impacts in Southern Africa'. *Geoforum* 126 (1 November 2021): 205–14. https://doi.org/10.1016/j.geoforum.2021.07.031.

Mautner, Gerlinde. 'A Research Note on Corpora and Discourse: Points to Ponder in Research Design'. *Journal of Corpora and Discourse Studies* 2 (3 July 2019): 2–13. https://doi.org/10.18573/jcads.32.

Mayr, Andrea. *Language and Power: An Introduction to Institutional Discourse*. London and New York, NY: Continuum International Publishing Group, 2008.

Mcallister, Edward. 'U.S. Climate Envoy Kerry Calls on African Nations to Help Curb Emissions'. *Reuters*, 15 September 2022, sec. Africa. https://www.reuters.com/world/africa/us-climate-envoy-kerry-calls-african-nations-help-curb-emissions-2022-09-15/.

McCarthy, James, and Scott Prudham. 'Neoliberal Nature and the Nature of Neoliberalism'. *Geoforum* 35, no. 3 (May 2004): 275–83. https://doi.org/10.1016/j.geoforum.2003.07.003.

McCrummen, Stephanie. 'This Will Finish Us'. *The Atlantic*, 8 April 2024. https://www.theatlantic.com/magazine/archive/2024/05/maasai-tribe-tanzania-forced-land-evictions-serengeti/677835/.

McIntosh, Janet. 'Land, Belonging and Structural Oblivion among Contemporary White Kenyans'. *Africa* 87, no. 4 (November 2017): 662–82. https://doi.org/10.1017/S0001972017000304.

McKay, Jackie. 'Native Hawaiians Fear Maui Wildfire Destruction Will Lead to Their Cultural Erasure'. *CBC News*, 17 August 2023. https://www.cbc.ca/news/indigenous/maui-lahaina-fire-indigenous-1.6939560.

McNally, David. 'Political Economy to the Fore: Burke, Malthus and the Whig Response to Popular Radicalism in the Age of the French Revolution'. *History of Political Thought* XXI, no. 3 (Autumn 2000): 427–47.

McWhorter, Ladelle. 'Enemy of the Species'. In *Queer Ecologies: Sex, Nature, Politics, Desire*, edited by Catriona Mortimer-Sandilands and Bruce Erickson, 149–72. Bloomington: Indiana University Press, 2010.

McWhorter, Ladelle. 'Foucault's Genealogy of Homosexuality'. *Journal of French and Francophone Philosophy* 6, no. 1–2 (1 March 1994): 44–58. https://doi.org/10.5195/jffp.1994.82.

Méndez, María José. '"The River Told Me": Rethinking Intersectionality from the World of Berta Cáceres'. *Capitalism Nature Socialism* 29, no. 1 (2 January 2018): 7–24. https://doi.org/10.1080/10455752.2017.1421981.

Mignolo, Walter D. 'The Geopolitics of Knowledge and the Colonial Difference'. *South Atlantic Quarterly* 101, no. 1 (1 January 2002): 57–96. https://doi.org/10.1215/00382876-101-1-57.

Mihai, Mihaela, and Camil Ungureanu. 'Far-Right Ecology and Geopolitical Resentment at Europe's Periphery: The Case of Romania's "Conservative Revolution"'. *Geoforum* 156 (1 November 2024): 104122. https://doi.org/10.1016/j.geoforum.2024.104122.

Milani, Tommaso M. 'Affect: Discourse, Politics, Intersectionality'. In *Handbook of Language and Youth Culture*, edited by Bente Svendsen and Rickard Jonsson, chapter 3. Abingdon: Routledge, 2023.

Milani, Tommaso M. 'Sexual Citizenship: Discourses, Spaces and Bodies at Joburg Pride 2012'. *Journal of Language and Politics* 14, no. 3 (21 August 2015): 431–54. https://doi.org/10.1075/jlp.14.3.06mil.

Milani, Tommaso M., and Erez Levon. 'Israel as Homotopia: Language, Space, and Vicious Belonging'. *Language in Society* 48, no. 4 (September 2019): 607–28. https://doi.org/10.1017/S0047404519000356.

Milani, Tommaso M., and Erez Levon. 'Sexing Diversity: Linguistic Landscapes of Homonationalism'. *Language & Communication*, Discourses of Diversity 51 (1 November 2016): 69–86. https://doi.org/10.1016/j.langcom.2016.07.002.

Milani, Tommaso M., and John E. Richardson. 'Discourse and Affect'. *Social Semiotics* 31, no. 5 (2021): 671–6. https://doi.org/10.1080/10350330.2020.1810553.

Milani, Tommaso M., and Rickard Jonsson. 'Who's Afraid of Rinkeby Swedish? Stylization, Complicity, Resistance'. *Journal of Linguistic Anthropology* 22, no. 1 (May 2012): 44–63. https://doi.org/10.1111/j.1548-1395.2012.01133.x.

Mildenberger, Matto. 'The Tragedy of "The Tragedy of the Commons"'. *Scientific American*, 23 April 2019. https://www.scientificamerican.com/blog/voices/the-tragedy-of-the-tragedy-of-the-commons/.

Mills, Charles W. 'Black Trash'. In *Faces of Environmental Racism: Confronting Issues of Global Justice*, edited by Laura Westra and Bill E Lawson, 2nd edition, 73–94. Lanham, MD: Rowman & Littlefield Publishers, 2001.

Mills, Charles W. 'Racial Liberalism'. *PMLA* 123, no. 5 (2008): 1380–97.

Millward-Hopkins, Joel. 'Inequality Can Double the Energy Required to Secure Universal Decent Living'. *Nature Communications* 13, no. 1 (26 August 2022): 5028. https://doi.org/10.1038/s41467-022-32729-8.

Milman, Oliver. 'Climate Denial Is Waning on the Right. What's Replacing It Might Be Just as Scary'. *The Guardian*, 21 November 2021, sec. Environment. https://www.theguardian.com/environment/2021/nov/21/climate-denial-far-right-immigration.

Milstein, Tema. 'Greening Communication'. In *Greening the Academy*, edited by Samuel Day Fassbinder, Anthony, J. Nocella II, and Richard Kahn, 161–73. Rotterdam: SensePublishers, 2012. https://doi.org/10.1007/978-94-6209-101-6_10.

Milstein, Tema. 'The Performer Metaphor: "Mother Nature Never Gives Us the Same Show Twice"'. *Environmental Communication* 10, no. 2 (3 March 2016): 227–48. https://doi.org/10.1080/17524032.2015.1018295.

Mirzoeff, Nicholas. 'It's Not the Anthropocene, It's the White Supremacy Scene: Or, the Geological Color Line'. In *After Extinction*, edited by Richard Grusin, 123–50. Minneapolis, MN: University of Minnesota Press, 2018. https://muse.jhu.edu/book/58563.

Mitchell, Brent A., Nigel Dudley, Sue Stolton, Jessica Campese, and Hannah L. Timmins, eds. '30×30: A Guide to Inclusive, Equitable and Effective Implementation of Target 3 of the Kunming-Montreal Global Biodiversity Framework'. WWF and IUCN World Commission on Protected Areas, August 2023. https://www.worldwildlife.org/publications/30x30-a-guide-to-inclusive-equitable-and-effective-implementation-of-target-3-of-the-kunming-montreal-global-biodiversity-framework.

Mittal, Anuradha. 'Justice Denied: East African Court of Justice Grants Tanzanian Government Impunity to Trample Human Rights of the Maasai'. *Oakland Institute* (blog), 30 September 2022. https://www.oaklandinstitute.org/east-african-court-justice-tanzanian-maasai-trample-human-rights.

Mittal, Anuradha, Zahra Moloo, and Frederic Mousseau. 'Stealth Game: "Community" Conservancies Devastate Land and Lives in Northern Kenya'. Oakland, CA: The Oakland Institute, 2021. https://www.oaklandinstitute.org/stealth-game-community-conservancies-devastate-northern-kenya.

Mix, Tamara L. 'The Greening of White Separatism: Use of Environmental Themes to Elaborate and Legitimize Extremist Discourse'. *Nature & Culture* 4, no. 2 (July 2009): 138–66. https://doi.org/10.3167/nc.2009.040203.

Mkutu Agade, Kennedy. '"We Will Not Watch Like Monkeys": Development Visions and Conflict Potentials in Northern Kenya'. In *African Futures*, edited by Clemens Greiner, Steven Van Wolputte, and Michael Bollig, 27:197–208. Leiden and Boston, MA: Brill, 2022. https://doi.org/10.1163/9789004471641.

Mohai, Paul, David N. Pellow, and J. Timmons Roberts. 'Environmental Justice'. *Annual Review of Environment and Resources* 34, no. 1 (November 2009): 405–30. https://doi.org/10.1146/annurev-environ-082508-094348.

Mohamed, Ashfak. 'No Thanks, KP, the Proteas Don't Need You'. *Independent Online*, 20 July 2017, sec. Sports. https://www.iol.co.za/sport/cricket/proteas/no-thanks-kp-the-proteas-dont-need-you-10381757.

Molek-Kozakowska, Katarzyna. 'The Hybrid Discourse of the "European Green Deal": Road-Mapping Economic Transition to Environmental Sustainability (Almost)

Seamlessly'. *Critical Discourse Studies* 21, no. 2 (3 March 2024): 182–99. https://doi.org/10.1080/17405904.2023.2197607.

Mondon, Aurelien, and Aaron Winter. *Reactionary Democracy: Racism, Populism, the Far Right and the People*. London: Verso, 2020.

Montesquieu, Charles de. *The Spirit of the Laws*. Translated by Thomas Nugent. 1748. Reprint, New York: D. Appleton, 1900.

Moore, Jason W. *Capitalism in the Web of Life: Ecology and the Accumulation of Capital*. London and New York: Verso, 2016. http://rbdigital.oneclickdigital.com.

Moore, Sam, and Alex Roberts. *The Rise of Ecofascism: Climate Change and the Far Right*. Newark: Polity Press, 2022.

Moreton-Robinson, Aileen. *The White Possessive: Property, Power, and Indigenous Sovereignty*. Minneapolis, MN: University of Minnesota Press, 2015.

Mortimer-Sandilands, Catriona, and Bruce Erickson. *Queer Ecologies: Sex, Nature, Politics, Desire*. Bloomington, IN: Indiana University Press, 2010.

Moses, A. Dirk. 'An Antipodean Genocide? The Origins of the Genocidal Moment in the Colonization of Australia'. *Journal of Genocide Research* 2, no. 1 (1 March 2000): 89–106. https://doi.org/10.1080/146235200112427.

Mouffe, Chantal. *For a Left Populism*. London and New York: Verso, 2018.

Mouffe, Chantal. *The Return of the Political*. 1993. Reprint, London and New York: Verso, 2005.

Mpanya, Mutombo. 'The Dumping of Toxic Waste in African Countries: A Case of Poverty and Racism'. In *Race and the Incidence of Environmental Hazards: A Time for Discourse*, edited by Bunyan I. Bryant and Paul Mohai, 204–14. Boulder, CO: Westview Press, 1992.

Mühlhäusler, Peter, and Adrian Peace. 'Environmental Discourses'. *Annual Review of Anthropology* 35, no. 1 (October 2006): 457–79. https://doi.org/10.1146/annurev.anthro.35.081705.123203.

Murdock, Esme G. 'Conserving Dispossession? A Genealogical Account of the Colonial Roots of Western Conservation'. *Ethics, Policy and Environment* 24, no. 3 (2021): 235–49. https://doi.org/10.1080/21550085.2021.2002625.

Murray, Joy, Manfred Lenzen, and John Murray. 'Forest Carbon – Questions of Indigenous Rights and Market Forces'. *Environmental Justice* 7, no. 2 (April 2014): 33–8. https://doi.org/10.1089/env.2013.0035.

Murray, Sally-Ann. 'The Idea of Gardening: Plants, Bewilderment, and Indigenous Identity in South Africa'. *English in Africa* 33, no. 2 (October 2006): 45–65.

Nadasdy, Paul. 'Transcending the Debate over the Ecologically Noble Indian: Indigenous Peoples and Environmentalism'. *Ethnohistory* 52, no. 2 (1 April 2005): 291–331. https://doi.org/10.1215/00141801-52-2-291.

Naess, Arne. 'The Shallow and the Deep, Long-Range Ecology Movement'. *Inquiry* 16, no. 1–4 (1 January 1973): 95–100. https://doi.org/10.1080/00201747308601682.

Narayan, John. 'The Wages of Whiteness in the Absence of Wages: Racial Capitalism, Reactionary Intercommunalism and the Rise of Trumpism'. *Third World Quarterly* 38, no. 11 (2017): 2482–500.

Nash, Joshua. 'Ecolinguistics and Placenames'. In *The Routledge Handbook of Ecolinguistics*, edited by Alwin F. Fill and Hermine Penz, 1st edition, 355–64. New York, NY : Routledge, 2018. | Series: *Routledge Handbooks in Linguistics*: Routledge, 2017. https://doi.org/10.4324/9781315687391-24.

Ndesanjo, Ronald B. 'Is Ngorongoro's Multiple Land-Use Model a Scam?' *Udadisi: Rethinking in Action* (blog), 3 March 2022. https://udadisi.com/is-ngorongoros-multiple-land-use-model-a-scam/.

Neely, Brooke, and Michelle Samura. 'Social Geographies of Race: Connecting Race and Space'. *Ethnic and Racial Studies* 34, no. 11 (November 2011): 1933–52. https://doi.org/10.1080/01419870.2011.559262.

Neil, Margaret Renata, and Federica M. Cerruti. 'South(s) of the South(s): Race, *Caporalato*, and the "Southern Question" Renewed in Contemporary Italian Border-Making'. *Modern Italy*, 20 September 2024, 1–15. https://doi.org/10.1017/mit.2024.18.

Neumann, Roderick P. *Imposing Wilderness: Struggles over Livelihood and Nature Preservation in Africa*. California Studies in Critical Human Geography 4. Berkeley, CA: University of California Press, 1998.

Neumann, Roderick P. *Making Political Ecology*. London: Routledge, 2005. https://doi.org/10.4324/9780203764206.

Neumann, Roderick P. 'Moral and Discursive Geographies in the War for Biodiversity in Africa'. *Political Geography*, Ethics in Political Ecology 23, no. 7 (September 2004): 813–37. https://doi.org/10.1016/j.polgeo.2004.05.011.

Nguyen, Li, and Kerry McCallum. 'Drowning in Our Own Home: A Metaphor-Led Discourse Analysis of Australian News Media Reporting on Maritime Asylum Seekers'. *Communication Research and Practice* 2, no. 2 (2 April 2016): 159–76. https://doi.org/10.1080/22041451.2016.1188229.

Niceforo, Marina. 'Hate Speech and Environmental Activist Discourse: Critical Analysis of User Comments to One Controversial Tweet by Just Stop Oil'. *Journal of Language and Discrimination* 8, no. 1 (16 February 2024): 1–23. https://doi.org/10.1558/jld.27262.

Nixon, Rob. *Slow Violence and the Environmentalism of the Poor*. Cambridge, MA: Harvard University Press, 2011. http://site.ebrary.com/id/10491781.

Norman, Wayne, and Chris MacDonald. 'Getting to the Bottom of "Triple Bottom Line"'. *Business Ethics Quarterly* 14, no. 2 (April 2004): 243–62. https://doi.org/10.5840/beq200414211.

Norval, Aletta J. *Deconstructing Apartheid Discourse*. New York: Verso, 1996.

Nouala, Simplice. 'Livestock Is a Form of Climate Justice in the Global South'. *Al Jazeera* (blog), 10 November 2022. https://www.aljazeera.com/opinions/2022/11/10/livestock-is-a-form-of-climate-justice-in-the-global-south.

Obertreis, Julia. 'Environmental and Climate Policies as the New Hobby Horse: The "Alternative for Germany" and the German Right Wing'. *Environmental History* 27, no. 4 (October 2022): 642–8. https://doi.org/10.1086/721412.

O'Boyle, Edward J. 'Classical Economics and the Great Irish Famine: A Study in Limits'. *Forum for Social Economics* 35, no. 2 (Fall 2006): 21–53. https://doi.org/10.1007/BF02746430.

*Ocean Garbage Patch Could Be Bigger than Mexico*. CNN, 2017. https://www.cnn.com/videos/world/2017/08/15/south-pacific-garbage-patch-kinkade-pkg.cnn.

Ogada, Mordecai. 'Marseille 2021: The 2nd Scramble for Africa'. *The Elephant* (blog), 24 September 2021. https://www.theelephant.info/reflections/2021/09/24/marseille-2021-the-2nd-scramble-for-africa/.

Ogada, Mordecai. 'Stealthy, Surreptitious Second Coming of Western Colonialism to Africa'. *The Elephant* (blog), 16 September 2022. https://www.theelephant.info/opinion/2022/09/16/stealthy-surreptitious-second-coming-of-western-colonialism-to-africa/.

Öhman, May-Britt. 'Embodied Vulnerability in Large-Scale Technical Systems: Vulnerable Dam Bodies, Water Bodies, and Human Bodies'. In *Bodies, Boundaries and Vulnerabilities: Interrogating Social, Cultural and Political Aspects of Embodiment*, edited by Lisa Folkmarson Käll, 47–79. Cham: Springer, 2016.

Ojala, Carl-Gösta, and Jonas M. Nordin. 'Mining Sápmi: Colonial Histories, Sámi Archaeology, and the Exploitation of Natural Resources in Northern Sweden'. *Arctic Anthropology* 52, no. 2 (2015): 6–21.

Olbermann, Zoe, Fabian Mayer, and Holger Schramm. 'Greenfluencing through the Power of Emotions? Impact of Message Frames and Emotionally Matching Background Music on the Effectiveness of Influencers' Environmental Communication'. *Social Media + Society* 10, no. 2 (1 April 2024): 20563051241254381. https://doi.org/10.1177/20563051241254381.

Olsen, Jonathan. *Nature and Nationalism: Right-Wing Ecology and the Politics of Identity in Contemporary Germany*. New York: St. Martin's Press, 1999.

O'Regan, John P., and John Gray. 'The Bureaucratic Distortion of Academic Work: A Transdisciplinary Analysis of the UK Research Excellence Framework in the Age of Neoliberalism'. In *Education and the Discourse of Global Neoliberalism*, edited by John Gray, John P. O'Regan, and Catherine Wallace, chapter 5. London: Routledge, 2021.

Oreskes, Naomi. 'Eight Billion People in the World Is a Crisis, Not an Achievement'. *Scientific American*, 1 March 2023. https://www.scientificamerican.com/article/eight-billion-people-in-the-world-is-a-crisis-not-an-achievement/.

Ormesher, Ellen, and Patrick Greenfield. 'Greenwashing Era Is over, Say Ad Agencies, as Regulators Get Tough'. *The Guardian*, 15 May 2023, sec. Environment. https://www.theguardian.com/environment/2023/may/15/greenwashing-era-is-over-say-ad-agencies-as-regulators-get-tough.

Össbo, Åsa, and Patrik Lantto. 'Colonial Tutelage and Industrial Colonialism: Reindeer Husbandry and Early 20th-Century Hydroelectric Development in Sweden'. *Scandinavian Journal of History* 36, no. 3 (1 July 2011): 324–48. https://doi.org/10.1080/03468755.2011.580077.

Ostrom, Elinor. *Governing the Commons: The Evolution of Institutions for Collective Action*. Cambridge: Cambridge University Press, 2015.

Oswald, Y., J. K. Steinberger, D. Ivanova, and J. Millward-Hopkins. 'Global Redistribution of Income and Household Energy Footprints: A Computational Thought Experiment'. *Global Sustainability* 4 (January 2021): e4. https://doi.org/10.1017/sus.2021.1.

Oumer, Abdella, Robbert Maseland, and Harry Garretsen. 'Was de Montesquieu (Only Half) Right? Evidence for a Stronger Work Ethic in Cold Climates'. *Journal of Economic Behavior & Organization* 173 (1 May 2020): 256–69. https://doi.org/10.1016/j.jebo.2020.03.010.

Oyěwùmí, Oyèrónké. 'The White Woman's Burden: African Women in Western Feminist Discourse'. In *African Women and Feminism: Reflecting on the Politics of Sisterhood*, edited by Oyèrónké Oyěwùmí, 25–43. Trenton: Africa World Press, 2004.

Pál, Viktor. 'Illiberal Environmentalism? The Case of Contemporary Hungary'. *Environmental History* 27, no. 4 (October 2022): 649–56. https://doi.org/10.1086/721414.

Paperson, La. 'A Ghetto Land Pedagogy: An Antidote for Settler Environmentalism'. In *Land Education: Rethinking Pedagogies of Place from Indigenous, Postcolonial, and Decolonizing Perspectives*, edited by Kate McCoy, Eve Tuck, and Marcia McKenzie, 115–30. Abingdon: Routledge, 2016.

Parenti, Michael. *Against Empire*. San Francisco: City Lights Books, 1995.

Park, Chang Sup, Qian Liu, and Barbara K. Kaye. 'Analysis of Ageism, Sexism, and Ableism in User Comments on YouTube Videos about Climate Activist Greta Thunberg'. *Social Media + Society* 7, no. 3 (July 2021): 20563051211036059. https://doi.org/10.1177/20563051211036059.

Pas, Annemiek, Elizabeth E. Watson, and Bilal Butt. 'Land Tenure Transformation: The Case of Community Conservancies in Northern Kenya'. *Political Geography* 106 (1 October 2023): 102950. https://doi.org/10.1016/j.polgeo.2023.102950.

Paul-Emile, Barbara. 'William Wordsworth: An English Romantic Poet's Response to Colonialism'. *History of European Ideas* 19, no. 4–6 (1994): 627–33.

Peluso, Nancy Lee. 'Coercing Conservation?: The Politics of State Resource Control'. *Global Environmental Change* 3, no. 2 (1 June 1993): 199–217. https://doi.org/10.1016/0959-3780(93)90006-7.

Peluso, Nancy Lee. 'Racial Fictions at Work in a Regional Plantation Complex'. *Annals of the American Association of Geographers* (3 July 2024): 1–5. https://doi.org/10.1080/24694452.2024.2354501.

Peluso, Nancy Lee, and Christian Lund. 'Introduction'. In *New Frontiers of Land Control*, edited by Nancy Lee Peluso and Christian Lund, 1–15. Critical Agrarian Studies. Abingdon and New York: Taylor & Francis (Routledge), 2013.

Peluso, Nancy Lee, and Michael Watts. 'Violent Environments'. In *Violent Environments*, edited by Nancy Lee Peluso and Michael Watts, 3–38. Ithaca: Cornell University Press, 2001.

Perret, Meg. '"Transgender Frogs Turn Your Son Gay": Endangered Amphibians, Estrogenic Pollution, and Male Extinction'. *Frontiers: A Journal of Women Studies* 44, no. 1 (2023): 27–52.

Perry, Keston K., and Leon Sealey-Huggins. 'Racial Capitalism and Climate Justice: White Redemptive Power and the Uneven Geographies of Eco-Imperial Crisis'. *Geoforum* (23 June 2023): 103772. https://doi.org/10.1016/j.geoforum.2023.103772.

Pickard, Sarah. 'Young Environmental Activists Are Doing It Themselves'. *Political Insight* 10, no. 4 (1 December 2019): 4–7. https://doi.org/10.1177/2041905819891364.

Pietiläinen, Sonja. '"They Will Not Survive Here": Bordering, Racialisation, and Nature in the Politics of the Finnish Populist Radical Right', 17 April 2024. https://doi.org/10.1075/jlp.23085.pie.

Pilles, Eric, Richard I. Nicklin, and Joshua M. Pearce. 'How We Can Mine Asteroids for Space Food'. *International Journal of Astrobiology* 23 (January 2024): e16. https://doi.org/10.1017/S1473550424000119.

Poole, Robert. 'A Corpus-Aided Ecological Discourse Analysis of the Rosemont Copper Mine Debate of Arizona, USA'. *Discourse & Communication* 10, no. 6 (1 December 2016): 576–95. https://doi.org/10.1177/1750481316674775.

Poole, Robert. *Corpus-Assisted Ecolinguistics*. London/New York/Dublin: Bloomsbury Academic, 2022.

Posocco, Lorenzo, and Iarfhlaith Watson. 'Nationalism and Environmentalism: The Case of Vauban'. *Nations and Nationalism* 28, no. 4 (2022): 1193–211. https://doi.org/10.1111/nana.12823.

Pouchet, Jessica. 'Participation, Inequality, and Conversations about Conservation'. *Journal of Sociolinguistics* 25, no. 3 (2021): 474–7. https://doi.org/10.1111/josl.12459.

Powell, Miles A. *Vanishing America: Species Extinction, Racial Peril, and the Origins of Conservation*. Cambridge: Harvard University Press, 2016.

Procházka, Ondřej, and Jan Blommaert. 'Ergoic Framing in New Right Online Groups: Q, the MAGA Kid, and the Deep State Theory'. *Australian Review of Applied Linguistics* 44, no. 1 (31 March 2021): 4–36. https://doi.org/10.1075/aral.19033.pro.

Pugliese, Joseph. 'More-than-Human Lifeworlds, Settler Modalities of Geno-Ecocide and Border Questions'. *Journal of Global Indigeneity* 5, no. 3 (2021): 1–34.

Pulido, Laura. 'Geographies of Race and Ethnicity II: Environmental Racism, Racial Capitalism and State-Sanctioned Violence'. *Progress in Human Geography* 41, no. 4 (1 August 2017): 524–33. https://doi.org/10.1177/0309132516646495.

Pulido, Laura, and Juan De Lara. 'Reimagining "Justice" in Environmental Justice: Radical Ecologies, Decolonial Thought, and the Black Radical Tradition'. *Environment and Planning E: Nature and Space* 1, no. 1–2 (1 March 2018): 76–98. https://doi.org/10.1177/2514848618770363.

Puwar, Nirmal. *Space Invaders: Race, Gender and Bodies out of Place*. Oxford: Berg, 2004.

Quijano, Aníbal. 'Coloniality of Power and Eurocentrism in Latin America'. *International Sociology* 15, no. 2 (1 June 2000): 215–32. https://doi.org/10.1177/0268580900015002005.

Quinton, Jessica, Lorien Nesbitt, and Daniel Sax. 'How Well Do We Know Green Gentrification? A Systematic Review of the Methods'. *Progress in Human Geography* 46, no. 4 (1 August 2022): 960–87. https://doi.org/10.1177/03091325221104478.

Rahman, Muhammad Habibur, Nejat Anbarci, and Mehmet A. Ulubaşoğlu. '"Storm Autocracies": Islands as Natural Experiments'. *Journal of Development Economics* 159 (1 November 2022): 102982. https://doi.org/10.1016/j.jdeveco.2022.102982.

Raidoo, Renugan. 'The Unruly in the Anodyne: Nature in Gated Communities'. In *Anxious Joburg: The Inner Lives of a Global South City*, edited by Nicky Falkof and Cobus van Staden, 132–51. New York: New York University Press, 2020.

Rajaram, Prem Kumar. 'Refugees as Surplus Population: Race, Migration and Capitalist Value Regimes'. *New Political Economy* 23, no. 5 (3 September 2018): 627–39. https://doi.org/10.1080/13563467.2017.1417372.

Rampton, Ben. 'Gumperz and Governmentality in the 21st Century: Interaction, Power and Subjectivity'. *Working Papers in Urban Language & Literacies*, no. 136 (2014). https://www.academia.edu/8114068/WP136_Rampton_2014_Gumperz_and_governmentality_in_the_21st_century_Interaction_power_and_subjectivity.

Ramutsindela, Maano, Frank Matose, and Tafadzwa Mushonga. *The Violence of Conservation in Africa: State, Militarization and Alternatives*. Cheltenham: Edward Elgar Publishing, 2022.

Rao, Mohan. 'An Imagined Reality: Malthusianism, Neo-Malthusianism and Population Myth'. *Economic and Political Weekly* 29, no. 5 (29 January 1994): PE40–52.

Rao, Smitha. 'Saffronisation of the Holy Cow: Unearthing Silent Communalism'. *Economic and Political Weekly* 46, no. 15 (2011): 80–7.

Raper, Peter E. *New Dictionary of South African Place Names*. Johannesburg and Cape Town: Jonathan Ball, 2004.

Ratcliffe, Rebecca. 'Prince William Accused of "White Saviour" Mentality in Africa Wildlife Film'. *The Guardian*, 12 October 2018, sec. Global development. https://www.theguardian.com/global-development/2018/oct/12/prince-william-accused-of-white-saviour-mentality-over-wildlife-conservation-video-tanzania.

Raworth, Kate. *Doughnut Economics: Seven Ways to Think Like a 21st Century Economist*. White River Junction, VT: Chelsea Green Publishing, 2018.

Rayner, Steve. 'How to Eat an Elephant: A Bottom-up Approach to Climate Policy'. *Climate Policy* 10, no. 6 (January 2010): 615–21. https://doi.org/10.3763/cpol.2010.0138.

Reid, Darren R. *Native American Racism in the Age of Donald Trump: Historical and Contemporary Perspectives*. Cham: Springer Nature, 2020.

ResearchGate. 'Scientists Announce the Existence of Another Ocean Garbage Patch on ResearchGate'. *ResearchGate Blog* (blog), 17 August 2018. https://www.researchgate.net/blog/another-ocean-garbage-patch.

Reynolds, Martin, Chris Blackmore, and Mark J. Smith, eds. *Environmental Responsibility Reader*. London and New York: Zed Books, 2010.

Richards, Imogen, Callum Jones, and Gearóid Brinn. 'Eco-Fascism Online: Conceptualizing Far-Right Actors' Response to Climate Change on Stormfront'. *Studies in Conflict & Terrorism* 48, no. 9 (18 December 2022): 982–1008. https://doi.org/10.1080/1057610X.2022.2156036.

Richardson, John E. 'Fascist Discourse'. In *The Routledge Handbook of Critical Discourse Studies*, edited by John Flowerdew and John E. Richardson, chapter 30. Abingdon: Routledge, 2017.

Richardson, John E. 'Racial Populism in British Fascist Discourse: The Case of COMBAT and the British National Party (1960–1967)'. In *Analysing Fascist Discourse*, edited by Ruth Wodak and John E. Richardson, 181–202. Abingdon: Routledge, 2012.

Richardson, John E., and Ruth Wodak. 'Anti-Sorosism: Reviving the "Jewish World Conspiracy"'. In *Conspiracy Theory Discourses*, edited by Massimiliano Demata, Virginia Zorzi, and Angela Zottola, 98:395–420. Discourse Approaches to Politics, Society and Culture. London: John Benjamins Publishing, 2022. http://benjamins.com/catalog/dapsac.98.17ric.

Rinat, Zafrir. 'Tree Hugging Settlers Make a Green Case for Seizing Palestinian Land'. *Haaretz*, 28 August 2022, sec. News. Access World News. https://infoweb.newsbank.com/apps/news/openurl?ctx_ver=z39.88-2004&rft_id=info%3Asid/infoweb.newsbank.com&svc_dat=AWNB&req_dat=66EC98D771734AC69572ABFB4998FDEF&rft_val_format=info%3Aofi/fmt%3Akev%3Amtx%3Actx&rft_dat=document_id%3Anews%252F18C2DADF70CE2BD0.

Riofrancos, Thea. *Resource Radicals: From Petro-Nationalism to Post-Extractivism in Ecuador*. Durham, NC, and London: Duke University Press Books, 2020.

Riofrancos, Thea. 'The Security–Sustainability Nexus: Lithium Onshoring in the Global North'. *Global Environmental Politics* 23, no. 1 (1 February 2023): 20–41. https://doi.org/10.1162/glep_a_00668.

Robbins, Paul. *Political Ecology: A Critical Introduction*. 3rd edition. Chichester: John Wiley & Sons, 2011.

Roediger, David R. *The Wages of Whiteness: Race and the Making of the American Working Class*. London; New York: Verso, 1991.

Rorty, Richard. *Contingency, Irony, and Solidarity*. 1st edition. Cambridge: Cambridge University Press, 1989.

Rush, Claire, Seth Borenstein, and Jennifer McDermott. 'Maui's Fire Became Deadly Fast. Climate Change, Flash Drought, Invasive Grass and More Fueled It'. *AP News*, 10 August 2023, sec. U.S. News. https://apnews.com/article/hawaii-wildfires-climate-change-92c0930be7c28ec9ac71392a83c87582.

Rushkoff, Douglas. 'How Tech's Richest Plan to Save Themselves after the Apocalypse'. *The Guardian*, 24 July 2018, sec. Technology. https://www.theguardian.com/technology/2018/jul/23/tech-industry-wealth-futurism-transhumanism-singularity.

Sachseder, Julia, Saskia Stachowitsch, and Clemens Binder. 'Gender, Race, and Crisis-Driven Institutional Growth: Discourses of "Migration Crisis" and the Expansion of Frontex'. *Journal of Ethnic and Migration Studies* 48, no. 19 (16 December 2022): 4670–93. https://doi.org/10.1080/1369183X.2022.2092461.

Said, Edward W. *Culture and Imperialism*. 1994. Reprint, London: Vintage, 2014.

Said, Edward W. *Orientalism*. 25th anniversary edition. 1978. Reprint, New York: Vintage Books, 2004.

Said, Edward W. 'Palestine: Memory, Invention and Space'. In *The Landscape of Palestine: Equivocal Poetry*, edited by Ibrahim Abu-Lughod, Roger Heacock, and Khaled Nashef, 3–22. Birzeit: Birzeit University Publications, 1999.

Saito, Kohei. *Marx in the Anthropocene: Towards the Idea of Degrowth Communism*. Cambridge: Cambridge University Press, 2023. https://doi.org/10.1017/9781108933544.

Salleh, Ariel. 'Class, Race, and Gender Discourse in the Ecofeminism/Deep Ecology Debate'. *Environmental Ethics* 15, no. 3 (1993): 225–44.

Sasa, Ghada. 'Oppressive Pines: Uprooting Israeli Green Colonialism and Implanting Palestinian A'wna'. *Politics* 43, no. 2 (1 May 2023): 219–35. https://doi.org/10.1177/02633957221122366.

Schetter, Conrad, Kennedy Mkutu Agade, and Marie Müller-Koné. 'Frontier NGOs: Conservancies, Control, and Violence in Northern Kenya'. *World Development* 151 (1 March 2022): 105735. https://doi.org/10.1016/j.worlddev.2021.105735.

Schmelzer, Matthias, Andrea Vetter, and Aaron Vansintjan. *The Future Is Degrowth: A Guide to a World beyond Capitalism*. New York: Verso, 2022.

Schmitt, Casey R. 'Anthropomorphism, Anthropocentrism, and Human-Orientation in Environmental Discourse', 4 July 2023. https://doi.org/10.1075/jlp.22123.sch.

Schultz, Susanne. 'Demographic Futurity: How Statistical Assumption Politics Shape Immigration Policy Rationales in Germany'. *Environment and Planning D: Society and Space* (26 April 2018): 0263775818772580. https://doi.org/10.1177/0263775818772580.

Schwartz, John. 'Two Climate Activists, 40 Years Apart in Age, on the Movement's Future'. *The New York Times*, 18 January 2023, sec. Climate. https://www.nytimes.com/2023/01/18/climate/climate-activists-bill-mckibben-xiye-bastida.html.

*Securing a Sustainable Future for the Amazon*. Davos, 2020. https://www.youtube.com/watch?v=9XKm0MUIJQs.

Seele, Peter, and Lucia Gatti. 'Greenwashing Revisited: In Search of a Typology and Accusation-Based Definition Incorporating Legitimacy Strategies'. *Business Strategy and the Environment* 26, no. 2 (2017): 239–52. https://doi.org/10.1002/bse.1912.

Selby, Jan, Gabrielle Daoust, and Clemens Hoffmann. *Divided Environments: An International Political Ecology of Climate Change, Water and Security*. Cambridge: Cambridge University Press, 2022. https://doi.org/10.1017/9781009106801.

Sène, Aby L. 'Against Wildlife Republics'. *The Republic*, 13 November 2022. https://republic.com.ng/october-november-2022/conservation-and-imperialist-expansion-in-africa/.

Sène, Aby L. 'Justice in Nature Conservation: Limits and Possibilities under Global Capitalism'. *Climate and Development* 16, no. 9 (20 October 2024): 838–47. https://doi.org/10.1080/17565529.2023.2274901.

Sène, Aby L. 'Land Grabs and Conservation Propaganda'. *Africa Is a Country* (blog), 17 June 2022. https://africasacountry.com/2022/06/the-propaganda-of-biodiversity-conservation.

Shahid, Waleed. 'The Left Didn't Sink Kamala Harris. Here's What Did.' *The Nation*, 18 November 2024. https://www.thenation.com/article/politics/corporate-democrats-not-woke-activists-doomed-kamala-harris/.

Shanaah, Sadi, Immo Fritsche, and Mathias Osmundsen. 'The Effect of Climate Change Threat on Public Attitudes towards Ethnic and Religious Minorities and Climate Refugees'. *Group Processes & Intergroup Relations* (23 July 2024): 13684302241262252. https://doi.org/10.1177/13684302241262252.

Sharma, Mukul. *Green and Saffron: The RSS, Modi, and Indian Environmental Politics*. 2nd edition. Ranikhet: Permanent Black, 2024.

Shaw, Christopher. *Liberalism and the Challenge of Climate Change*. Abingdon; New York: Routledge, 2024. https://doi.org/10.4324/9780429463488.

Shotter, John, and Michael Billig. 'A Bakhtinian Psychology: From out of the Heads of Individuals and into the Dialogues between Them'. In *Bakhtin and the Human Sciences: No Last Words*, edited by Michael Bell and Michael Gardiner, 13–30. London; Thousand Oaks: SAGE, 1998. http://public.eblib.com/choice/publicfullrecord.aspx?p=483326.

Siddiqui, Sophia. 'Racing the Nation: Towards a Theory of Reproductive Racism'. *Race & Class* 63, no. 2 (1 October 2021): 3–20. https://doi.org/10.1177/03063968211037219.

Simpson, Leanne Betasamosake. *As We Have Always Done: Indigenous Freedom through Radical Resistance*. 3rd edition. Minneapolis: University of Minnesota Press, 2017.

Singh, David. 'A New Regime of Dispossession in Neo-Liberal India? Wind Energy, Hindutva, and Land Politics in Western Gujarat'. *Journal of Contemporary Asia* (15 July 2024): 1–22. https://doi.org/10.1080/00472336.2024.2376595.

Singo, Leiyo. 'When Maasaiphobia Became Policy'. *The Republic*, 3 August 2022. https://republic.com.ng/august-september-2022/maasai-predicament-ngorongoro/.

Slovic, Scott, Swarnalatha Rangarajan, and Vidya Sarveswaran. *Routledge Handbook of Ecocriticism and Environmental Communication*. Abingdon and New York: Routledge, 2019.

Smith, Anthony D. *Myths and Memories of the Nation*. Oxford: Oxford University Press, 1999.

Solnit, Rebecca. 'Why Climate Despair Is a Luxury'. *New Statesman*, 19 October 2022. https://www.newstatesman.com/environment/2023/07/rebecca-solnit-climate-despair-hope.

Soloway, Richard A. *Demography and Degeneration: Eugenics and the Declining Birthrate in Twentieth-Century Britain*. Chapel Hill and London: UNC Press Books, 1995.

Spence, Mark David. *Dispossessing the Wilderness: Indian Removal and the Making of the National Parks*. New York: Oxford University Press, 1999. http://site.ebrary.com/id/10085416.

Staff Reporter. 'Prince Harry Visits Rwanda on a Wildlife Conservation Trip'. *Rwanda Dispatch News Agency*, 22 August 2022, sec. Regional. https://rwandadispatch.com/prince-harry-visits-rwanda-on-a-wildlife-conservation-trip/,https://rwandadispatch.com/prince-harry-visits-rwanda-on-a-wildlife-conservation-trip/.

Stanley, Brooke, and Walter Dana Phillips. 'South African Ecocriticism'. *Oxford Handbooks Online: Scholarly Research Reviews*, 6 February 2017. https://doi.org/10.1093/oxfordhb/9780199935338.013.154.

Staudenmaier, Peter. 'Organic Farming in Nazi Germany: The Politics of Biodynamic Agriculture, 1933–1945'. *Environmental History* 18, no. 2 (1 April 2013): 383–411. https://doi.org/10.1093/envhis/ems154.

Staudenmeier, Peter. *Ecology Contested: Environmental Politics between Left and Right*. Porsgrunn: New Compass Press, 2022.

Stavrakakis, Yannis. 'Green Ideology: A Discursive Reading'. *Journal of Political Ideologies* 2, no. 3 (October 1997): 259–80.

Stavrakakis, Yannis. *The Lacanian Left: Psychoanalysis, Theory, Politics*. Edinburgh: Edinburgh University Press, 2007. https://doi.org/10.1515/9780748629077.

Stephens, Murdoch. *Critical Environmental Communication: How Does Critique Respond to the Urgency of Climate Change?* Lanham: Lexington Books, 2018.

Steyn, Melissa E. *Whiteness Just Isn't What It Used to Be: White Identity in a Changing South Africa*. Albany: State University of New York Press, 2001.

Stibbe, Arran. 'An Ecolinguistic Approach to Critical Discourse Studies'. *Critical Discourse Studies* 11, no. 1 (2 January 2014): 117–28. https://doi.org/10.1080/17405904.2013.845789.

Stibbe, Arran. *Ecolinguistics: Language, Ecology and the Stories We Live By*. 2nd edition. London: Routledge, 2021.

Stibbe, Arran. 'Positive Discourse Analysis: Re-Thinking Human Ecological Relationships'. In *The Routledge Handbook of Ecolinguistics*, edited by Alwin Fill and Hermine Penz, 165–78. Routledge Handbooks in Linguistics. New York: Routledge, 2018.

Stoegner, Karin, and Ruth Wodak. '"The Man Who Hated Britain" – the Discursive Construction of "National Unity" in the Daily Mail.' *Critical Discourse Studies* 13, no. 2 (1 April 2016): 193–209. https://doi.org/10.1080/17405904.2015.1103764.

Strick, Simon. 'The Alternative Right, Masculinities, and Ordinary Affect'. In *Right-Wing Populism and Gender: European Perspectives and Beyond*, edited by Gabriele Dietze and Julia Roth, 207–30. Bielefeld: transcript Verlag, 2020.

Stroud, Christopher. 'Turbulent Linguistic Landscapes and the Semiotics of Citizenship'. In *Negotiating and Contesting Identities in Linguistic Landscapes*, edited by Robert Blackwood, Elizabeth Lanza, and Hirut Woldemariam, 3–18. London and New York: Bloomsbury, 2016.

Sturgeon, Noel. *Ecofeminist Natures: Race, Gender, Feminist Theory and Political Action*. Oxford: Taylor & Francis Group, 1997.

Sturgeon, Noël. *Environmentalism in Popular Culture: Gender, Race, Sexuality, and the Politics of the Natural*. Tucson: University of Arizona Press, 2009.

Sultana, Farhana, ed. *Confronting Climate Coloniality: Decolonizing Pathways for Climate Justice*. London: Routledge, 2024. https://doi.org/10.4324/9781003465973.

Sturgeon, Noël. 'The Unbearable Heaviness of Climate Coloniality'. *Political Geography* 99 (1 November 2022): 102638. https://doi.org/10.1016/j.polgeo.2022.102638.

Suzuki, Yuka. *The Nature of Whiteness: Race, Animals, and Nation in Zimbabwe*. Seattle: University of Washington Press, 2017.

Swaffield, Joanne. 'After a Decade of Critique: Neoliberal Environmentalism, Discourse Analysis and the Promotion of Climate-Protecting Behaviour in the Workplace'. *Geoforum* 70 (March 2016): 119–29. https://doi.org/10.1016/j.geoforum.2016.02.014.

Táíwò, Olúfẹ́mi O. *Elite Capture: How the Powerful Took over Identity Politics*. Chicago: Haymarket Books, 2022.

Táíwò, Olúfhemi O. *Reconsidering Reparations*. New York: Oxford University Press, 2022.

Takaki, Ronald. '"An Entering Wedge": The Origins of the Sugar Plantation and a Multi-ethnic Working Class in Hawaii'. *Labor History* 23, no. 1 (January 1982): 32–46. https://doi.org/10.1080/00236568208584643.

Tănăsescu, Mihnea. 'Rights of Nature, Legal Personality, and Indigenous Philosophies'. *Transnational Environmental Law* 9, no. 3 (November 2020): 429–53. https://doi.org/10.1017/S2047102520000217.

Taylor, Charlotte. 'Metaphors of Migration over Time'. *Discourse & Society* 32, no. 4 (1 July 2021): 463–81. https://doi.org/10.1177/0957926521992156.

Taylor, Dorceta E. 'American Environmentalism: The Role of Race, Class and Gender in Shaping Activism 1820–1995'. *Race, Gender & Class* 5, no. 1 (1 January 1997): 16–62.

Taylor, Dorceta E. 'The Rise of the Environmental Justice Paradigm Injustice Framing and the Social Construction of Environmental Discourses'. *American Behavioral Scientist* 43, no. 4 (1 January 2000): 508–80. https://doi.org/10.1177/0002764200043004003.

Taylor, Julia. 'Just an Energy Transition? A Gendered Analysis of Energy Transition in Northern Cape, South Africa'. *Agenda*, 7 August 2023, 1–14. https://doi.org/10.1080/10130950.2023.2240855.

Taylor, Sarah McFarland. '"F*ck Earth": Unmasking Mars Colonization Marketing, from Planetary Perceived Obsolescence to Apocalyptic "New Earth" Rhetoric'. *Journal of Religion, Media and Digital Culture* 11, no. 1 (22 November 2022): 54–84. https://doi.org/10.1163/21659214-bja10067.

Tebaldi, Catherine. 'Metapolitical Seduction: Women's Language and White Nationalism'. *Journal of Linguistic Anthropology* 34, no. 1 (2024): 84–106. https://doi.org/10.1111/jola.12418.

Tebaldi, Catherine, and Scott Burnett. 'The Science of Desire: Beauty, Masculinity, and Ideology on the Far Right'. *Journal of Right-Wing Studies* 2, no. 2 (18 January 2025). https://doi.org/10.5070/RW3.1604.

Thakholi, Lerato. 'Conservation Labour Geographies: Subsuming Regional Labour into Private Conservation Spaces in South Africa'. *Geoforum* 123 (1 July 2021): 1–11. https://doi.org/10.1016/j.geoforum.2021.04.019.

The Economist. 'Why the Fires in Hawaii Have Been so Bad'. *The Economist*, 11 August 2023. https://www.economist.com/united-states/2023/08/11/why-the-fires-in-hawaii-have-been-so-bad.

'The Green Energy Scandal Exposed'. *Panorama*. BBC One, 3 October 2022. https://www.bbc.co.uk/programmes/m001cw6z.

'The Not-So-Benevolent Billionaire (Part II) – Bill Gates in Africa'. Podcast. *Citations Needed*. Medium, 1 August 2018. https://citationsneeded.medium.com/episode-46-the-not-so-benevolent-billionaire-part-ii-bill-gates-in-africa-4329389dd4a3.

The Red Nation. *The Red Deal: Indigenous Action to Save Our Earth*. Brooklyn: Common Notions, 2021.

Thing, Sudeep Jana, Roy Jones, and Christina Birdsall Jones. 'The Politics of Conservation: Sonaha, Riverscape in the Bardia National Park and Buffer Zone, Nepal'. *Conservation and Society* 15, no. 3 (September 2017): 292. https://doi.org/10.4103/cs.cs_15_2.

Thompson, Kim-Ly, and Natalie C. Ban. '"Turning to the Territory": A Gitga'at Nation Case Study of Indigenous Climate Imaginaries and Actions'. *Geoforum* 137 (1 December 2022): 230–36. https://doi.org/10.1016/j.geoforum.2021.11.006.

Thurlow, Crispin. 'Queering Critical Discourse Studies or/and Performing "Post-Class" Ideologies'. *Critical Discourse Studies* 13, no. 5 (19 October 2016): 485–514. https://doi.org/10.1080/17405904.2015.1122646.

Thurlow, Crispin, Alessandro Pellanda, and Laura Wohlgemuth. 'The Discursive Chronotopes of Waste: Temporal Laminations and Linguistic Hauntings'. *Language in Society* 51, no. 5 (November 2022): 819–37. https://doi.org/10.1017/S0047404522000616.

Tilley, Lisa, and Max Ajl. 'Eco-Socialism Will Be Anti-Eugenic or It Will Be Nothing: Towards Equal Exchange and the End of Population'. *Politics* 43, no. 2 (1 May 2023): 201–18. https://doi.org/10.1177/02633957221075323.

Tilley, Lisa, Anupama M. Ranawana, Andrew Baldwin, and Tyler M. Tully. 'Race and Climate Change: Towards Anti-Racist Ecologies'. *Politics* 43, no. 2 (29 September 2022): 141–52. https://doi.org/10.1177/02633957221127166.

Todorov, Tzvetan. 'The 2 Principles of Narrative'. *Diacritics* 1, no. 1 (1971): 37–44. https://doi.org/10.2307/464558.

Tornel, Carlos. 'Decolonizing Energy Justice from the Ground Up: Political Ecology, Ontology, and Energy Landscapes'. *Progress in Human Geography* 47, no. 1 (10 October 2022): 43–65. https://doi.org/10.1177/03091325221132561.

Torres, Émile P. 'The New Optimist Movement Has a Numbers Problem'. *Truthdig* (blog), 20 June 2024. https://www.truthdig.com/articles/the-new-optimist-movement-has-a-numbers-problem/.

Torres, Émile P. 'Do We Live in Hell? A Comprehensive Empirical Survey of the World's Vast Suffering.' *SSRN Electronic Journal*, (27 June 2025). http://dx.doi.org/10.2139/ssrn.5327670.

Tran, Mark. 'David Attenborough: Trying to Tackle Famine with Bags of Flour Is "Barmy"'. *The Guardian*, 18 September 2013, sec. Global development. https://www.theguardian.com/global-development/2013/sep/18/david-attenborough-famine-population.

Trask, Haunani-Kay. *From a Native Daughter: Colonialism and Sovereignty in Hawaii*. Revised 2nd edition. Honolulu: University of Hawai'i Press, 1999.

Trotsky, Leon. *The History of the Russian Revolution*. Translated by Max Eastman. New York: Monad Press, 1980.

Trucost. 'Natural Capital at Risk: The Top 100 Externalities of Business'. London: Trucost, 2013. http://naturalcapitalcoalition.org/wp-content/uploads/2016/07/Trucost-Nat-Cap-at-Risk-Final-Report-web.pdf.

Tuana, Nancy. 'Populationism and Climate Change: Controlling Images, Ecointersectional Alternatives'. In *Patricia Hill Collins*, edited by Amy Allen and Eduardo Mendieta. State College, Penn State University Press, forthcoming.

Tuana, Nancy. *Racial Climates, Ecological Indifference: An Ecointersectional Analysis*. New York: Oxford University Press, 2023.

Tuck, Eve, and K. Wayne Yang. 'Decolonization Is Not a Metaphor'. *Decolonization: Indigeneity, Education & Society* 1, no. 1 (2012): 1–40.

Tuck, Eve, and Marcia McKenzie. *Place in Research: Theory, Methodology, and Methods*. Routledge Advances in Research Methods 9. New York; London: Routledge, Taylor & Francis Group, 2015.

Turner, Joe, and Dan Bailey. '"Ecobordering": Casting Immigration Control as Environmental Protection'. *Environmental Politics* 31, no. 1 (2 January 2022): 110–31. https://doi.org/10.1080/09644016.2021.1916197.

Tutt, Owen, and Harald Heubaum. 'Nationalism and the Energy Transition: The Case of the SNP'. *Nations and Nationalism* 30, no. 2 (April 2024): 339–55. https://doi.org/10.1111/nana.12993.

Tyerman, Thom. 'Everyday Borders in Calais: The Globally Intimate Injustices of Segregation'. *Geopolitics* 26, no. 2 (3 March 2021): 464–85. https://doi.org/10.1080/14650045.2019.1631807.

Uhl, Isabella, Johannes Klackl, Nina Hansen, and Eva Jonas. 'Undesirable Effects of Threatening Climate Change Information: A Cross-Cultural Study'. *Group Processes & Intergroup Relations* 21, no. 3 (1 April 2018): 513–29. https://doi.org/10.1177/1368430217735577.

Valle, Gaby Del. 'When Environmentalism Meets Xenophobia'. *The Nation*, 8 November 2018. https://www.thenation.com/article/archive/environment-climate-eugenics-immigration/.

Van Valkenburgh, Shawn P. 'Digesting the Red Pill: Masculinity and Neoliberalism in the Manosphere'. *Men and Masculinities* 24, no. 1 (2021): 84–103. https://doi.org/10.1177/1097184X18816118.

Vanclay, Frank. 'Principles to Gain a Social Licence to Operate for Green Initiatives and Biodiversity Projects'. *Current Opinion in Environmental Sustainability* 29 (December 2017): 48–56. https://doi.org/10.1016/j.cosust.2017.11.003.

Varco, Matt. 'Volk Utopia: Racial Futures and Ecological Politics on the German Far-Right'. *Geoforum* (July 2023): 103823. https://doi.org/10.1016/j.geoforum.2023.103823.

Vettese, Troy, and Drew Pendergrass. *Half-Earth Socialism: A Plan to Save the Future from Extinction, Climate Change and Pandemics*. London and Brooklyn, NY: Verso, 2024.

Vidal, John. 'It Should Not Be Controversial to Say a Population of 8 Billion Will Have a Grave Impact on the Climate'. *The Guardian*, 15 November 2022, sec. Opinion. https://www.theguardian.com/commentisfree/2022/nov/15/population-8-billion-climate.

Wainaina, Binyavanga. 'How to Write about Africa'. Literary. *Granta* (blog), 2005. https://granta.com/how-to-write-about-africa/.

Walenta, Jayme. 'The Making of the Corporate Carbon Footprint: The Politics behind Emission Scoping'. *Journal of Cultural Economy* 14, no. 5 (3 September 2021): 533–48. https://doi.org/10.1080/17530350.2021.1935297.

Wang, Claire. 'Lahaina, Site of Incalculable Native Hawaiian Importance, Reels from Cultural Losses'. *The Guardian*, 15 August 2023, sec. US news. https://www.theguardian.com/us-news/2023/aug/15/maui-wildfires-lahaina-hawaii-rebuild.

Wang, Claire. '"Occupied by the US": Wildfires Renew Native Hawaiian Call for Sovereignty'. *The Guardian*, 7 September 2023, sec. US news. https://www.theguardian.com/us-news/2023/sep/07/native-hawaiian-sovereignty-maui-wildfires.

Warren, Charles. 'Nativeness and Nationhood: What Species "Belong" in Post-Devolution Scotland?' In *Invasive and Introduced Plants and Animals: Human Perceptions, Attitudes and Approaches to Management*, edited by Ian D. Rotherham and Robert A. Lambert, 67–80. Abingdon: Routledge, 2011.

Watts, Eric King. *Postracial Fantasies and Zombies: On the Racist Apocalyptic Politics Devouring the World*. Oakland: University of California Press, 2024.

Weldemichel, Teklehaymanot G. 'Inventing Hell: How the Ethiopian and Eritrean Regimes Produced Famine in Tigray'. *Human Geography* 15, no. 3 (1 November 2022): 290–4. https://doi.org/10.1177/19427786211061431.

Westberg, Gustav, and Henning Årman. 'Family Politics in Contemporary Fascist Propaganda: Multimodal Entanglements of National Socialist Ideals, Populist Rhetoric, and Image Bank Semiotics'. *Journal of Right-Wing Studies* 2, no. 2 (18 January 2025): 103–35. https://doi.org/10.5070/RW3.1619.

Westhuizen, Christi van der. 'Afrikaners in Post-Apartheid South Africa: Inward Migration and Enclave Nationalism'. *HTS Teologiese Studies / Theological Studies* 72, no. 4 (26 August 2016): 9. https://doi.org/10.4102/hts.v72i4.3351.

Wetherell, Margaret. *Affect and Emotion: A New Social Science Understanding*. Los Angeles: SAGE, 2012.

Wetherell, Margaret. 'Trends in the Turn to Affect: A Social Psychological Critique'. *Body & Society* 21, no. 2 (1 June 2015): 139–66. https://doi.org/10.1177/1357034X14539020.

Whyte, Kyle Powys. 'Indigenous Science (Fiction) for the Anthropocene: Ancestral Dystopias and Fantasies of Climate Change Crises'. *Environment and Planning E: Nature and Space* 1, no. 1–2 (March 2018): 224–42. https://doi.org/10.1177/2514848618777621.

Whyte, Kyle Powys. 'Too Late for Indigenous Climate Justice: Ecological and Relational Tipping Points'. *WIREs Climate Change* 11, no. 1 (2020): e603. https://doi.org/10.1002/wcc.603.

Widengård, Marie. 'The Mining-Conservation Nexus in Jamaica's Cockpit Country'. *Edge Effects* (blog), 15 June 2023. https://edgeeffects.net/cockpit-country-mining-conservation/.

Wilson, Edward O. *Half-Earth: Our Planet's Fight for Life*. New York: Liveright, 2017.

Wilton, Robert D. 'Colouring Special Needs: Locating Whiteness in NIMBY Conflicts'. *Social & Cultural Geography* 3, no. 3 (1 January 2002): 303–21. https://doi.org/10.1080/1464936022000003541.

Wittmer, Josie. 'Dirty Work in the Clean City: An Embodied Urban Political Ecology of Women Informal Recyclers' Work in the "Clean City"'. *Environment and Planning E: Nature and Space* 6, no. 2 (1 June 2023): 1343–65. https://doi.org/10.1177/25148486221102374.

Wodak, Ruth. 'The Discourse-Historical Approach'. In *Methods of Critical Discourse Analysis*, edited by Ruth Wodak and Michael Meyer, 63–93. Introducing Qualitative Methods. London and Thousand Oaks: SAGE, 2001.

Wodak, Ruth. *The Politics of Fear: The Shameless Normalization of Far-Right Discourse*. 2nd edition. London: SAGE, 2021. https://us.sagepub.com/en-us/nam/the-politics-of-fear/book265617.

Wolf, Martin. 'Market Forces Are Not Enough to Halt Climate Change'. *Financial Times*, 2 July 2024. https://www.ft.com/content/b2b6fb7a-9477-4485-a9e3-435b5e9c987e.

Wynter, Sylvia. *Sylvia Wynter: On Being Human as Praxis*. Edited by Katherine McKittrick. Durham: Duke University Press, 2015.

Wynter, Sylvia. 'Unsettling the Coloniality of Being/Power/Truth/Freedom: Towards the Human, after Man, Its Overrepresentation – An Argument'. *CR: The New Centennial Review* 3, no. 3 (2003): 257–337. https://doi.org/10.1353/ncr.2004.0015.

Yam, Kimmy. 'Why Hawaiian Sovereignty Has Undeniable Context for the Maui Fires'. *NBC News*, 24 August 2023. https://www.nbcnews.com/news/asian-america/hawaiian-sovereignty-undeniable-context-maui-fires-rcna100972.

Yamada, Shoko, Lav Kanoi, Vanessa Koh, Al Lim, and Michael R. Dove. 'Sustainability as a Moral Discourse: Its Shifting Meanings, Exclusions, and Anxieties'. *Sustainability* 14, no. 5 (January 2022): 3095. https://doi.org/10.3390/su14053095.

Yeginsu, Ceylan, and Christine Chung. 'Travel in 2023: 12 Months That "Took Chaos to a New Level"'. *The New York Times*, 20 December 2023, sec. Travel. https://www.nytimes.com/2023/12/20/travel/wildfires-floods-weather-travel.html.

Zabalegui, Maite Arraiza. 'The Tale of EDCs and Trans Identities'. *Diogenes* 65, no. 1 (February 2024): 110–30. https://doi.org/10.1017/S0392192123000287.

# Index

www.ingramcontent.com/pod-product-compliance
Ingram Content Group UK Ltd.
Pitfield, Milton Keynes, MK11 3LW, UK
UKHW020240290926
15470UKWH00002B/654

9 781350 360297